The Fourth Circle

EAST-WEST CENTER
SERIES ON

CONTEMPORARY ISSUES IN ASIA AND THE PACIFIC

Series Editor, Muthiah Alagappa

The Fourth Circle

A POLITICAL ECOLOGY OF SUMATRA'S
RAINFOREST FRONTIER

John F. McCarthy

Stanford University Press

Stanford, California

2006

Library of Congress Cataloging-in-Publication Data

McCarthy, John F. (John Fitzgerald), 1964–
The fourth circle : a political ecology of Sumatra's rainforest frontier /
John F. McCarthy.
p.   cm. — (Contemporary issues in Asia and the Pacific)
Includes bibliographical references and index.
ISBN 0-8047-5211-7 (cloth : alk. paper)
ISBN 0-8047-5212-5 (pbk. : alk. paper)
1. Forest policy—Indonesia—Nanggroe Aceh Darussalam.
2. Forest management—Indonesia—Nanggroe Aceh Darussalam.   3. Biological
diversity conservation—Indonesia—Nanggroe Aceh Darussalam.   4. Farms, Small—
Indonesia—Nanggroe Aceh Darussalam.   5. Adat law—Indonesia—Nanggroe Aceh
Darussalam. I. Title. II. Series.

SD657.I5M37 2006
333.7509598'1—dc22
2005025804

Original Printing 2006

Last figure below indicates year of this printing:

15   14   13   12   11   10   09   08   07   06

A Series from
Stanford University Press and the East-West Center

CONTEMPORARY ISSUES IN ASIA AND THE PACIFIC

Muthiah Alagappa, Series Editor

A collaborative effort by Stanford University Press and the East-West Center, this series addresses contemporary issues of policy and scholarly concerns in Asia and the Pacific. The series focuses on political, social, economic, cultural, demographic, environmental, and technological change and the problems related to such change. A select group of East-West Center senior fellows—representing the fields of political science, economic development, population, and environmental studies—serves as the advisory board for the series. The decision to publish is made by Stanford.

Preference is given to comparative or regional studies that are conceptual in orientation and emphasize underlying processes and to work on a single country that addresses issues in a comparative or regional context. Although concerned with policy-relevant issues and written to be accessible to a relatively broad audience, books in the series are scholarly in character. We are pleased to offer here the latest book in the series.

The East-West Center is an education and research organization established by the U.S. Congress in 1960 to strengthen relations and understanding among the peoples and nations of Asia, the Pacific, and the United States. The Center contributes to a peaceful, prosperous, and just Asia Pacific community by serving as a vigorous hub for cooperative research, education, and dialogue on critical issues of common concern to the Asia Pacific region and the United States. Funding for the Center comes from the U.S. government, with additional support provided by private agencies, individuals, foundations, corporations, and the governments of the region.

To Henny Dewaty Purba

And then we descended the dark scarp of Hell . . .
into the Fourth Great Circle
and ledge of the abyss . . .

More shades were here than anywhere above,
and from both sides, and with mad howls,
straining their chests, they rolled enormous weights.

And when they met and clashed against each other
they turned to push the other way, one side
screaming, "Why hoard?," the other side, "Why waste?"

And so they moved back round the gloomy circle,
returning on both sides to opposite poles
to scream this shameful tune another time;

Again they came to clash, turn and roll
forever in their semicircle joust.

—Dante Alighieri, *Inferno*, Canto VII

Sulitnya mengatur kayu, sama seperti menghadapi lingkaran setan. Katanya, daripada terbentur dan menjadi pahlawan kesiangan, lebih baik ikut terlibat di dalamnya. Maka, mulailah setiap kutipan tidak resmi dilakukan.

—*Kompas*, 11 February 1997

Lingkaran itu tidak mempunyai ujung (berputar terus) sehingga sulit menentukan mana ujung dan mana buntutnya. Semuanya setan, tapi yang mana yang jadi boss nggak tahu.

—Official, Ministry of Environment, 4 April 2000

Contents

Tables, Maps, and Figures

Acknowledgments

I owe a special debt of gratitude to the many people in South and Southeast Aceh who shared their houses, meals, and lives with me. Unfortunately, for reasons of confidentiality, and in light of the continuing conflict occurring in Aceh, I am unable to acknowledge fully the villagers, officials, and NGO field workers in Southeast Aceh and South Aceh who contributed to this research. May their beautiful homeland find peace.

My supervisors at Murdoch University, Carol Warren and Susan Moore, offered priceless support, advice, and suggestions over the course of this research. Carol Warren's Australia Research Centre grant and a research support grant from the School of Environmental Science also provided partial funding for fieldwork.

The Indonesian Institute of Sciences (LIPI) granted me a visa and research permit to undertake this research. Pak Effendi Sumardja from the Ministry of Environment, Prof. Dr. Sediono Tjondronegoro from Institut Pertanian Bogor, and Dr. Mummamad Ismail Gade from Universitas Syiah Kuala sponsored my research application and gave valuable counsel in the early stages. Tuti Herndarawati Mintarsih, from the Ministry of Environment, processed letters of support for visa applications at short notice and offered valued suggestions.

The Asia Research Centre, Murdoch University, provided a research grant, travel funding, and office facilities. Thanks to colleagues and staff associated with the Centre including Sidney Adams, Mark Besson, Del Blakeway, David Bourchier, Vedi Hadiz, Christopher McNeely, Teuku

Rezasyah, Richard Robison, Robert Roche, Andrew Rosser, Cisca Spenser, Jasmin Sungkar, and Zian Zhiang. Thanks also to Gaynor Dawson and Sean Foley for help during the early phase of the research; to Esther Velthoen for encouraging me to look into Dutch historical material, suggesting several useful sources, and patiently translating the first of these; to Karel Eringa for spending many long evenings providing working translations of both turgid and exotic colonial texts concerning Aceh; to Voss Patoir, who grew up in Banda Aceh during the colonial period, for helping with translations and giving insights into the lost colonial world from whence he came; to Gitte Heij, who provided invaluable leads into legal sociology and patiently translated colonial taxation laws from Aceh and Sumatra; and to Hidayat Sunarsyah, who drew the extensive maps of South and Southeast Aceh.

Thanks also to Dr. Kathryn Monk from the Leuser Management Unit (LMU), who initially encouraged me to work in the Leuser area; to Pak Zainal Abidin Pian and Dolly Priatna, who, with Kathryn, provided me with invaluable access to LMU's extensive archive on the Leuser area and also generously took the time to discuss the issues I was studying; to Mike Griffiths, Pak Ikram Syarif Sangadji, and many other LMU staff members who discussed the challenges facing resource management in Leuser; to the field staff of local NGOs active in South Aceh and the WWF Indonesia Programme, who encouraged me to include Menggamat as a field site, providing precious insights and otherwise explaining many aspects of the area. I found the people involved in these projects to be highly dedicated and motivated. My intention in interpreting the problems that these projects faced is not to pass judgment on their well-intentioned efforts, but rather to draw out lessons to be learnt for the future. I ask for their forbearance for any shortcomings in my analysis.

Thanks also to the faculty at the Pusat Kajian Antropologi Pembangunan (Puskap), Fakultas Ilmu Sosial dan Ilmu Politik, Universitas Sumatra Utara, who hosted a seminar discussion regarding my research; to Zulkifli Lubis, who accompanied me on a fieldtrip to Aceh Tenggara in late 1998, helping me to understand Alas tenure and territoriality as well as to deepen and recheck the conclusions suggested by earlier fieldwork; to Libby and Fraser Caghil, then at the Australia Centre in Medan, who extended their friendship and support, particularly during the protracted crisis of 1997–99.

I am grateful to the Centre for International Forestry Research (CIFOR), which provided support funding for fieldwork and office facilities during several visits to Bogor; to Lini Wollenberg, who enthusiastically helped shape the early stages of the research; to David Edmunds, who gave astute suggestions with respect to a paper based on the Menggamat case; to Chris Barr, Anne Casson, Daju Pradnja Resosudarmo, and Katrina Brown, who provided helpful editorial comments as this particular case study was being prepared to be published as a CIFOR occasional paper; to Keebet and Franz von Benda Beckmann, who gave valued comments on an earlier version of the Alas case that I presented to a Seminar on Legal Complexity, Natural Resource Management and Social (in)Security in Indonesia held in Padang in 1999; to the Resource Management in Asia Pacific Project at the Australian National University's Research School of Pacific and Asian Studies, which hosted a seminar that provided valuable feedback on a paper based on the Sama Dua material; to the anonymous reviewers for the journals *Development and Change, Human Ecology,* and the *Journal of Southeast Asian Studies*, who also offered important suggestions in the subsequent development of this study; to the participants at the Indonesian-Netherlands Study on Environmental Law and Administration (INSELA) workshop on Decentralisation and Environmental Management in Indonesia, held by Leiden University's Van Vollenhoven Institute and the Indonesian Centre for Environmental Law (ICEL) in Jakarta in June 2000, who helped refine the discussion of decentralization issues found in the final chapter.

Heartfelt thanks go also to Keebet von Benda Beckmann, Lesley Potter, Peter Brosius, and two anonymous reviewers for their perceptive readings of the completed manuscript and suggestions for revision. I revised the manuscript for this book during a fellowship at the Van Vollenhoven Institute, Leiden University, and an Australian Research Council fellowship at the Asia Research Centre. I am grateful for my colleagues at the Van Vollenhoven Institute, particularly Adriaan Bedner and Barbara Oomen, for their helpful comments on the manuscript and for their engagement with the issues under consideration here.

My parents supported my shift to Perth and patiently encouraged me during the trials of prolonged fieldwork in remote places. My brother, (Dr.) James McCarthy, shared the experience in many ways, from meeting me at the airport at odd hours to receiving telephone calls from remote villages asking for advice regarding assorted tropical ailments. Henny Dewaty Purba, whom I met along the way, gave her love and emotional support during the roller-coaster ride that was fieldwork at a time of crisis.

Abbreviations

ABRI	*Angkatan Bersenjata Republic Indonesia.* Armed Forces of the Indonesian Republic
APBD	*Anggaran Pembangunan dan Belanja Daerah.* The total district government budget
BAP	Biodiversity action plan
Bappenas	*Badan Perencanaan Pembangunan Nasional.* The National Development Planning Agency
Bappeda	*Badan Perencanaan Pembangunan Daerah.* Regional Development Planning Board
BPN	*Badan Pertanahan Nasional.* National Lands Agency
CBC	Community-Based Conservation
CBNRM	Community-based natural resource management
CCF	Community Conservation Forest
Dephutbun	*Departmen Kehutanan dan Perkebunan.* Department of Forestry and Estate Crops, formerly the Department of Forestry (*Dephut*)
DINAS	Decentralized provincial office under authority of governor
DPRD	*Dewan Perwakilan Rakyat Daerah.* District Legislative Assembly

EU	European Union
HGU	*Hak Guna Usaha.* Enterprise Use Rights
ICDP	Integrated Conservation and Development Program
IPK	*Ijin Pemanfaatan Kehutanan.* Timber Use Permit (a permit to extract timber from an area)
IPKH	*Industri Pengolahan Kayu Hulu.* Timber Industry Processing Permit (a license for a sawmill)
GLNP	Gunung Leuser National Park, or *Taman Nasional Gunung Leuser* (TNGL), national park founded in 1980
HPH	*Hak Pengusaha Hutan.* Forest Concession
ICDP	Integrated Conservation and Development Project
IUCN	World Conservation Union
Kanwil	*Kantor Wilayah.* Provincial office of a central government ministry
KKN	*Korrupsi, kolusi dan nepotisme.* Corruption, collusion, and nepotism
KLH	*Menteri Negara Kependudukan dan Lingkungan Hidup.* State Ministry for Population and Environment
Krismon	*Krisis moniter.* Monetary crisis, the East Asia economic crisis
LDP	Leuser Development Programme
LKMD	*Lembaga Ketahanan Masyarakat Desa.* Village Community Security Institution (village council)
LMU	Leuser Management Unit. The organization entrusted with the implementation of the Leuser Development Programme
LMD	*Lembaga Musyawarah Desa.* Village Deliberation Institution, the village assembly
NGO	Non-government Organization
PAD	*Pendapatan Asli Daerah.* Revenues generated directly by the district
Pemda	*Pemerintah daerah.* "Local government": this term refers to both the provincial government (*Pemerintah Daerah Tingkat I*) and the district government (*Pemerintah Daerah Tingkat II*)

PKA	Directorate General of Nature Conservation and Protection within the Ministry of Forestry (formerly PHPA)
PHPA	*Perlindungan Hutan dan Lelestarian Alam*. Directorate General for Forest Protection and Nature Conservation
Polda	*Kepolisian Daerah*. Provincial Police Command
Polres	*Kepolisian Resort*. District Police Command
Polsek	*Kepolisian Sektor*. Sub-district Police Command
RePPProt	Regional Physical Planning Programme for Transmigration
SAKB	*Surat Angkutan Kayu Bulat*. Permit for transporting unprocessed logs
SAKO	*Surat Angkutan Kayu Olahan*. Permit for transporting sawn timber
TGHK	*Tata guna hutan kesepakatan*. Forest Land Use Agreement
TNGL	*Taman National Gunung Leuser*. Gunung Leuser National Park
TPHT	*Tim Pengamaan Hutan Terpadu*. Integrated Team for Forestry Protection
UNCED	United Nations Conference on Environment and Development, Rio de Janeiro, June 1992.
WWF-LP	World Wide Fund for Nature's Leuser Project
YLI	*Yayasan Leuser International*. Jakarta-based foundation formed by influential Acehnese to support the Leuser Development Programme and conserve the Leuser Ecosystem
YPPAMAM	*Yayasan Perwalian Pelestarian Alam Masyarakat Adat Menggamat*. Menggamat Community Representative Body for Conservation Forests

Note on Terminology

In this work the term "regional government" (*pemerintah daerah*) is used for the four main levels of government structure outside Jakarta:

Province (Region Level 1)	Propinsi (Daerah Tingkat 1)
District (Region Level 2)	Kabupaten (Daerah Tingkat 2)
Sub-district	Kecamatan
Village	Desa

At the time of this research, central government departments, such as the Department of Forestry, had representative offices—that is, *Kantor Wilayah* (*Kanwil*)—in each province that looked after centrally administered (deconcentrated) responsibilities. In addition, each province had its own autonomous offices (*Dinas*) that, with technical guidance from the relevant central government department, provided services within a province but reported directly to the governor. Regional government offices of different sectorial agencies might be present at each of these levels (Freeman 1993; MacAndrews 1986). To avoid confusion, I will specify whether the *Dinas* or the *Kanwil* office is involved.

The state forestry agency as a totality (including its *Dinas* and *Kanwil* line agencies in the provinces) is referred to as the Department of Forestry (*Departmen Kehutanan,* or *Dephut*). It became known as the Department of Forestry and Plantation Estates (*Dephutbun*) in 1998. The term "Ministry of Forestry" refers to the central office headquarters of the department under the Minister of Forestry (and Estate Crops) in Jakarta.

The Fourth Circle

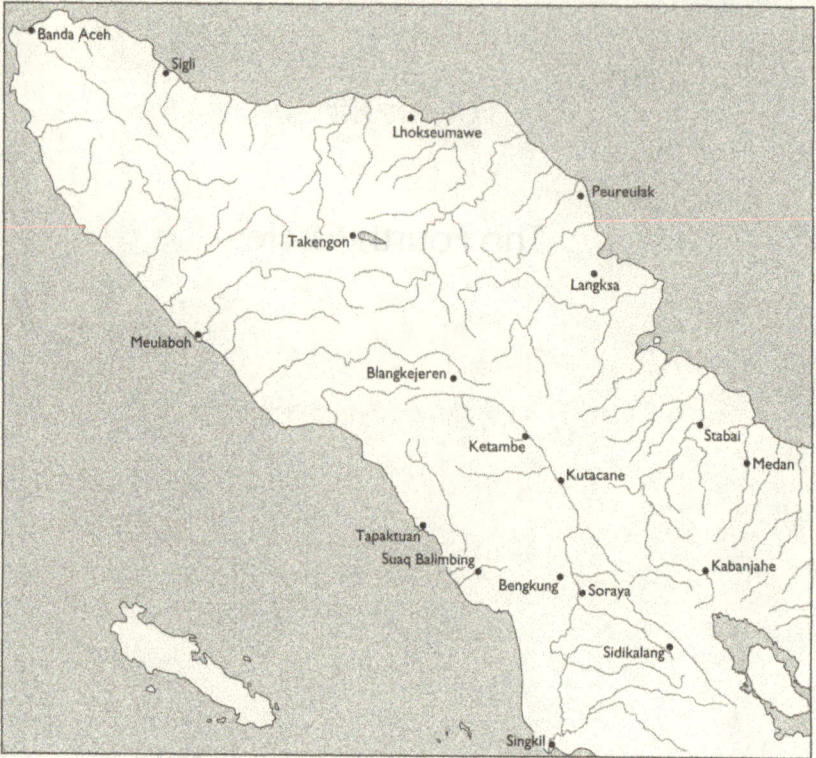

MAP 1. Northern Sumatra.

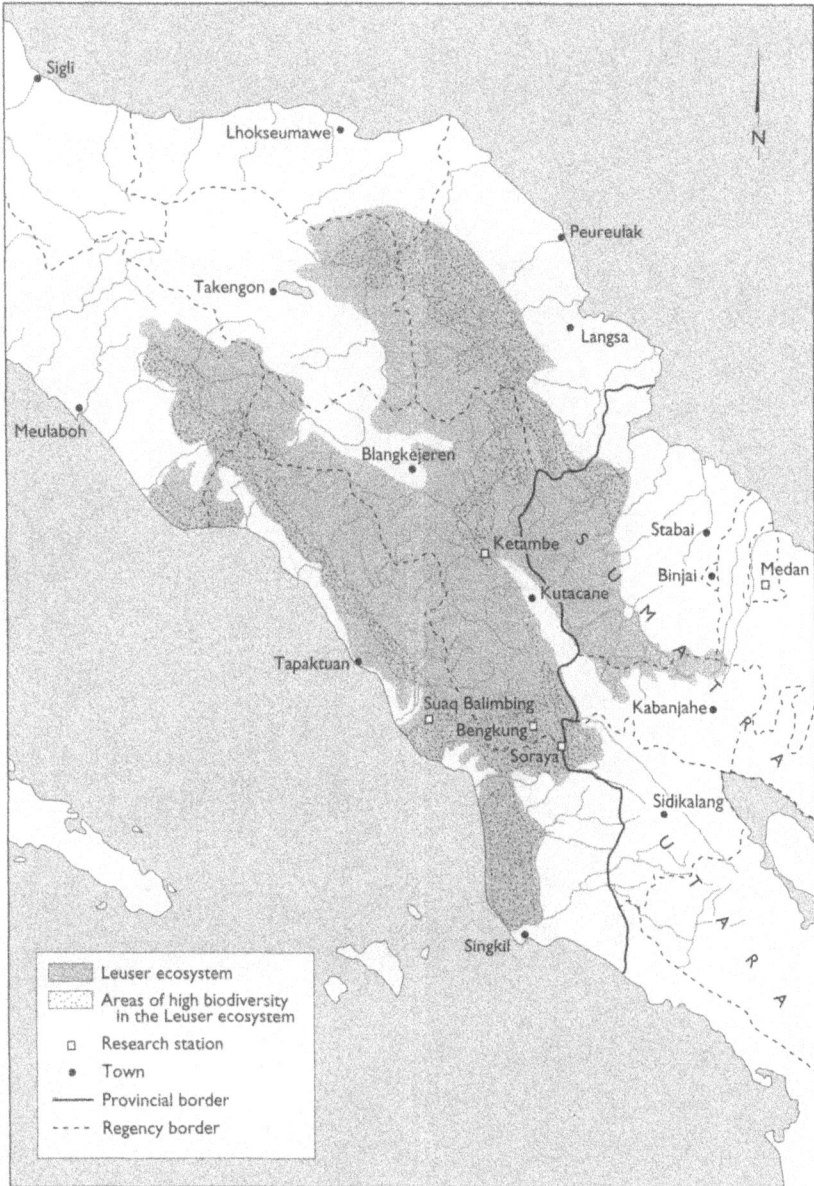

MAP 2. The Leuser Ecosystem. Courtesy of Leuser Management Unit.

MAP 3. Indonesia.

1

Introduction: Institutional Arrangements and Forest Regimes

In October 1997, in the midst of the forest fires that all but consumed Sumatra that year, I took an old village Land Rover west from the town of Terangon in the highlands of Gayo Lues in Southeast Aceh down through the wilderness to the west coast. In several places the monsoon rains had washed away the road. In other places there had been serious landslides, and the road clung precariously to the mountainside. Following the example of an elderly Acehnese trader who sported the white cap worn only by those who have made the pilgrimage to Mecca, I concentrated on the road behind, avoiding the view of the precipitous drop to the side or the dangerously steep and eroded track ahead. Nevertheless, through the thick smoke from the fires, occasionally I glimpsed a series of mountains completely covered in natural forest. Halfway down the mountain pass, I asked Pak Haji who owned the forest. "God owns it," he said. "Or Suharto," he added with a laugh, "but the letter [i.e., the exploitation permit] has not yet come down."

On reflection, it became clear that this statement was more than a casual joke. The Acehnese traditionally have considered wilderness that has not yet come into human possession as *tanah Allah*, "God's land" (Soekirman 1994: 303). At the same time, the state (or Suharto) could allocate exploitation rights in "forest estate." As Article 33 of the 1945 Constitution states, "land and water and the natural riches contained therein shall be controlled by the State and shall be made use of for the people." The Forestry Law (UU 11/1967) extended the state's sovereignty over "forests." Article 5 of the law established that "all forests within the territory of the Republic of Indonesia, including the natural resources they contain, are administered by the State."

By the 1990s, what might once have been considered *tanah Allah* remains only in remote places. Here the forest exists in all its primeval glory—tall *meranti* stretch towards the sky and the calls of gibbons and monkeys echo from deep green forest groves. Human interlopers in this foreign domain advance cautiously through the understory, wary of the green pit-vipers (ground-dwelling snakes that do not flee when sensing human presence) and leeches and various stinging insects, and stepping carefully over fallen logs covered in thick carpets of moss or slashing away at vines and the enveloping barbs of wild rattan. Villagers tell frightening stories of the Sumatran tiger, and those taking to the forest hope to avoid meeting the feared *raja hutan* ("king of the jungle"). But usually it is the intruder who causes the forest to shake, as squads of monkeys take flight, flinging themselves through quivering trees. Just occasionally the visitor glimpses the dark red plumage of a wizened orangutan sitting wearily in the fork of a tree or adroitly fleeing through the treetops. At other times the forest resounds with the brilliant cantata of the hornbill, accompanied by the helicopter beating of its blazing wings.

In other places, villagers' activity has rolled over the landscape to make room for gardens—clusters of nutmeg or candlenut trees, interspersed between other plants, including fleshy-leafed coffee plants, chili shrubs with their iridescent fruit, or stunted green *nilam* plants. In other places, especially alongside the highway, clear felling has occurred. The forest lays supine, like some vision of the holocaust. In other places the desolation has already been converted into neat rows of oil-palm trees, industrial and uninteresting worker trees that supply the prosperous oil palm industry. In yet other places, the marking of the landscape is unclear. Observers see neither the deep hew of undisturbed forest nor the forest-like gardens, neither the destruction of clear-cut forests nor the neat rows of industrial trees. Here the forest is thin. All the large trees have been combed out, and the resulting secondary forest is bisected with roads through which giant logging trucks carry the huge tropical logs. In yet other places, "invisible" or "illegal" hands have intruded to mine the valuable tropical hardwood. Here the forest is pockmarked with buffalo tracks and other paths on which loggers—usually local villagers—have pulled their loads.

This variability results from the way various actors—government, industry, village—have marked the landscape and used it for their ends. Where these actors have interests in resources, advancing claims over resources in a certain area, they may act to secure those claims or to contend against the claims advanced by other actors. Institutional arrangements exist in which specific actors have secured socially recognized rules and conventions that clarify who "owns" a resource or, in other words, who can otherwise constrain how others behave in respect to that resource (Bromley 1989: 870). Institutional arrangements involve binding rules and conventions concerning who has the right to access and to use natural resources and under what conditions. These arrangements establish a particular resource regime—a structure of rights and duties characterizing the relationship of one individual to another with respect to a resource (Bromley 1989: 871; Feeny et al. 1990: 6).[1]

As institutional arrangements are played out in space, they can also take the form of maps or spatial plans that exist in government offices or in human minds, or are inscribed on the landscape. As I noted earlier, these spatial plans are written on the natural landscape, with its rivers, mountains, and thick tropical jungle. In most areas, the land has long since become the territory of a village. In the years, decades, or even centuries of local use, villagers have collected forest products or opened areas for cultivation. Consequently, communities have long asserted a claim over the surrounding land and territory vis-à-vis other groups. Access to the land and the forest within such territories has long been a right of members of a village, a right mediated by the customary institutional arrangements of local communities. Since the time of Dutch colonial scholarship, the customary institutional arrangements of local communities have been known as *adat*. *Adat* amounts to understandings and institutional arrangements that inform village practice and facilitate the settlement of village disputes.[2]

This book is about the political, legal, and economic dynamics shaping environmental outcomes across southern Aceh, one of the richest and most expansive areas of tropical rainforest in Southeast Asia. The setting, remote southern Aceh, now the site of a separatist conflict, is unusual, given the paucity of detailed studies of this area.[3] The study examines how local customary (*adat*) regimes, *de facto* district authority systems, and the state in its various manifestations together create the particular institutional arrangements governing resource use in three different sites. The central theme is that the present cycle of ecological decline can be understood in terms of the specific institutional patterns operating at the district level.

This volume also considers how local agro-ecological systems and customary practices have adjusted to radical political change and uncertainty, economic collapse, and globalization in these particular sites. In the process, it explores the nature and role of local customary arrangements and the place of "customary law" (*adat*) within the unitary

Indonesian state. It also considers the effects of project and policy innovations that attempt to address environmental justice issues, whether
by reforming or taking the place of inept state agencies, by incorporating
local people in resource governance regimes, or by revitalizing customary
systems.

In addition, I examine two situations where extra-legal logging has
been rampant, areas also subject to high-profile biodiversity conservation projects. Consequently, my study analyses both the dynamics driving
resource degradation in rural Indonesia and the complex problems facing biodiversity conservation projects. A major theme here concerns the
connection between district patterns of governance and environmental
decline in the two districts concerned. Since I made this study the Indonesian government has implemented decentralization reforms, which have
handed sweeping new powers to this administrative level. It is likely that
the institutional patterns identified in this study will affect the outcome
of the current decentralization reforms on resource use.

By way of introduction, this first section substantiates the central
theme regarding resource management in Indonesia. The second section
presents a framework for analyzing the complex institutional arrangements governing resource use at the local level. Here I draw on several
thus far quite distinct sets of theoretical perspectives, including the new
institutional economics, political science, political ecology, and legal anthropology. Finally, the last section of this chapter sets out the structure
of this book.

Institutional Arrangements in Indonesia's "Forestry Estate"

In an earlier period the state had yet to carry out extensive cadastral
surveys or create land titles over extensive areas in the remote "outer
islands" of Indonesia. In these areas the Indonesian state has embarked
on a strategy to control natural resources by designating these areas as
"forest" irrespective of existing agrarian uses or the variety of plant life
found there (Li 1999). Doing so involved a process of "territorialization"
to set geographic boundaries, and attempt to control the activities of
people—including their access to natural resources—within these boundaries. The state set about this resource-control strategy through a mapping exercise that divided national territory into complex zones, creating
regulations that delineate how and by whom resources in these zones
may be used. Agencies whose jurisdictions are territorial as well as func-
tional then administered these zones (Vandergeest and Peluso 1995).[4]

During the 1980s, government agencies embarked on a mapping exercise that had created the so-called Consensus Forest Land Use Plan (*Tata Guna Hutan Kesepakatan,* or TGHK), what might be called a "forestry map."[5] This mapping exercise divided the geography of outer-island Indonesia in accordance with the categories established in the Forestry Law (UU 11/1967), designating areas for conversion, production, limited production, protection forest, and nature reserves, or leaving them as unclassified areas.[6] Thus, on paper, at least, the authorities had marked the acceptable boundaries within which local village and cultivation areas would legitimately exist. Inside the boundaries of the vast state "forest zone" (*kawasan hutan*), the forestry department maintained the right to allocate use (or non-use, for protection areas) of forest under state control. This took the form of either direct or indirect control. In areas mapped as "protection forest" and "nature reserve," in theory at least the forestry agency maintained direct control. In practice, the forest authorities attempted to forbid local access to "forest" and land located within national park boundaries. In areas mapped as "conversion forest," the forestry department received requests for conversion of the area to plantation agriculture, granting "timber use permits" (*Ijin Pemanfaatan Kayu,* or IPK) and "enterprise use rights" (*Hak Guna Usaha,* or HGU) to lessees over what remained state land. In areas mapped as "production forest," the forest authorities granted long-term leases (*Hak Pengusaha Hutan,* or HPH) to timber concessionaires, who log under forestry department supervision.

Another official mapping process took place later, in accordance with the Spatial Planning Law (UU 24/1992), which incorporated improved environmental planning procedures and involved a planning process at the district and provincial levels. This process, carried out according to a new set of criteria that divided the land area into "cultivation" and "non-cultivation" areas, revised the earlier territorial categories, which had resulted in huge areas being allocated for production and conversion under the TGHK Forest Maps.[7] This process led to the creation of "regional spatial plans" (*Rencana Umum Tata Ruang Daerah,* or RUTRD), in provincial government offices, producing maps in which territorial classifications in many areas contradicted the Forest Map.[8]

Using the state's spatial planning process as a metaphor for a village's ways of arranging its activities in space, Indonesian writers have referred to "community spatial plans" (Abdullah 1994; Soekirman 1994). Yet, in the oral culture of the villages, no one has ever seen the need or taken the time to translate this "Community Plan" onto paper, or create a "Community Map." This "Community Map" has only taken physical form after long deliberation, in those few areas where NGOs—through a community consultation process—have carried out a "counter-mapping" exercise, to use Peluso's term (Peluso, 1995; Momberg et al. 1996). In South Aceh, such a map would show villages, hamlets, rice paddy, gardens, fishing and bathing spots, and paths leading to areas where local people have in the past collected or still continue to collect non-timber forest products. In the minds of villagers, it would also be marked with events in local history, showing where the Dutch colonial forces used to camp; the paths across the mountains that local *Muslimin* forces used to flee from the Dutch; where various guerrilla forces hid during the insurrections of recent Acehnese history; where villagers fish or collect forest products; where they hold the *kenduri* feasts; and where old men pray to the spirit of the place, known as the *aulia*.

In many cases, the state carried out its mapping exercises without taking into account the locally evolved and long-practiced "Community Spatial Plan."[9] As a result, local communities may not accept the formal status of a state-designated "forest" because *adat* property rights (known in legal language as *hak ulayat,* or "right of avail") are still operative in these areas.[10] This can have two consequences. On the one hand, when land management authorities implement their spatial plans, there may be conflict with local communities—either taking an open form or the form that Scott (1985) calls "everyday forms of resistance"—unheralded sabotage, clandestine theft, and passive resistance—the disorganized, unplanned guerrilla-style activities through which peasants make their resistance felt. On the other hand, where the state does not implement its plans (or implements them unevenly), to a large degree local communities are able to continue their land management practices according to local understandings.

Adding up the effects of overlapping local and government institutional arrangements could lead to what might be called a Composite Map—the sum of the layers of untidily overlaying and contradictory spatial plans. Of course, as it plays itself out in the field, in various sections, one plan predominates or is imposed while the others are left, like invisible plastic overlays or imaginary dotted lines. In some places, the spatial plans of different categories of actors come into conflict. In yet other places, like the *tanah Allah* that awaited the letter from Suharto, the spatial plan of a particular actor remains virtual, a space awaiting the assertion of a claim. For in many places, the *de jure* status of state "forest estate" (*kawasan hutan*) may have little relation to the *de facto* situation in the field. Alternatively, if the area has already been converted to other uses, this formal status may no longer have any meaning, unless someone wants to make use of national law to make a claim over the "forest" in the "national interest." Where *tanah Allah* persists, the *hutan negara* (state forest) status may not be enforced, leaving resources largely unprotected. In this fashion, complex configurations of state and local regimes overlap in very localized ways.

In a real sense, then, the fate of a given area is the outcome of how the configuration of these various claims is played out in space. Depending on the outcome, in areas mapped as state "forest" it is possible to observe:

1. Logging of "production forest" by a concessionaire operating with the official permission of the Forestry Department

2. Rapid conversion of an area into plantation agriculture by corporate interests with a lease over an area

3. Conversion of an area for government-sponsored transmigration programs

4. Retention of forest within the boundaries of a protected area

5. Maintenance of village gardens for food and tree crop production, perhaps combined with swidden agriculture, and/or the conversion of new areas for agricultural use

6. Retention of forest areas within village territory

7. "Uncontrolled" logging by "illegal" logging networks.

The possibility of overlapping and conflicting claims over the same place allows for yet more variation. Nonetheless, such variation suggests that types of land uses can be loosely associated with the presence (or absence) of a particular kind of institutional arrangement. In other words, the institutional arrangements operating in a particular area are intimately related to a particular outcome.

From the 1980s on, increased international attention has focused on the fate of tropical rainforests. Apart from the unquantifiable values inherent in these corners of the natural world that are associated with the biodiversity they harbor, these forests supply essential ecological services. These include the hydrological functions of the forest—protection from floods, erosion and landslides and provision of water for industry, agriculture, and domestic use—as well as carbon sequestration and provision of resources for ecotourism and timber and non-timber products (Effendi 1998). However, if forests are to continue to provide these goods, surrounding communities or other actors need to devise institutional arrangements to enforce restraint on resource users who might otherwise deplete local resources.

In theory, through institutional arrangements a community of users determines who has access to resources and under what conditions. Consequently, the problems of resource depletion, deforestation, and biodiversity loss are also institutional problems. Indeed, as Bromley and Cernea (1989) have argued, resource degradation in developing countries is often a consequence of institutional problems. Contemporary events and a broad literature point to the critical nature of these problems.

Institutional Problems

A variety of social, economic, demographic, and ecological factors are involved in deforestation in Indonesia.[11] In addition, these problems are also institutional problems: socio-economic structures and power relations underlie the problems facing the management of Indonesia's resources, particularly in connection with the changing structure of property rights operating in the vast "forest" domain.

In 1997 and 1998, extensive forest fires ravaged the Indonesian islands of Kalimantan and Sumatra, burning some seven to ten million hectares of forest and agricultural land.[12] Because political turmoil erupted in the wake of a currency meltdown, the environmental devastation combined into a wider economic and political crisis. As in other sectors of the economy, during the New Order (1966–98) powerful conglomerates and politico-business families were able to dominate the forestry sector at the expense of sound environmental management. Critics then directed the same charges that had been leveled against the regime's economic management against its environmental management: precarious policies, "the capricious exercise of power, cronyism and incompetent systems of governance" (Robison and Rosser 1998: 1593).

In contrast, the official discourse of the New Order period tended to ascribe the failing of the state forest institutions either to the managerial and logistical difficulties facing a state forestry agency's managing such an extensive area of "forest" with few resources, or to difficulties of law enforcement. In this view, management problems could be remedied by making available greater resources with the help of international donors and by a more thoroughgoing implementation of the law. However, though these two factors were clearly at play, this policy discourse avoided discussion of a third issue—the allocation of property rights over areas of the "forestry estate" to politico-business interests.

If one considers environmental policy as the imposition of a particular structure of property rights, it is apparent that difficulties arise from the way co-existing property regimes—associated with the state or with *adat* customary orders—are linked to overlapping claims over access. In constructing a regulatory order for resource extraction, the New Order followed the colonial example and overlaid indigenous notions of tenure and territoriality with the concepts and regulatory institutions of a state property regime. Subsequently, to generate state revenue and further capital accumulation by clients of the regime, the state allocated use rights to commercial interests in rich forest areas. By failing to institute a state property regime in more than name, a *de jure* state property regime slipped into the *de facto* control of private interests. These consisted of the conglomerates and politico-business families at the apex of the state, together with local networks of power and interest that dominated illegal timber operations at the district level. The end result was a complex layering of competing property claims covered with the normative discourses of state forestry.

At the field level, the New Order's forest policy entailed a transfer of the flow of benefits arising from resource extraction away from villagers.[13] Irrespective of long-standing local uses, the state allocated areas of the state "forest estate" to private interests, usually urban elites with close ties to key political figures. When these claims overlapped with customary land rights, local communities became "squatters" in "forest" territory. External land claims helped stimulate local farmers, afraid of future land shortages, to gain *de facto* control of available land by planting trees, including in poorly protected "protection forests" and national parks (Angelsen 1995a). Meanwhile, underpaid local officials—including police, army, and forestry department staff paid to protect conservation areas—joined in networks of power and interest that benefited from uncontrolled logging. As a consequence, the forest authorities failed to implement the strictures of the state forest regime.

To take the analysis of these institutional dynamics further in field investigations, this study necessarily raises a theoretical problem: How can we understand and analyze institutional arrangements? In the next section I will briefly discuss the approach used in this study.

Orientations: An Approach to Understanding Institutional Arrangements

The Institutional Turn

Over the last few decades economists and social scientists using explanatory models derived from rational choice, game theory, and organizational theory have focused on how institutions evolve in response to individual choices and strategies, and how institutions affect the performance of political and economic systems (Acheson 1994). Within this field of "institutional economics," North (1990: 3) has defined institutions as "the humanly devised constraints that shape human interaction," noting that "if institutions are the rules of the game, organisations and their entrepreneurs are the players." Ostrom (1992a: 19), a thinker more concerned with the management of natural resources, has defined an institution "as a set of rules of rules actually used (the working rules or rules-in-use) by a set of individuals to organize repetitive activities that produce outcomes affecting those individuals and potentially affecting others."[14]

Within this framework, property rights have been considered important institutional arrangements. A property right can be understood to be "a secure claim or expectation over a future stream of benefits arising from a thing or situation" (Bromley 1989: 870). A property right entails the existence of an authority that will enforce that right. This is a formal or informal structure able to exclude those without legitimate property rights and regulate access to and use of those with the rights to use a resource. Seen from this perspective, property is not a thing, but "a network of social relations that governs the conduct of people with respect to the use and disposition of things."[15] A "property regime" is a structure of rights and duties characterizing the relationship of one individual to another with respect to a resource (Bromley 1989: 870). A particular property regime will gain its character from the way in which those controlling it decide to allocate rights and duties. How they do so in turn depends upon what the decision-making group believes to be scarce and valuable, and hence what needs to be protected with rights (Bromley 1989: 870). Thus, a political process determines property rights by assigning rights to specific resource users. Consequently, the way authority works and the way those with discretionary power grant access rights—whether wielding authority within the state or within the village—are critical to resource outcomes.

Institutional arrangements encompass more than property rights, however; they also include the socially recognized rules governing labor relations, the attainment of permits and licenses, and modes of transport and access to market. Because commodities jump in value as they circulate through sites associated with these various institutional arrangements, actors who have privileged access to or control over strategic nodes in the network of exchange are able to constrain the conditions under which others benefit from resources and gain disproportionate benefits for themselves (Ribot 2000; Zerner 2000).

The complexity of the research situation complicates the problem of understanding institutional arrangements as "rules-in-use" governing access to common pool resources. In complex rural settings there may be competing views concerning what should constitute the rules-in-use. Since defining and applying rules involves political acts, particular parties invest in a specific institutional order while other parties may contest this order. The rules-in-use—which differ from laws laid down in a distant capital—tend to reflect certain power relations. For instance, depending on the situation, they may reflect either customary local notions of rights, clientelist patterns of exchange, state laws or an overlapping, and complicated, crossing over of the three. Moreover, this reality may not be obvious: the powerful may forbid alternative claims over resources from being openly articulated. Dissenting parties may only express their views "off stage" or in covert acts of resistance (Scott 1985). Where normative arrangements governing access to resources are subject to on-going conflict, the interpretation of rules will differ according to the perspective of the speaking subject and his or her particular interests. This can make locating the rules-in-use particularly problematic. Rather than attempting to conceptualize the essence of a particular set of rules-in-use, this suggests that a given normative order needs to be understood in terms of its evolving relations to different legal and normative orders operating in a changing political and socio-economic context.[16]

Legal Anthropology and History

In contrast with the generalized and rather ahistorical institutionalist meta-theory suggested by institutionalist accounts influenced by rational choice theory, legal anthropologists have developed a nuanced understanding of local institutional arrangements. This emerged from the study of how colonial states imposed on and attempted to reconfigure local arrangements as "customary law." While the term customary law "sounds as if it designates a straightforward set of traditional rules," the view that emerged—particularly from African legal studies—is of a concept with its own history (Moore 1986). This concept bears the traces of colonial policy towards indigenous polities and the social and historical context in which different claims regarding what should constitute "customary law" were formulated and advanced. While the histories of different colonial circumstances varied widely, in many cases, colonial administrations constructed formal structures of indirect rule and labeled them "traditional," even though they were largely new entities. Creating "traditional authorities" for indirect rule involved a process of defining and perpetuating positions of power and advantage, in the process investing selected chiefs and traditional heads with a greater range and different sorts of authority than they had previously enjoyed. Indirect rule also involved constructing "customary law" in accordance with European approaches to law. This required turning fluid accounts of relationships at work and selective presentations of moral claims by particular actors into "customary rules" (Chanock 1982, 1985).

The colonial system of indirect rule in the Dutch East Indies also involved the development of a large variety of "native municipalities." Here indigenous rulers and headmen approved by the colonial administration were authorized to manage the "internal affairs" of their "jural communities" under the paternalistic guidance of colonial administrators (Sonius 1981). During the nineteenth century, in accordance with the doctrine that all property rights over "waste lands" could be decided by the colonial state, in many respects *adat* property rights over forest lands had been systematically neglected. In response, from the early twentieth century on, a group of legal scholars associated with Van Vollenhoven ("the Leiden school") set out to create the "science" of *adat* law. This involved systematically studying aspects of indigenous custom (*adat*) that "had the power of compulsion" and hence could be considered to have "legal consequences" (Otto and Pompe 1989: 239). Given that anything designated as law became the concern of lawyers and the state, by creating this "*adat* law," these scholars were able to draw part of *adat* into the sphere of law and thereby attempt to preserve "*adat* law" from the incursions of European law. In this fashion Van Vollenhoven and his colleagues endeavored to find a way to recognize indigenous rights over land and prevent the thorough-going alienation of Indonesian land to European interests, especially the rubber and sugar industries. With the renewed interest in the *adat* order that this entailed, the attempt to abolish the plural system of colonial law and replace it with a uniform European system failed. To a significant degree, *adat* authorities were allowed to continue to dispense "indigenous justice" in accordance with the norms and procedures of long-standing *adat* systems until the end of the colonial period (Sonius 1981).

In a similar fashion to other new states, post-independence, the Indonesian government dismantled the indigenous polities they inherited from the colonial past. Still, in the rural hamlets of Indonesia, what might be considered a customary socio-legal order has often endured in domestic levels of organization, in kinship relations, in property relations, and in hamlet and village politics. This customary system, where it still holds sway, has subtly altered over time, ensuring that this is neither the customary order of the pre-colonial past nor the colonially constructed "*adat* law." Yet, as villagers have reshaped their traditions to face a changing world, remnants of this past can still be discerned.[17]

From this literature emerges a historically contextualized view of the dialectical mutually constitutive relation between indigenous normative systems and the European legal systems imported by colonial powers that produced "customary law" (Merry 1988). The picture of customary systems that emerged is not of "internally consistent codes of action analogous to Western written law," but rather of "negotiable and internally contradictory repertories that were applied with discretion" (Merry 1992).

In viewing contemporary situations, it is necessary to distinguish how co-existing legal and institutional arrangements pertain to the same set of activities and relationships, each with different underlying conceptual structures (Benda-Beckmann and Benda-Beckmann 1999). Developing this perspective further, more recent theorizations have pointed out that the earlier dualistic approach—that saw the "legal" and the "societal" as coherent orders with their own distinct logics—tended to overlook the real connections between local socio-legal orders and state law. Rather than assuming that there is a dialectical relation between a dominant state political-legal order and a resistant local order, this contrasting perspective seeks to understand how, in attempting to control the direction of social change, actors make use of various institutions and competing value orientations. In attempting to understand "the complexities of a historically produced political-legal context," this analytical framework turns its attention "to shifting patterns of dominance, resistance, and acquiescence, which occur simultaneously" (Wilson 2000).

Seeing the State from Below

In addition to evolving "customary" orders, this research inevitably needs to focus on the state. Earlier Abrams (1988) observed that political sociology had reified "the idea of the State" as an integrated institution that we can understand in terms of cohesion, purpose, independence, common interest and morality (Abrams 1988: 68). He argued that, rather than seeing the state as an ultimate point of reference, the focus should shift to the "state system"—"the internal and external relations of political and governmental intuitions" (Abrams 1988; 75). The state system appears as a set of bureaucracies—"multiple agencies, organisations, levels, agendas, and centres that resists straightforward analytical closure" (Gupta 1995: 392). Often such state systems appear "riven by internal contradictions, incomplete consciousness of interest, incorrect implementation of projects aimed at furthering its interests, and conflict between individual officials and the organisation."[18] As different state actors and levels of administration create laws and administrative rules according to different aims and agendas, we see the coexistence of different logics of regulation. This leads to state heterogeneity or "internal pluralism" (Santos 1992: 134). This heterogeneity has significant implications for the development of local socio-legal configurations.

Viewed from below, rather than appearing like a large-scale, unitary structure, the state is visible in a more disaggregated form. Villagers deal with sub-district officials, village heads, the local offices (*dinas*) of government line agencies. When villagers need to carry out official business, rather than traveling to the agency office in town, they can seek out lower-level officials via kinship or neighborhood links, people with whom they have long-standing personal ties. As they seek their advice directly through village household links, the official order can be somewhat domesticated (Gupta 1995).

Members of the village, including active or pensioned civil servants, soldiers, and policemen and others who have some kind of relationship to the state system, are able to mediate relations with the state. They can provide villagers with the favors they most urgently need. For instance, they can control access to the plethora of licenses and letters of authority required to conduct business, they can obtain or help validate official rights to land, and they can intervene with the army and the police, "who both officially and unofficially exercise a good deal of local control" (Kahn 1999: 94). In upland areas of Indonesia where land is more equally distributed—and consequently it is difficult to identify a discrete class of landlords—those with this capacity occupy a privileged position within the village elite (Kahn 1999). In many cases it is these actors who broker state services, control seasonal employment, or perhaps connect the village with other produce and labor markets (Peluso 1992c: 213–14). Their contacts, confidence, knowledge, and access to capital give them the capacity to act as patron, broker, or intermediary between local villagers and entrepreneurs, outside customers, and the bureaucracy.[19]

As local agents of the state act in this context, a significant gap emerges between the practices of local-level bureaucracies and the abstract idea of a translocal state institution (Gupta 1995). It is also hardly surprising that state agencies face significant problems implementing policy. While local officials face "domestic" pressures and temptations, state policy prescriptions often reflect the perceptions and interests of policy elites in a distant capital. They often are inappropriate, lack local legitimacy, or even contradict indigenous social values and long-standing local understandings. Therefore, in remote areas—such as South and Southeast Aceh—power may be highly localized while the official institutional frameworks of state policy may be weakly established (cf. Scott 1977; Oi 1989: 9).

Here a meticulous implementation of existing or new policies would involve changing the local rules of the game. Because often doing so would endanger entrenched local interests who receive disproportionate benefits from existing arrangements, a local official who attempts to implement official policy may face vehement opposition from powerful local leaders and entrepreneurs. In many cases those charged with implementing policy have limited support from their superiors, and face retaliation from powerful local actors who object to the application of a particular policy or law. Officials proceeding against the tide could face significant risks to their career or other unpleasant consequences. Meanwhile, actors with an interest in seeing state policy implemented and laws enforced may be too weak politically to apply pressure to this end. At the same time, as superiors within the state hierarchy may lack the capacity to supervise local officials effectively, there are also few outside controls on local officials. Field-level officials have considerable incentives to act contrary to policy or legal prescriptions (Migdal 1988, 1994). Consequently, depending on the context and the exigencies of the moment, field-level officials face conflicting demands, constraints, and enticements connected with their different roles. These include the role of loyal agent of the state, the role of local village patron who responds to local expectations and reciprocities, and the unrequited needs of their own households (Peluso 1992c: 223). The behavior of officials as they attempt to balance the "self-contradictory perceptions" associated with these different roles generates the inconsistency for which state agencies are known in the local domain.

Yet, even if official laws cannot be implemented, state law continues to affect social realities. Law maintains "modes of conceptualizing and categorizing the world," structuring the discourse within which issues are framed. These modes are implicated in processes of power (Merry 1995). If the state law is to retain a degree of authority, it has to maintain a gloss of legitimacy: to do so, it should appear just and free of gross manipulation. Legal forms can impose inhibitions upon state actors interested in maintaining the legitimacy of their rule. They can also provide space for opposition: social movements tend to mobilize legal language and categories to provide a language and locale for resistance (Merry 1995). In a similar fashion, as discussed later in this study, official law provides resources that village and *adat* leaders can use to defend the village order from outside incursions.

But state law is a double-edged sword: those with the power to do so often abuse it. In so many cases, official law helps authorize flawed state policy agendas or the accumulative strategies of private interests in ways that systematically work against villagers. As the key actor in the allocation of state resources or with discretionary authority over how state regulations are implemented, state-based actors can control resources that other actors depend upon to pursue their interests. Officials can mobilize state rules, or threaten to do so, in their business dealings within the local social order. In this way officials can occupy a privileged position within local systems of exchange. For while an official may not enforce the law, "it is his legal ability to do so that gives him something to exchange" (Moore 1973: 728).

Clientelist Networks of Exchange and Accommodation

In addition to the state and evolving "customary" orders, institutional arrangements also emerge through the interplay of actors. Anthropologists have long observed that reciprocities can create the regularities that maintain conforming behavior (Malinowski 1922). In Indonesia, as in rural Thailand, the logic of the gift remains important to institutional arrangements:

A strong cultural norm concerning reciprocity and more specifically gratitude and moral indebtedness, underpins almost all social relationships, and patron-client relations in particular. Giving creates an obligation to reciprocate on behalf of the recipient . . . No act of giving . . . is performed without expectation of future return in some form or other, and the morality of reciprocal obligation is present in all relationships—with peers as with subordinates and superiors. (Arghiros 2001)

To defy this norm is to put oneself outside the sphere of normal social interaction: it is these "regular reciprocities and exchanges" between "mutually dependent parties" that generate the norms and values of particular "social fields" (Moore 1973: 728). Each such social field can be distinguished by its "processual characteristic"—its ability to generate its "own custom and rules and the means of coercing or inducing compliance" (Moore 1973: 720–21).

The implication then is that, in addition to emanating from government administrative and judicial decisions, from custom mediated by the deliberations of customary authorities, or from the conscious act of individuals crafting working rules in collective choice arenas, institutional arrangements also emerge in the course of transactions and interactions between actors. Reciprocity and shared but tacit understanding can regulate open social networks, creating the *de facto* "rules of the game."[20] While such binding obligations may not be legally enforceable, they can structure the processes of competition, negotiation, and exchange within a particular setting.

In addition to "rogue officials" (referred to in Indonesian as *oknum*) engaged in extra-legal or quasi-legal transactions, there are other important parties in the district networks of exchange and accommodation. Local leaders wishing to maintain their position need to extend control over sources of patronage and influence—including the means of accessing resources—that can offer them the means of accumulation and political advancement as well as strategies of survival for those dependent upon them. In addition, regional politicians have an interest in using their discretionary power over budgetary allocations, contracts at the center and other assets at their disposal for their political and economic purposes. To do so, they enter into exchanges with bureaucratic peers and colleagues as well as with local leaders on whom they depend to mobilize resources and control groups of clients (Migdal 1988). Entrepreneurs wishing to remain engaged in business also face a strong pressure to conform to a system of exchange that involves extra-legal gifts and favors. They use these gifts, inducements, and coercions "to induce or ease the allocation of scarce resources" (Moore 1973: 722). While none of these obligations are legally enforceable, they are required of entrepreneurs to stay in the game.

These reach down to involve village elites, and as I described earlier, leads to "the development of chains or matrices of power relations that extend into peasant villages linking 'patrons' and 'clients' into reciprocal relations involving the flow of money, services, favours and support." These matrices extend from within the state, to the bottom of its hierarchy, and then to the clients who lie outside the state (Kahn 1999: 94).[21]

Some writers have criticized this way of seeing societies as being dominated by clientelist systems comprised of a multiple of interlocking, asymmetrical patron-client relations: they argue that this analytical framework tends to obscure horizontal differentiation, underplays class differences and social conflict, and fails to differentiate effectively the different types of social and political relations (Kemp 1984). Given the increased importance of the market, we would expect capital to become the most important source of power, increasing the salience of short-term relationships that are mediated by cash rather than by highly personalistic relations. Applying the concept of clientelism more selectively, it has been suggested that clientelist relations exist in the middle of a spectrum between, on the one hand, relations of naked power and, on the other, highly personalistic kinship relations (Kemp 1984).[22]

However, given the comparative lack of social differentiation in upland areas of Indonesia, it can also be argued that clientelist systems remain significant in these areas (Kahn 1999).[23] Here clientelist patterns emerge because there is a need for "the informal system of politics that circumvents the contradictions inadvertently created by the formal political and economic systems" (Oi 1989: 229), for "the formal channels for meaningful political participation and interest articulation are weak" and individuals continue to need informal networks built upon personal clientelist ties to secure their interests (Oi 1989: 9).

Studies of corruption avoid the conceptual problems associated with the patron-client concept while offering an explanation of the type of networks of exchange and accommodation discussed in this study.[24] Corrupt transactions begin when the instigator of a transaction gains access to someone with the resources that he or she requires. As a relationship develops over time, the parties may carry out their transactions through an intermediary and thereby hide their relationship. However, due to the secret and sanctionable nature of corrupt transactions, parties to such transactions need to trust one another, and this typically involves face-to-face interaction (Warburton 2001). As the two parties develop a history of shared secrets (*tau sama tau*), they also build a mutual interest in protecting their transactions and sharing the benefits of continued resource access. Thus, the personalistic aspect of the relationship will tend to expand, with the two parties socializing—for instance, inviting each other to their children's weddings. Once such a relationship is established, the parties involved become "locked into a complex social interaction in which trust must be nurtured and maintained" (Warburton 2001: 226). Consequently, such relationships typically combine a highly personalistic relationship with a strong element of economic exchange.

Such simple dyadic relationships typically occur within wider social networks of exchange and accommodation. To begin with, a person wishing to gain access to a resource may make payments, make a promise, or do a favor to an official with power over a resource—whether it be permits, licenses, information, or law enforcement activities.[25] In many cases, officials—acting as *oknum*—also initiate transactions acting either directly or through business partners, and in this fashion generating rent for self-enrichment and/or for political and bureaucratic objectives. If an actor begins with access to an official who has a high degree of discretionary authority within a regime, corruption can flow downwards through a hierarchy as lower-level officials become involved and earn their shares. Alternatively, if an actor requires access to officials with greater discretionary authority than the initial point of contact, corruption can flow upwards through the hierarchy of the state and party.[26] In such cases, higher officials may even expect a percentage of all the funds generated by the extra-legal activities of their subordinates within a regime. If authority is scattered among a variety of state agencies, an actor seeking to gain access to particular resources needs to work across bureaucratic agencies. As a multiplicity of actors become involved in gaining access to those actors within different agencies who have discretionary powers over the allocation of state resources, permits, and the application of law enforcement, the network of exchange and accommodation can become expansive. Given the constraints in the formal system that work against local understandings and interests, there are also considerable incentives for state-based actors, local politicians, leaders, and entrepreneurs to enter into exchange relationships of this sort. Accordingly, a network can involve exchanges and accommodations between a wide range of organizational and social actors. As network members act to protect and extend their interests by entering into exchanges and accommodations with new actors and by using the power and resources in their control, networks can develop a "processual characteristic": the capacity to generate its own rules and to coerce or induce compliance. This occurs as those involved attempt to ensure that actors within the network gain promotion while non-members are isolated and ejected. In this fashion a network can extend towards "places in the organisation where discretionary power resides and can be utilised to the network's advantage" (Warburton 2001: 232). When the accumulated power of a network reaches a critical point, it can be said that a corrupt network has "captured" a particular organization or setting.

Bringing together the usually separate conceptual approaches suggested by the anthropology of law, political science, property rights, and institutional analysis discussed so far suggests a particular understanding of institutional arrangements. There are parallel fields of control associated with the state in its various local manifestations, with village-based customary (*adat*) orders, and with the institutional orders that emerge through the interactions and reciprocities of actors working within and across a range of sites. At times these orders appear to work in parallel as discrete sets of institutional arrangements. Yet, as actors seek their objectives in rapidly changing political and economic circumstances, frequently these orders stand in complex, overlapping, and dynamic relations to each other. In such situations, the emergent institutional arrangements governing a particular situation are multiple, fluid, ambiguous, and contested, subject to manipulation, negotiation, and redefinition. In such situations, rather than existing as simple sets of rules and enforcement mechanisms, institutional arrangements are an on-going process.[27]

This perspective grants that the effective institutional arrangements are constituted through processes of negotiation, exchange, conflict, and compromise involving a multiplicity of institutional orders. As actors conduct their affairs within these processes, they are continuously engaged in securing their objectives, experimenting with strategies, making claims supported by particular institutional orders, and contesting or renegotiating rules associated with others. Understood in this way, institutional arrangements are shaped by power and involve continuous,

multiple, and changing processes of negotiation, exchange, accommodation, and dispute that tend to be open-ended and resistant to closure.
Such processes may be episodic, or they may be unsynchronized and lack
consistency, occurring simultaneously or in close succession. Yet, as they
occur over long periods of time and develop diachronically, they tend to
accrue into institutionalized patterns. These institutionalized patterns are
dynamic, shifting over time as actors adjust to changing social, political,
or ecological circumstances (Berry 1997; Leach et al. 1999).

Approach

This approach provides the means to analyze the institutional patterns
associated with resource outcomes in complex contexts, facilitating the
understanding of:

> How interactions—conflicts, accommodations, and exchanges—between
> different categories of actors in changing circumstances affect institutional
> arrangements over time.

> How different categories of actors in specific sites and at different social and
> economic scales have different capacities to affect specific levels of decision-
> making and rule-enforcement activities, thereby attempting to shape the
> institutional matrix to their ends.

> How political and economic changes affect the capacity of different actors to
> influence institutional arrangements and environmental outcomes.

Through field investigations this study seeks to understand how institutional arrangements affect resource outcomes. It considers the following
questions: How do local institutional arrangements emerge and evolve
under various conditions? How do particular institutional patterns lead
to certain environmental outcomes? These inquiries also raise a number
of subsidiary questions: Are villagers the primary villains of the piece?
Or, are other actors (such as the state or entrenched local networks of
exchange and accommodation) primarily responsible for deforestation
and over-exploitation? Can a revitalized customary (*adat*) order serve
as the basis for the more sustainable management of natural resources?
Despite the extensive resources involved, why do outside project interventions so often fail? What role do the state and its legal system play
in the management of natural resources? How can *adat* institutions be
nested in the wider state regime? How can we best study or conceive of
village institutions with respect to natural resource management? What
has been the effect of the economic and political crisis that occurred following the fall of Suharto? Does decentralization provide possibilities for

putting natural resource management on a more sustainable basis? While it is not possible to directly answer all of these questions, this study is relevant to them all.

Earlier I identified at least seven configurations of state and local regimes associated with various environmental outcomes. As this study could focus on only a small number of study sites within a range of complex and diverse situations, I decided to work in areas that had been set aside by the official spatial plan for forest protection, and eliminate areas zoned for logging concessions and conversion to plantations.[28] From the original list of institutional configurations identified earlier, I thus restricted the investigation to institutional arrangements related to

> retention of forest within a protected area;
>
> management of forest-like gardens by villagers, coupled with the gradual extension of agricultural activities into surrounding forest areas;
>
> retention of natural forest under village regimes;
>
> unrestrained logging of forests by logging networks.

To maximize the relevance of the research, I decided to focus on an area of high value for biodiversity conservation that was subject to intensive outside interventions. After various investigations, I decided to carry out the research in Aceh (northern Sumatra), an area that subsequent to this research became the site for a separatist conflict.[29] Here, as I describe in the Appendix, the European Union and the Indonesian government had chosen to implement an Integrated Conservation and Development Project (ICDP). Over the same period, the World Wide Fund for Nature (WWF-FN) had chosen to carry out a community-based natural resource management (CBNRM) project.

Structure of This Study

From the outset this research involved keeping in view a number of ethnographically conceived "sites": the actual village sites selected as the direct subjects of the study; the state associated spheres of political action at district, provincial, and national scales; and the ICDP and CBNRM projects and the interpenetrating transnational and state policy frameworks that both supported and worked against them.[30] To make the study of the last-mentioned more compelling, I decided to carry out the main body of research in more than one village complex within the extensive South and Southeast Aceh area. Doing so allowed me to make comparisons across cases, facilitating the recognition of the unique characteristics of each situation and the connections, parallels, and contrasts across cases.

During this preliminary stage, three particular characteristics of the area affected the process of selecting field sites. First, it was important to study as much institutional diversity as possible. The Leuser area is characterized by a high degree of variability in its people and their institutional arrangements, even across neighboring sites. For instance, there are five ethnic groups within the Badar, Sama Dua, and North Kluet sub-districts (*kecamatan*) included in this study. Within these sub-districts, moving several villages in any direction often involves interacting with a second ethnic group, with its own language and *adat* system. Neighboring communities also differed in the length of time since the first settlement of the area; the age and type of farming system; the steepness of the terrain; the history of integration into the wider cash economy; the existence of accessible and valuable timber; the degree of activity of NGO, ICDP, or the state forestry agency; the extent of illegal logging in the area; and the degree of maturity of *adat* institutions. Sites were chosen to capture this diversity.

Second, the study needed to face the difficulty of either pinning down the influences affecting institutional arrangements or indeed at times grappling with the imprecise nature of the arrangements themselves.

Given the complexity of the situation, at the outset of this study I could not specify the precise institutional arrangements and influences under investigation. As such, it was important to select case studies that included as many different influences as possible. For example, if possible the study should include an area free of logging influences as well as a site dominated by logging.

Third, the question of access remained decisive. As writings on ethnographic methods suggest, the amount and quality of information gathered depends very much on the rapport established with informants (Ely 1991). Thus, I selected sites where it was possible to build relationships with people, especially key informants. It was only after investing long periods of time building up contacts and investigating a site that its characteristics became apparent. Also, in the course of the research, it became obvious that the picture drawn from preliminary fieldwork differed considerably from that provided by prolonged fieldwork.

As the characteristics of the research area made it difficult to select sites according to a predetermined pattern of variables, the research warranted a qualitative or ethnographic approach. Qualitative research tends to be exploratory and open ended. Such an approach does not involve developing an *a priori* research design but rather moves from a general area of focus and develops its categories inductively in the course of field research (Strauss and Corbin 1990: 23).[31]

Initially I choose to work in two areas (Sama Dua and Badar) where

I had established contacts within local villages. These two sites met the main criteria mentioned earlier: they were adjacent to "forest," were not currently zoned for logging concessions or agribusiness development, and were situated within the Leuser Ecosystem, an area of high priority for biodiversity conservation. As I describe in the Appendix, I subsequently included a third case study (Menggamat).

Proceeding from the less complicated to the most complex, these sites represented a range of situations.

1. The Sama Dua Case (Chapter 2):

At Sama Dua the native tropical forest lay at least three hours' walk inland from the coastal villages. Compared to the other areas, it was virtually inaccessible to both logging networks and conservationist interests, which both focused their attentions elsewhere. Here local farmers opening plots on the forest frontier constituted "transformative users": that is, they principally depended on agricultural practices that involved opening land and transforming it into forest-like gardens.[32] At the same time, villagers also extracted resources such as timber, other forest products, and animal prey for their own use or for petty trade. In this sense, the villagers were also non-extractive users: their existence depended on the hydrological functions, heavy rainfall, and rich soils provided by the surrounding forest ecosystem. In this site, the study focused on these complex socio-ecological relationships, considering the agricultural activities affecting resource outcomes and how the latter were mediated by village institutional arrangements and the unstable economic situation.

2. The Menggamat Case (Chapter 3):

In this case, logging networks could gain access to the forest via the road and the Menggamat River; therefore, in addition to local villagers, loggers with outside backing were also significant actors. These "secondary appropriators" were actors with an instrumental interest in appropriation of resources in terms of "exchange value": they held "no intrinsic interest in the sustainability of the resource system over the long term" (Selsky and Creahan 1996: 354–55). To the extent that conservation agencies—initially WWF-LP and later LMU—had activities in Menggamat, it was also possible to trace the effect of these "non-extractive users" interests in the area. They included those who derived benefit from the resources system but do not enter into or use the natural area in any direct way. In addition to conservationist NGOs, they included ecologists, researchers, tourists, and a national and international policy elite who have an interest in retention of tropical rainforest habitat (Edwards and Steins 1998: 357).

3. The Alas Case (Chapter 4):

This was the most complex case. Here resource outcomes were subject to the activities of the complete set of actors: villagers, loggers, and non-extractive actors were present and attempted to ensure that the institutional arrangements reflected their interests. In addition, as the area is at the heart of Gunung Leuser National Park, *jagawana* (forest police) were also active, operating in both official and extra-legal capacities.

As this discussion suggests, from the beginning this research project necessarily entailed using historical, political, and social analysis to understand an ecological problem. In other words, the problem would be examined from the perspective of political ecology. This research agenda "combines the concerns of ecology and a broadly defined political economy," encompassing "the constantly shifting dialectic between society and land-based resources and also within classes and groups within society itself" (Blaikie and Brookfield 1987: 17).[33]

Earlier approaches to political ecology assumed the unproblematic existence of a "material/ecological base and a series of actors, differentially empowered but with clear interests, contesting the claims of others to resources in a particular context" (Brosius 1999). Post-structuralist approaches to political ecology, however, have focused on how the identities and interests of various agents are constituted, suggesting that both are "contingent and problematic" (Brosius 1999).[34] As this study will show, the interests of villagers are constantly negotiated in light of a range of factors that include severe fluctuations in the price of the commodities that they sell and upon which they depend, the advent of timber entrepreneurs backed by powerful local interests, and the presence or absence of forestry department and conservationist actors. Villagers and other actors are caught between short-term needs (feeding their families) and long-term considerations of ecological integrity and economic security. As we will see, this situation affected how actors conceived of their identities and interests, the meaning of *adat*, the use of state law, and the resultant patterns of resource use.[35]

To begin, the next chapter will discuss the nature of customary *adat* arrangements in Sama Dua, and consider how *adat* has changed over time as villagers have adjusted to shifting ecological and economic conditions.

2

Local Institutions in Sama Dua

Introduction

After leaving Tapaktuan, the capital of South Aceh district, a few kilometers north the winding road skirts Gunung Kambil, a mountain that
marks the border between Sama Dua and Tapaktuan sub-districts (see
maps 4 and 5). Here a coffee shop (*warung*) perched strategically over
a cliff offers a spectacular view of swaying palm trees and a turquoise
sea dancing in the tropical light. The road then descends and for a few
kilometers crosses the flat coastal plain of the Sama Dua River, a narrow band of intensely green wet rice fields. Most of Sama Dua's twenty-
seven villages lie here on a small coastal plain facing the sea to the west.
Arid, denuded foothills surround this plain on three sides and in turn are
overshadowed by steeper slopes covered by leafy nutmeg gardens (*kebun
pala*). Further to the north, the valley narrows and the steep hills come
down almost to the beach. Here, at a place known as Air Dingin, day
trippers from Tapaktuan come to bathe beneath a waterfall cascading
out of the nutmeg forests beyond. Shortly beyond Air Dingin, the road
climbs over the buttress of Gunung Lhok Pawoh, a mountain that marks
the northern end of the sub-district (*kecamatan*) of Sama Dua.

In the context of this study, Sama Dua is remarkable in a number of
ways. First, there are better historical records for this area than for the
other two more remote field sites. For some two centuries local farmers
here have produced cash crops for export. Foreign sailing ships visited

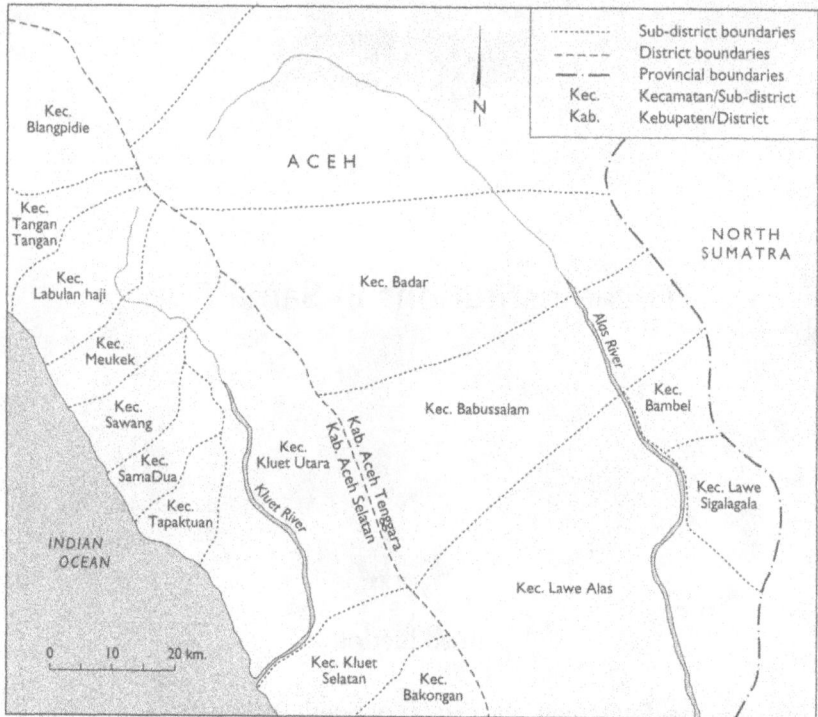

MAP 4. South and Southeast Aceh: sub-district boundaries (Kecamatan).

what was once known as the "pepper coast" in the late eighteenth and early nineteenth century to buy pepper. In addition, historically Tapaktuan's port exported substantial amounts of forest products, and foreign sailors had significant contact with this coast over some two centuries. Moreover, after the Dutch conquest at the turn of the twentieth century, Sama Dua lay near the seat of colonial administration at Tapaktuan. This meant that the colonial records regarding local institutions and agriculture and forest use are comparatively complete.[1] These historical records deepen the analysis of local institutions and help clarify the direction of institutional change by setting the contemporary situation in a historical context. These records disclose the outlines of present-day *adat* institutional arrangements. They suggest that the *adat* institutional arrangements operating in Sama Dua have subtly shifted with significant political and economic changes. Even so, there are many characteristics, such as *adat* assumptions regarding property rights, that have been particularly persistent. This suggests that, though *adat* in Sama Dua is not the same today as it was during the colonial period, nevertheless there remains something that has recognizably been inherited from the past.[2]

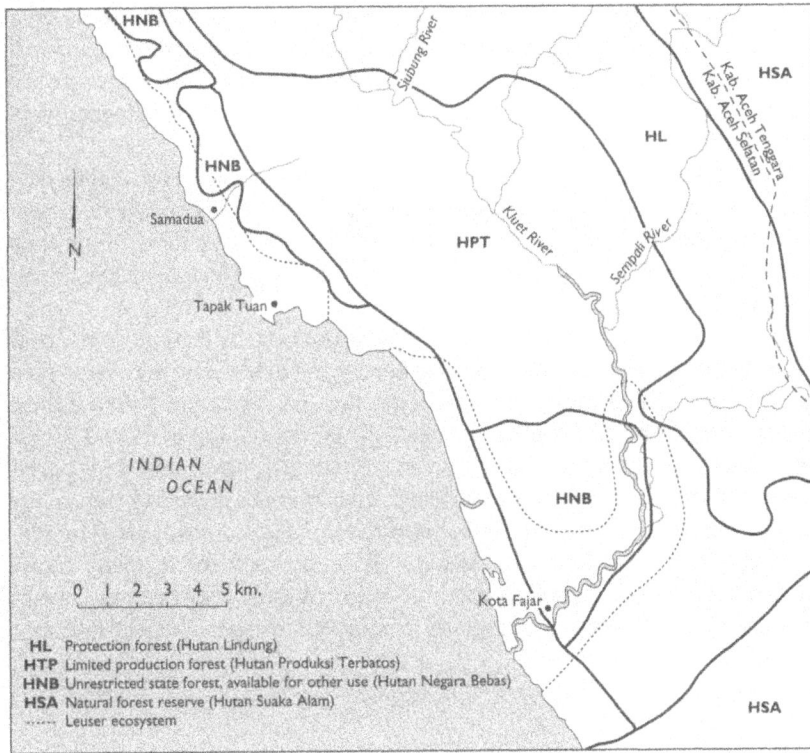

Map labels: HNB, HNB, HNB, HNB, HSA, HSA, HL, HL, HPT, HSA
Samadua, Tapak Tuan, Kota Fajar
Selburag River, Kluet River, Sempali River
Kab. Aceh Tenggara / Kab. Aceh Selatan
INDIAN OCEAN
N
0 1 2 3 4 5 km.

HL   Protection forest (Hutan Lindung)
HTP Limited production forest (Hutan Produksi Terbatos)
HNB Unrestricted state forest, available for other use (Hutan Negara Bebas)
HSA Natural forest reserve (Hutan Suaka Alam)
...... Leuser ecosystem

MAP 5. South Aceh: forest boundaries. (*RePPProt.*)

A second feature of Sama Dua also makes the area apposite to this study. Due to the ease of access and the suitability of the area for tree crops, the hills immediately behind Sama Dua had long ago been converted into gardens. In the absence of roads other than the north-south coastal road, the natural forest lies approximately two to three hours' walk to the east of the coastal villages. Due to the difficulty of gaining access to natural forest, loggers and the clientelist networks in which they were embedded have bypassed Sama Dua. Moreover, the area immediately behind Sama Dua was neither national park nor subject to an operational logging concession. Therefore researchers, ecologists and conservationist NGOs, and the state forestry agency have focused their attentions elsewhere. Consequently, like many other areas in Indonesia, the fate of the forest has largely lain in the collective hands of villagers living in the immediate vicinity. That local institutions have been less complicated by outside actors in Sama Dua makes the study of local institutions and the fundamental direction of institutional change much easier. Accordingly, in the context of this research project the Sama Dua case acted somewhat like a control; the understanding of local institutions—including property rights—provided by Sama Dua served to inform the other case studies, where local institutional arrangements were more difficult to unravel.

This chapter investigates the nature of local institutions in Sama Dua and how they have evolved over time. This discussion focuses on how *adat* institutions have interacted with wider state institutions, how they have responded to economic fluctuations, and what has been the corresponding impact on forest use.

In the course of this discussion it is important to bear in mind that Sama Dua's villagers have primarily been agriculturalists: they have pursued their interests within a social order that has principally served their needs as farmers. Local *adat* regimes have governed access to and use of land and forestry resources within what villagers have long considered the territory of Sama Dua. Historically, *adat* regimes governed the collection of forest products and timber extraction by local residents. The major local farming institution, the *seuneubok*, protects the property rights of farmers in their hillside gardens, also providing for the transformative use of the forest—clearing and conversion of native forest into tree crops.

Sama Dua

Older residents of Sama Dua recount various legends about the origins of the community and its name. One version holds that the first person who opened land here was a person from Padang who followed the Minangkabau tradition of roaming in search of opportunity (*merantau*).[3] After opening land here, he invited others to follow, forming the first settlement at Suak. The first person to open the land became the headman (*datuk*), a title that has been inherited by his descendants.[4] Upon their arrival, the first two settlers saw two seagulls flying toward the shore. Comparing the two seagulls with their own situation, they said, "It is just like us"; thereafter, villagers have referred to the area as "Sama Dua," which means "two together."[5] Another legend about the origin of the name reflects the relation between Minangkabau (or Padang) people from West Sumatra and the wider Acehnese population. This legend holds that a woman from East Aceh married a man from Padang; this couple were the first settlers of what became known as Sama Dua. In this version, "Sama Dua" refers to the coming together of the two ethnic groups.[6] As one village elder explained, "Sama Dua refers to the fact that *Adat Padang* and *Adat Aceh* were both here and both were used."[7]

For the most part, the inhabitants of Sama Dua are of Aneuk Jamee ethnicity. In 1997, there were approximately 14,000 Aneuk Jamee living in the sub-districts of Tapaktuan, Sama Dua, and Manggeng (Hidayah 1997).[8] The Aneuk Jamee are descendants of Minangkabau (or Padang) settlers from West Sumatra who migrated to South Aceh during the nineteenth century.[9] For some time after settling the area, the Minangkabau people—now known as *Aneuk Jamee* (Acehnese for *anak tamu*, or "children of guests")—retained the distinctive matriarchal social organization of their homeland. The headmen bore the Minang title of *datuk* rather than the Acehnese title of *teuku*. However, the Aneuk Jamee lived closely with the Acehnese and learned to accommodate *adat* Aceh; sometimes the headmen combined both titles, becoming known as *teuku datuk* (Maksum 1983: 13).[10] By the early twentieth century, according to colonial sources, the Aneuk Jamee had subsumed Minangkabau *adat* under Acehnese forms of social organization that more closely reflected Islamic precepts (Adatrechtbundels 1938b). These included Acehnese tenurial and inheritance traditions. While today Sama Dua residents report that they follow *adat* Aceh, they maintain that their *adat* still has a "Minang aroma." This is symbolized in Sama Dua marriage ceremonies; the man wears a hat known as a *peci* in the Aceh style, while the woman wears the elaborate Minang-style headdress.[11] This adaptive yet persistent sense of ethnic identity continues in contemporary times.[12] Up to the present day the Aneuk Jamee around Tapaktuan and Sama Dua continue to speak a dialect of the Padang language.

Agro-ecological Context

In Sama Dua a very limited amount of land was suited to wet rice cultivation, because Sama Dua has only 306 hectares of irrigated land. The population increased from approximately 2,650 in 1901 to 13,111 in 1995 (BKI 1903; Kantor Statistik Kabupaten Aceh Selatan 1995). This represented a five-fold increase in ninety-four years.[13] Accordingly, by the 1960s, villagers had long brought all the land suited to wet rice agriculture (*sawah*) under cultivation.[14] With the increasing population, each passing generation's holdings of *sawah* became increasingly fragmented. Today, many families have little or no *sawah*. Those wishing to work as farmers have had to open dry land plots in the hills. However, where possible, villagers supplement farming incomes by working as fishers, drivers, or petty traders. Due to the high cash value of products derived from dry land agriculture in the hills, the predominant land use in Sama Dua has been cultivation of perennial tree crops.

Forest History

Besides working *sawah* and *ladang* (swidden plots), in the past the people of Sama Dua exploited *damar* (resin) and *getah* (native forest rubber) from the forest. Indeed, a colonial report noted that the best *getah* collectors came from Sama Dua (BKI 1912: 413). Older villagers remember that during the colonial period the people of Sama Dua used to range freely across the forest collecting non-timber forest products. There were footpaths through areas of forest, and villagers would pack rice and make week-long expeditions into the forest to gather *damar*, forest rubber, gutta percha, and rattan. Colonial sources also tell us that the sale of forest products was an important source of local income, and that forest products such as *damar*, rattan, forest rubber, and gutta percha were major exports from the Tapaktuan port, alongside cultivated cash crops such as nutmeg and cloves (Paulus and Stibbe 1921: 272).[15]

At the end of the nineteenth century and in the first years of the twentieth there was a boom in demand for gutta percha, a latex obtained after felling large adult trees (*Palaquium* and *Payena* genera) found in the Dipterocarp forests across Southeast Asia. According to colonial records, gutta percha was widely collected around Tapaktuan at that time. In addition, villagers also harvested rubber (*getah*) from a naturally occurring fig species (*Ficus* spp), including that known as *rambung* (*Ficus elastica*). When opening a plot of land, forest farmers used to leave *rambung* trees for tapping and also to mark a farmer's property rights over a plot of land or to distinguish the border between adjacent plots (WWF n.d.). In addition, they also planted extensive areas of *rambung*, developing plantations for rubber production. However, as the price of natural rubbers declined sharply in the early decades of the twentieth century, farmers in this district replaced the *rambung* plantations with the widely cultivated *para* rubber (Potter 1997: 300; WWF n.d.).

After independence, the role of non-timber forest products experienced a further decline. According to local informants, villagers became reluctant to make expeditions into the forest; they preferred to stay in the *kampung* (village). Previously rattan had been used for a wide range of everyday uses such as for packing food. However, with the advent of plastic, the demand for rattan fell dramatically. People still collect it from forest paths, on fishing expeditions, or from uncleared areas around their gardens (*kebun*), but nowadays it is mainly used for making baskets. At the same time, with the invention of synthetic resins, the price of *damar* also fell: according to village informants, because *damar* was only fetching Rp 250 per kilogram in 1997, it was not worth collecting.[16] A *seuneubok* head explained that economic reasons led to the decline of non-timber forest products (NTFP)[17]: "The important thing was not that the forest was far, people always had to travel a long way, but that it wasn't worth it anymore."[18] During 1996–99, collection of *damar* and rattan was no longer a major activity.

From Pepper to Nutmeg

The history of agriculture in this area can roughly be divided into three periods, each relating to a specific land use. The first period was that of the pepper boom, the period of settlement. The sloping lands of South Aceh were well suited for the cultivation of pepper. At the same time, the nearby bay of Tapaktuan offered an excellent anchorage; in the 1870s, Veth reported that Tapaktuan was the most important port in this area for the export of pepper (Veth 1873). During the nineteenth century the coastal districts of Southern Aceh (known as "the pepper coast") played a primary role in the world pepper trade. Pepper cultivation declined after World War I. In 1920, pepper exports virtually ceased, due to a sudden price fall that "played havoc with the plans of pepper farmers"; subsequently, farmers in South Aceh turned to clove and nutmeg production (Kreemer 1922: 477; 1923: 19).

During the second period, stretching up to 1965, the farmers of Sama Dua integrated the cultivation of cloves (*Szygium aromaticum*) with shifting agriculture. They opened new dry land plots (*ladang*) in the hills just behind Sama Dua. The first year they would open the forest, cutting and burning the vegetation. They would plant dry land rice (*padi ladang*), crops such as onions, garlic, sweet potatoes, chili, and bananas and other fruits and food crops, primarily for their own use. The next year, they then planted clove trees, and this enrichment planting ensured that the initial swidden area would be succeeded by clove gardens.[19] Because clove trees took five years to mature, to meet their everyday needs farmers would continue to plant *padi ladang* and seasonal crops, or even cash crops such as the plant producing patchouli oil (*Pogostemon cablin*, known locally as *nilam*); coffee (*Coffea robusta*) might be grown in adjacent areas, extending their plots or even opening new *ladang* somewhere else.[20]

People would plant something, and when the soil became tired they would open *ladang* somewhere else and plant rice there. Some planted cloves—or whatever had a price, they would plant. When the price fell they would abandon it and plant something else in another place. Those with skill and experience would plant say ten things . . . if the price of one product failed, they had something else.[21]

When the clove trees began to bear the nail-shaped flower buds known as cloves, there was less need to open new land. In this way, farmers avoided dependence on a single crop. Nevertheless, the thick forest was thinned out and was largely replaced by clove gardens in the hills immediately behind Sama Dua.

At this time, swidden agriculture was integrated with cash crop production from permanent clove gardens. Clove production from permanent gardens and the production of rice and seasonal crops from shifting plots formed mutually supporting elements of an agricultural practice.[22] According to Conklin's (1975) typology of swidden agriculture, this was a "partial supplementary" swidden system, in that the agriculturist "devotes only part of his agricultural efforts to the cultivation of the swidden." Conklin contrasts such "partial systems" with more environmentally benign "established integral systems," which were "largely self-contained and ritually sanctioned ways of life" and involved the clearance of very little or no climax forest each year (1975: 3).[23]

Tigers and other wild animals were still abundant then, and villagers would usually go into the hills only in parties of three or more. As the opening of *ladang* was very labor intensive, farmers would exchange labor, coordinating their activities with others and opening land in groups. Forest farmers tended to build huts on their hillside plots, where they mostly stayed with their families. Because wild boar, deer, goats, and other animals coming out of the forest would eat the rice and other crops, they had to guard their crops. According to an elder informant, people also liked to live in the hills partly because it was one way to avoid the experience of Dutch colonial rule and the forced labor the colonial regime imposed: "People preferred to live in the mountains . . . My mother was born in the mountains, people referred to her as *nenek ladang rimba* ("forest farm grandmother") . . . We felt safe in the hills, and there was plenty of food . . . The land was very fertile, and things grew well."[24] Another older villager remembered how many members of his extended family lived in their forest gardens (*kebun*): "They planted many crops just for eating rather than for sale. We would visit them, and they would give us all these products to carry down and sell."[25]

In the early 1960s, the hills around Sama Dua were covered in clove trees, and Sama Dua was the center of clove production in South Aceh. However, around 1963 a clove pest struck the clove gardens, and most of the trees died. In response, the farmers abandoned the clove gardens on the foothills behind the coast; initiating a third agricultural regime, they began cultivating nutmeg, which grew well in the fertile and cooler forest land on steeper hills beyond the first foothills. The abandoned foothills became unused *Imperata* (*alang-alang*) grasslands. These grasslands tend to catch fire spontaneously in the dry season, and to this day the foothills of the Bukit Barisan Mountains behind Sama Dua remain a testimonial to the era of clove cultivation.

Agricultural scientists have listed the Sama Dua district as one of the seven major nutmeg-producing centers of Indonesia (Lubis 1992). Nutmeg agriculture fit into a complex agroforest succession that began with opening a *ladang* and ended with the existence of a productive garden. Because seedlings require protection from the sun for the first years and need to be well spaced, when they opened a new plot farmers would leave other, larger trees in their gardens, including towering forest durian and *rambung* trees.[26] The first year, farmers would open a plot and plant dry land rice (*padi ladang*). The second year, depending on the farmer's preference and the market prices, the land would be enriched with other crops such as *nilam*, coffee, *pinang*, and tropical fruits including rambutan, *salak*, and papaya as well as sugar-producing palm (*gula merah*). According to an older farmer, "In the past people plant *padi*, bananas (*pisang*), cassavas, or sweet potatoes—there was no general pattern: nutmeg was generally planted six by six meters, coffee two by two meters, as well as durian and other things."[27] Farmers usually planted the nutmeg the second year or even a year later. As the nutmeg begins to fruit only after seven to ten years, farmers would then open a second plot further out. This strategy enabled them to harvest other crops while waiting for the nutmeg trees to bear fruit. After a nutmeg tree matured, fruit production would then continue to rise until the tree reached peak production at around twenty-five years. Trees would then bear fruit twice a year until they reached sixty to seventy years of age. After this time, fruit production begins to decline. However, unlike clove production, nutmeg cultivation did not form an element of a "partial supplementary" swidden system. Today, the practice of opening temporary, shifting plots to meet everyday food needs has virtually disappeared. In the 1970s, nutmeg cultivation was integrated into a more complex system of agroforestry.[28] Subsequently, *ladang* became a transitory stage in the preparation of nutmeg gardens, which yielded perennial crops that provided cash income.[29]

For at least three reasons, since the 1970s villagers have shifted away from tree crop cultivation. First, there were wild fluctuations in the price of nutmeg from year to year, and even over the course of a single year. In 1997, for instance nutmeg farmers began the year obtaining only 800 Rp per *bambu*; by June prices had risen to Rp 3,200 per *bambu*, but by October they had fallen to 2,300 per *bambu*.[30] With such large fluctuations in nutmeg prices common, farmers have needed to look for other sources of cash income.

After the price of nutmeg fell in Sama Dua during the 1970s, people began to pay less attention to their *kebun*. When nutmeg gardens were weeded and well tended, the trees did not have to compete with weeds or colonizing tree species, and nutmeg production increased. However, with prices low, farmers were reluctant to invest labor maintaining a productive nutmeg garden for negligible returns. Many villagers abandoned their nutmeg gardens and let them run to weed. When nutmeg prices increased, however, they would return to their gardens, weed the area, harvest the fruit, or even open new plots. Consequently, farmers have been unable to manage their nutmeg gardens efficiently.

This dynamic was exacerbated by a second factor: the increased importance of the cash economy. When the road between Sama Dua and Tapaktuan was paved in 1984, Sama Dua was better connected to outside markets and job opportunities. As the effects of Indonesia's long economic boom began to reach South Aceh, many people found paid work in other sectors—in construction, as drivers, as day laborers, as fishers, or in the timber industry. Once young farmers began moving to other areas seeking cash work, they could buy motorbikes and televisions. Rather than wait long months for an uncertain nutmeg harvest, young farmers preferred to see the direct results from their work in the form of cash, and over time work in the forest gardens came to have lower status than wage labor.

The paving of the road between Sama Dua and Tapaktuan, along with the improvement of road transport to Medan and Banda Aceh, also allowed farmers to find distant markets for their durians. Accordingly, the cultivation of durians became increasingly significant. Indeed, durian became one of the top eight income earners for farmers in South Aceh (Jordan and Anwar 1997). Villagers have long tended groves of these highly valued trees in gardens located in the valleys and on the lower slopes of the hills just behind Sama Dua. They preferred to plant durian there because they could more easily carry the heavy, thorny fruits back to the village from nearby gardens. The smaller nutmeg fruits could readily be carried back from more distant forest gardens up to four hours' walk from the village. Durians were exported to as far as Banda Aceh, and Sama Dua was proud of its reputation for delicious durians.[31] Durian yields fluctuated significantly from season to season, however a single tree could yield forty to fifty durians over a few nights. In 1996 durians sold for 1,000 to 1,500 Rp per fruit (US $0.40 to 0.65), and a farmer could earn 50,000 Rp or more for a few nights work. Some farmers had groves of durian trees, and in the durian season farmers formed teams and would take turns sleeping by their durian plots. They would wait for durians to fall, collect them, and then bear them down to market.

Third, the nature of nutmeg cultivation supported the changing patterns. As noted earlier, the opening of new nutmeg gardens is a long-term investment that entails significant opportunity costs. To reduce the time taken away from other activities, farmers usually sought the help of other farmers to clear a new *kebun* quickly. As wage labor largely replaced labor exchanges between farmers, they paid for this labor. Besides the opportunity cost invested in this work, farmers had to buy and prepare the seedlings. During the seven years before nutmeg trees bore fruit, they could partly offset this unproductive time by intercropping nutmeg with short-term cash crops such as chili pepper and *nilam*. Nonetheless, they generally needed capital to establish a viable nutmeg garden, capital that could be obtained through cash jobs. However, once established, nutmeg gardens were low maintenance, compared with other crops; they only required intermittent weeding. Therefore villagers employed in other sectors could visit their nutmeg gardens on holidays or on a Sunday to weed the garden or harvest the nutmeg fruit. These considerations meant that, for many farmers, agroforestry moved from being the main activity to being a side activity. As a nutmeg wholesaler explained in 1997, "a farmer cultivating 50 trees can harvest 300 bambu every 3 months. This amounts to enough to live on for one month. So the farmer will need to find other income for two months."[32] In this way some farmers shifted away from supplementing the cultivation of clove cultivation with swidden agriculture to supplementing nutmeg cultivation with income from other cash jobs.

Informants also noted that farmers opening far-flung *kebun* had to waste long hours trekking through the hills. If they wished to sleep in the village, they needed to make the journey twice a day. For those with far away *kebun*, this was well nigh impossible. Farmers with remote *kebun* left the village on a Saturday and did not return to the village until Thursday. They also faced the difficult labor of carrying seedlings, agricultural supplies, and food up to the *kebun* and hauling big sacks of nutmeg back from the mountains. The time and energy expended on these long journeys were large disincentives to opening new gardens. "This life is very difficult for them," an informant reported, "especially during the time when the nutmeg garden has not yet come into production. Many of these people are poor, but what other choice is there?"[33]

Ecological Change and Forest Outcomes

Over the last decades, the story of agricultural expansion in Sama Dua became one of the progressive conversion of natural forests in the hills behind the coast into nutmeg gardens—a story of the marching agrofor-

est frontier. There are no maps of the forest paths behind Sama Dua;
nor are there official measurements of this conversion of the forest or
accurate records regarding ecological change in the Leuser Ecosystem, let
alone in Sama Dua. However, there are some indicators of the degree of
ecological change and the rate of forest conversion.

On many occasions between 1996 and 1999, I discussed nutmeg cultivation with farmers, village heads, and *adat* heads. In the course of
this research I also interviewed eleven *seuneubok* heads. These interviews
revealed the pattern of forest conversion. As forest farmers opened new
kebun at the frontier end of the *seuneubok* territory, over time this frontier had become increasingly distant.[34] When they remembered the year
that they opened a particular *kebun*, and the time it took to reach it,
the pace of the march of nutmeg gardens into the forested hills became
clearer. Farmers revealed that, when deciding to open a new plot, generally they would choose the closest areas of land available for opening
while attempting to avoid steep terrain. Consequently, if the frontier of a
particular *seuneubok* could only be reached after a steep climb, farmers
would attempt to avoid the effort involved by finding more accessible
land in an adjacent *seuneubok*. This meant that some farmers have plots
in more than one *seuneubok*.

Contrary to this trend, the head of one particular *seuneubok* had continued to open successive plots in his *seuneubok*. By tracing the year he
opened a particular plot and the time taken to reach it, it was possible
to gain a rough measure of the pace that this particular *seuneubok* had
expanded. He described how he walked for a half hour to reach his first
kebun, a plot he carved out of the forest in 1956. In 1966, he opened a
second *kebun*, a site reached after one and a half hours walking up the
seuneubok path. That year he planted *padi ladang*, and then in 1967
he planted the plot with nutmeg trees. The following year, 1968, he decided to extend his *kebun*. However, in the meantime, many farmers had
opened plots behind him and the unopened forest was now considerably
farther out: it now took two and a half hours to walk to the forest frontier where he opened new *ladang*. However, by 1998 the forest frontier
had moved even further out, and farmers wishing to open new *kebun*
now needed to walk around three and a half hours.[35] Thus the time taken
to reach the forest frontier in this *seuneubok* had tripled between 1956
and 1966. Then, in just two years, the time taken to reach the frontier
increased by the same measure again. However, after this time agricultural expansion slowed: it took thirty years (1968–98) to increase by
this amount once more. These figures suggested that the nutmeg gardens
expanded rapidly in the 1950s. In the 1960s, when the clove gardens
were abandoned, growth was greatest. Then the expansion slowed until

the East Asian economic crisis of 1998, when large numbers of farmers again opened new plots. Consequently, the patterns of expansion related closely to the patterns of economic and ecological crisis.

Unlike this *kepala seuneubok*, other farmers have neither continued to open forest gardens over such a sustained period of time nor opened them in the same *seuneubok*. Yet, they confirmed the trends established by his experience: before 1965, farmers opened *ladang* close to the village, converting these areas into clove gardens. A major increase in forest clearance occurred in the few years after 1965, when farmers abandoned the clove gardens to open nutmeg gardens farther out. Then, especially since the 1970s, villagers found other sources of income outside nutmeg cultivation. As nutmeg cash crop production (to be discussed later) became increasingly unreliable, the rate of forest conversion subsequently slowed until the recent crisis.

A survey carried out by Leuser Development Programme (LDP) fieldworkers in 1998 offers another measure of ecological change. A survey of older informants indicated that, on average, the water levels in the streams of Sama Dua have decreased by some 18 percent from 1988 to 1998. These informants also estimated that water levels in the Sama Dua River had decreased by 20 percent in this same period. In January 1999, an older villager in a *warung* next to the Sama Dua River offered an explanation of these changes. He recalled that in 1966 and again in 1990 there had been two large floods. Generally he noted that the river had widened significantly. In his opinion, this happened because the Sama Dua watershed no longer held water as it had in the past. As farmers cleared the forest and replanted it with nutmeg, large forest trees disappeared.[36] The nutmeg trees were much smaller than the forest giants, with less foliage and smaller root systems. Thus, water ran off the hills more quickly, and the hydrological functions of the forests had changed.

A significant change began to affect nutmeg agroforestry during the 1980s, suggesting the shift to a fourth period of agroforestry. In the mid 1980s, a caterpillar pest known as "trunk driller" (*penggerek batang*) and identified as *Batocera hector*, began to attack nutmeg trees, boring into the trunks of trees and causing them slowly to die. Farmers and agricultural department staff interviewed during the course of this research were perplexed by this plague and offered a number of theories to explain the infestation. Previously, two species of *Copsycus* birds had preyed on the *Batocera* pest and controlled its numbers.[37] Local villagers said that in previous decades bird hunters from Medan had visited the South Aceh area, capturing large numbers of the beautiful *Copsycus* birds for sale in bird markets. In addition, villagers hunting with air guns further reduced the population of these birds. Another theory held that these birds lived in the *rambung* (*ficus*) trees that local farmers had previously retained in their forest gardens.

As the *ficus* trees no longer had any economic function, nutmeg farmers had increased the intensity of nutmeg cultivation by cutting them down, incidentally decreasing the habitat for the birds (Barber 1997: 23).

To complicate things further, in the late 1990s a second, smaller insect pest appeared. Known locally as *bubuh cabang*, this predator attacks nutmeg trees from the end of the branches.[38] Officials from the agricultural department believe that *bubuh cabang* was an insect resident in the forest. This infestation indicates the environmental consequences of forest management practices of South Aceh, most notably the unrestrained logging of other areas (see Chapter 3). According to an official in the district agricultural office, heavy logging had disturbed the insect from its natural habitat in the Kluet area, and some displaced insects found a new niche in the nutmeg *kebun* below the forest. As the adult insect flew during its reproductive cycle, depositing larvae over a large area, the pest spread readily, quickly moving up the hillsides of South Aceh.[39] The combination of the two pests proved extremely deadly; nutmeg trees could die within three days, and a whole *kebun* that had taken years to reach maturity could be destroyed in a matter of months. With every hectare of nutmeg producing at least one ton of undried nutmeg annually, as of October 1997 it was estimated that the farmers of South Aceh were losing 2,202 tons per year due to the nutmeg attack (*Kompas* 1997b). The following year the situation continued to deteriorate: according to an agriculture official, in 1994 nutmeg covered an estimated 11,300 hectares, but by 1998 this was reduced to 8,906 hectares.[40] As Indonesia slipped into a deep economic crisis, nutmeg fetched higher prices and farmers became increasingly dependent on income from this crop. Then, both the nutmeg agroforestry system and the outside economic system collapsed together. The attack of the pest could not have occurred at a worse time.

Crisis

In 1997, the East Asian economic crisis struck Indonesia. Across Indonesia, the crisis increased commodity prices (in rupiah terms) and made their production more attractive, leading to a temporary increase in forest clearing (Sunderlin et al. 2000).[41] In South Aceh, as prices increased dramatically, opportunities to earn cash in the towns diminished. With a falling rupiah, prices of export crops increased dramatically, and export-oriented farmers could enjoy a short-term gain.[42] In the past, *nilam*, a crop introduced by the Dutch and long associated with shifting agriculture, was a significant cash crop in South Aceh's hilly soils.[43] Just as the value of the rupiah sank, the U.S. dollar value of the patchouli oil produced from *nilam* skyrocketed. These twin influences led to a drastic increase in the local price of patchouli oil from around 35,000 Rp/kg in 1995, first to 150,000 Rp/kg in early 1997 and then to around 1,080,000 Rp/kg at the beginning of 1998. If a one-hectare crop of *nilam* yielded 64 kilograms of oil, as *nilam* farmers suggested, a single hectare would produce a profit of around Rp 20 million, enough to build a house or buy land.[44] Villagers cultivating *nilam* cultivation could reap windfall profits. With spiraling prices and the collapse of many cash jobs due to the economic crisis, the contagion of "*nilam* fever" (*demam nilam*) spread across South Aceh. Shopkeepers, public servants, and even forestry officials began to cultivate unused areas of land, even behind government offices.

In Sama Dua villagers began seeking plots of land to plant cash crops and farmers enthusiastically began opening dry *ladang* plots in the distant forest.[45] *Nilam* grew particularly well in the cool and fertile environment offered by newly opened ground in the distant hills. Here the land was better and the chance of the *nilam* being cross-infected by agricultural pests associated with other cultivated crops was low. As the income earned from this cash crop far outstripped what could be earned from other crops, farmers concentrated on *nilam*.[46] Groups of five to thirty people proceeded into new forest to open their own plots.[47] The new *ladang* lay up to five hours from the coastal villages, and many farmers built huts to live alongside their hillside plots, only returning to the village once a month. However, the *nilam* boom was based on a commodity price fluctuation. The first *nilam* crops could only be harvested seven months after planting. By mid-1998, *nilam* prices began to fall, just as many of the crops planted during the previous wet season began to be harvested.[48] As so many farmers had opened plots, *nilam* production increased just as demand slackened. According to farmers, in 1998 values *nilam* was worth growing only if prices were higher than 500,000 Rp/kg; by January 1999, *nilam* had fallen to 160,000 Rp/kg. Villagers unable to secure their livelihoods in other ways even begun to log areas of remnant forest behind Sama Dua.

Before the economic crisis, given the steepness of the terrain, the labor of taking timber down to the road made logging Sama Dua's hillsides an unattractive activity. Except for periods in the agricultural cycle when villagers could find little, the labor involved compared poorly with the price of timber: it was hardly worth the effort. In 1996–98, I interviewed furniture manufacturers and others with significant commercial needs who related that they obtained timber from other more accessible areas where wide-scale "illegal logging" was occurring.[49] Yet, by January 1999, it was clear that logging was now occurring on a much wider scale.

With the collapse of *nilam* prices and the devastation of nutmeg gardens, many villagers needed to find other sources of income quickly. One *ketua seuneubok* explained that if people did not earn income from logging the forest, they would have had "to eat rocks" (*makan batu*). With many villagers in dire circumstances and with so few other job opportunities, and with higher rupiah values for timber (due to the currency crisis), by early 1998 timber now "had a price." For those impoverished by the crisis and with few alternatives, it was now worth the serious effort of carting timber down from the hills. Villagers participated in logging of Sama Dua's forests.[50] Previously, trade in wood had occurred on a small scale within the village. For instance, if a villager wanted to build a house he might buy wood from someone else in the *village*. Now it was sold to wood traders from outside the area, who in turn sold it on to Medan. Thus, a large amount of the profit fell to outside businessmen.[51]

The logging was taking place in diverse ways. For instance, in some cases a person with capital supplied costs for a logging team involving a chainsaw operator. Chainsaw operators negotiated with farmers with trees standing on their land, or took timber from abandoned land or from the forest frontier. After a tree was felled, a chainsaw operator cut it into planks. Villagers in need of cash left the village before 7 A.M., walking up and then carrying planks down on their shoulders into the village. After returning to the village around 11 A.M., they then would make a second trip into the hills.[52] In other cases, farmers wishing to open new plots supported the logging; in the past, however, when a farmer opened a new plot, often a lot of wood was just left to rot. But now they were obtaining up to four cubic meters of wood from one tree.

Alternatively, farmers gave loggers permission to take trees remaining on their land. Under such arrangements, farmers would sell the wood to a trader and divide the profits, perhaps paying a group of villagers to help carry the wood down to the village. In yet other cases, loggers were just taking the wood.[53]

This account of changing agricultural regimes in Sama Dua demonstrates how the agroforestry system in Sama Dua has evolved over time. With ecological dynamics and widely fluctuating markets affecting the abundance and value of the agricultural products and forest products on which local farmers depend, they had to rapidly change their livelihood strategies to adapt. At the same time, these local livelihood strategies affect the local ecology. This in turn suggests that the rate at which villagers convert forest to agroforestry, extract timber, and collect non-timber forest products not only depends on the ebbs and flows of the local ecology and widely fluctuating markets, it in turn also feeds back into and helps create an intertwined pattern of ecological and economic change.

Adat and Village Institutions

Sama Dua and Its Territory

"Territoriality" is best understood as "the attempt by an individual or group to affect, influence, or control people, phenomena, and relationships by delimiting and asserting control over a geographical area" (Sack 1986: 19). Territoriality is based on a claim over a territory that is maintained vis-à-vis other groups that tends to be "justified by some ideological, moral, legal or political reason" (Dijk 1996: 19).[54] As the following discussion will indicate, the people of Sama Dua have long-established concepts of territoriality.

Following the colonial conquest at the turn of the century, Dutch colonial administrators surveyed the indigenous political order in preparation for setting up a system of indirect rule. Their reports reveal that the Sama Dua River divided the residents of Sama Dua.[55] According to these reports, those to the south of the river were predominantly descendants of settlers from the Sao area, while those on the north bank were descendants of settlers from the Pariaman area of West Sumatra. Each group was originally organized as a clan (*suku*) and each had its own headman or *datuk*.[56] Formerly there had been two *datuk*, one for each side of the river. The Dutch report reveals that in Tapaktuan the *datuk* appointed assistants, known as *panglima* (commander) or *keucik* (Acehnese for village head), to act as their spokesmen in the villages, and it was likely that the *datuk* of Sama Dua also played the decisive role in Sama Dua's government.[57] While officially the two *datuk* ruled the area together, the Colonial report notes that actually there was little cooperation between the two and each ruled his own area. Eventually, in the course of a dispute within the villages, the southern settlement established a second *datuk* (BKI 1912).

As I discussed earlier, historically the people of Sama Dua ranged widely across the mountains behind the coast gathering forest products, fishing in the streams and rivers, and hunting prey in the forest. Within this area, at times farmers also opened *padi ladang* and planted cash crops on suitable land. Older members of the Sama Dua relate that men of ascetic disposition used to withdraw into the forest to practice religious austerities. Over time, all these practices came to mark the features of the forest behind Sama Dua in the local sense of place. Natural geographic boundaries, such as river watersheds and mountain ridges, and the presence of neighboring communities—such as the Menggamat people along the Kluet River on the other side of the mountains (see next chapter)—all helped consolidate local notions of territoriality.

In setting up a colonial state, the Dutch colonial government preferred to govern through indigenous elites, particularly in areas not closely tied to Dutch colonial interests. As a part of this system, as in other parts of the world, the colonial administration constructed a formal structure of indirect rule, which was considered to be customary or traditional, although it involved constructing new entities or restructuring existing indigenous forms of organization.[58] Colonial scholarly and administrative practices involved the identification and creation of "jural communities" or "*adat* law communities." Indigenous institutions were adapted and co-opted to maintain social control.[59] As the colonial government developed a structure of "traditional government" for indirect rule above the village level, it is possible to see customary law (*adatrecht*) as having been created and used as a tool of the colonial legal-administrative system, and accordingly as having been implicated in colonial domination. However, especially in more remote areas (such as Aceh), village communities continued to make and enforce their own rules in accordance with long-standing customary practices also referred to as *adat*. While the colonial presence undoubtedly affected these localities, detailed information tends to be lacking. Nonetheless, it appears that at this level, *adat* continued to develop and adapt to change without help from courts or legislators (Holleman 1981). In short, the colonial state presided over a system of overt institutional pluralism.

As a part of this process, the colonial state embarked on a process of "territoralization," setting out "to control people and their actions by drawing boundaries around a geographic space . . . and proscribing or prescribing specific activities within these boundaries" (Vandergeest and Peluso 1995: 159). The first step in this process involved establishing territorial administrations in the newly conquered areas. Doing so entailed consolidating the population into definite groups under a centralized, hierarchical leadership through whom colonial rule could be exercised.

Dutch colonial reports from the turn of the nineteenth century reveal how, after conquering the South Aceh area in 1901, this process took place in Sama Dua. As noted earlier, originally Sama Dua were organized as clans (*suku*), each with its own headman (*datuk*). While formally the *datuk*s ruled the area together, the colonial report notes that actually there was little cooperation between them and each ruled his own domain (BKI 1912).[60] One report noted how the process of identifying the "self-governing head" (*Zelfbestuurder*) occurred: "In 1903 these *datuk* gave control over the whole territory or district (*Landschap*) to a certain Teukoe Paneu who as next of kin to one of the *datuk* stood above the heads in Sama Dua. The intention was that he would bear responsibility to the European government" (Kreemer 1923). In the colonial system, Teukoe Paneu became the "territorial head" (*landschapshoofd*); older residents of Sama Dua refer to the position as that of *raja*. This account suggests that local leaders did not remain passive in this process of centralizing local authority but rather actively negotiated the process.[61] From this time on, Sama Dua constituted what the Dutch considered to be an "*adat* jural community."

As a clearly defined *adat* community needed to be associated with a circumscribed territory within the colonial schema, this territorialization process involved the mapping of the administrative territory (*Landschap*) under each "territorial head." By negotiating *landschap* boundaries with local groups and then fixing them on maps, the colonial government seems to have further fixed notions of local group identity.[62] Today these form the administrative boundaries with the neighboring sub-district (*kecamatan*) within the South Aceh region.[63]

A village elder (*pakar adat*) offered some insights into how this territorialization process consolidated local territoriality. He explained that Sama Dua's territory extends to a place known as Tanah Hitam (see Map 6):

Tanah Hitam is the boundary of Sama Dua. It is a half day walk from here. Our ancestors had already opened land here, and the Dutch put the boundary here. A half day further from Tanah Hitam lies Sarah Baru [a hamlet of neighboring Menggamat in the sub-district directly to the east].[64]

The head of a *seuneubok* confirmed this, explaining that, in a cave here, Tengku Dagang, a Padang holy man took up a contemplative life (*bertapa*) and became an *Aulia* (protective spirit of the area).

This cave is a most pure place, and those wishing to pass—to fish in *Pucuk Kluet* which is the best fishing spot—must ask permission [from the *Aulia*] and burn incense (*kemenyan*).[65] If we don't, it can be dangerous.[66]

The head of another *seuneubok* remembered how, during the colonial period, groups of up to thirty villagers collecting forest products in the forest would sleep in this cave, despite the fact that a tiger was also known to stay there.[67]

State territorialization extended to the demarcation of forest areas which were mapped and placed under state control; this process facilitated the control of natural resources within the newly constructed

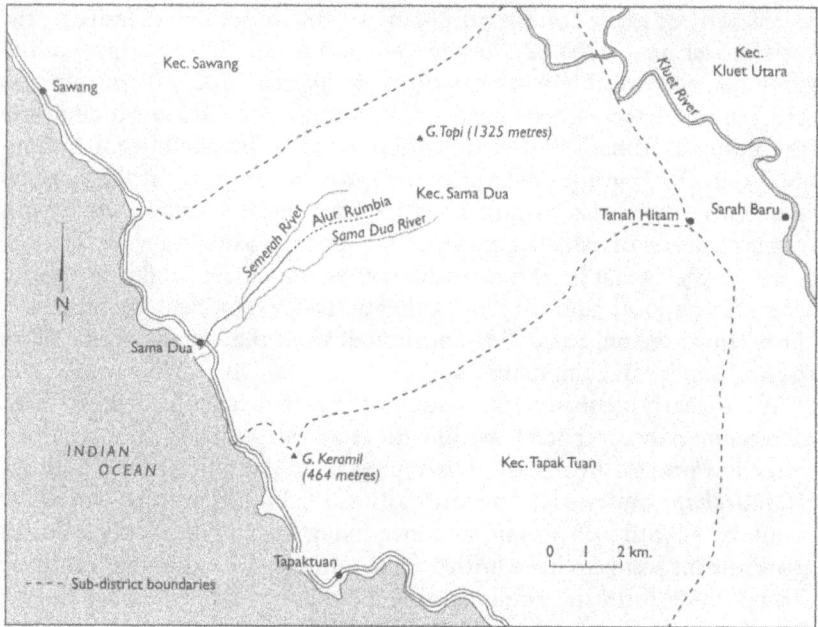

MAP 6. Sama Dua Sub-district (Kecamatan): sub-district boundaries.

boundaries (Vandergeest and Peluso 1995: 388). In accordance with the "domain principle," the colonial authorities allocated leases over such areas for developing rubber plantations or for logging operations or set areas aside as forest reserves. However, Sama Dua's geography was unsuited to plantation agriculture. In contrast to the Alas valley (see Chapter 6), the colonial territorialization did not proceed any further here. When the Leuser Reserve was created, it also did not extend to Sama Dua: its western boundary lies some fifteen kilometers east of the Kluet River (see "Protection Forest" (HL) boundaries in Map 5) beyond the eastern administrative boundary set at Tanah Hitam. In respect to colonial mapping, local informants mention that the Dutch put small cairns on the tops of mountains to mark the peaks. However, villagers did not recall the existence of colonial forest boundaries. In contemporary Sama Dua, when questioned on the issue, villagers would say that there were no forestry boundary markers in the Sama Dua territory.[68]

Postindependence

After independence, the administrative boundaries of Sama Dua were extended to include several Acehnese villages just to the north of the Sama Dua valley. Moreover, the structure of local government also changed: the model of village administration found in northern Aceh was now to be followed throughout the Special Region of Aceh (*Daerah Istimewa Aceh*) including the areas inhabited by the Aneuk Jamee.

The traditional structure of village government in Aceh is two tiered. The first tier consists of the village or *gampong*. From historical times, in what Mattugengkeng calls the tradition of *gampong* government, religion, *adat*, and government were not separated. This meant that the *gampong* head (*keucik*) and the religious leader (*tengku meunasah*) were considered a dyad, the "parents of the *gampong*." However, these village leaders did not make decisions on their own: decision making was "guided and advised by intelligent (*cerdik-pandai*) elders of experience and wisdom" (Mattugengkeng 1987). These advisers formed a coordinating body or council of elders known as the *petuhapet* for the management of village affairs.[69] Mattugengkeng recalls the traditional metaphor used to describe the status of the *petuhapet* in relation to the *keucik* and *teungku meunasah*: it stood above the *keucik* and *teungku menuasah* in the role of parents (Mattugengkeng 1987). In addition to the *keucik* and *teungku menusasah*, there were also other *adat* officials, such as the *kejuren blang*, who is responsible for regulating the supply of irrigation water and protecting the forest along the river course.[70] The second tier of village government was the village league or *mukim*. Historically in Aceh the *mukim* consisted of the villages and hamlets that shared a mosque and that over time had come to consider themselves to be a single community under the leadership of a figure known as the *imam mukim*.[71]

To bring Sama Dua's governance structure into accord with the Acehnese pattern, the positions of local heads were now renamed: the *datuk* were replaced by *kepala mukim*, who were now elected (Table 2.1). At the same time the *landschapshoofd/raja* was replaced by the *camat* (or sub-district head), a government appointed official.[72] Henceforth, the *kepala mukim* became the *adat* heads in Sama Dua under the *camat*. Village heads (*keucik*) were elected and village decision making was guided by a permanent council of elders (*cerdik pandai*). As elsewhere in Aceh, this council was known as the *petuhapet*—a coordinating body for the management of village affairs.[73] The *petuhapet* consisted of six to eight people recognized for their wisdom and knowledge of *adat*, including a *petuhapet* head (*ketua petuhapet*), the religious leader (*imam mesjid*),

TABLE 2.1
Administrative Sturctures in Sama Dua

Colonial period (1923)[i]	Existing (1999)
Keresidenan Aceh	Governor of Aceh Province
Asisten Residen stationed at Meulaboh administered Division of West coast of Aceh (*Afdeeling Westkust Van Aceh*).	
Dutch Controleur administered Tapaktuan Subdivision (*Onderafdeeling Tapaktuan*)	*Bupati,* District (*Kabupaten*) of South Aceh District administered from Tapaktuan
Raja or *Landschapshoofd* head of territorial administrative area	*Camat* administers sub-district (*Kecamatan*)
Datuk clan (suku) head	*Kepala mukim* (ceremonial adat role only), head of village league (*kemukiman*).
Keucik (*Panglima*) with decision making in consultation with council of village elders (*petuhapet*).	Village Head (*kepala desa*)
	After village government law, LKMD/LMD village councils (under leadership of village head) replaces *petuhapet*.

[i] I offer this rather schematic rendering of administrative structures as an aid to understanding. A full history of various administrative reorganizations in this area is beyond the scope of this study.

and other village elders. The *petuhapet* was independent of the village head (*keucik*): the *petuhapet* would make a decision and the *ketua petuhapet* would hand this over to the *keucik* for consideration. As the membership of the *petuhapet* could not be changed by the *keucik*, and as the balance of power rested with this council, the *keucik* had to work with the *petuhapet*.[74]

A village elder, who had served as village head (*keucik*) of a Sama Dua village during the 1980s, at a time when the structure of village government changed, described the previous pattern of village government. As he recalled, in the past many villagers had never left the village: communications were very poor and "every village was like its own country":

Before everything was tied up with tradition. The *petuhapet* might say we have to do it this way because it has always been this way . . . Old people of limited vision and education dominated the *petuhapet*. You would have to approach them slowly and skillfully suggest some innovation. But people would say this is how our ancestors did it and so why do you want to change it. We received it from our parents and it will be the same until we die. For example, Acehnese all wore black, and only certain rich or special people could wear other colours. You would be fined or face sanctions for departing from village norms.[75]

However, during his period of office, village government was reformed in accordance with the Village Government Law (Act No. 5/1979). The law aimed to make the structure of village government across Indonesia uniform. In essence, as Mattugengkeng has argued, the reform cut off the head of the previous structure of *adat* government (Mattugengkeng 1987). Under the new law the village head (*kepala desa*) was to be responsible to the sub-district head (*camat*). This meant that in Aceh the *kepala mukim* were to lose their official position in the structure of government and be reduced to symbolic *adat* leaders. The implementation of this law meant that the *petuhapet* was also replaced. The new law prescribed that each village would have two village councils—the Village Assembly (*lembaga musyawarah desa*, or LMD) and Village Community Resilience Council (*lembaga ketahanan masyarakat desa*, or LKMD). Both these bodies would now be under the leadership of the village head.[76] However, a former village head remembered how the situation changed after he became village head in 1981:

The LMD and LKMD existed in name but hardly functioned. Slowly after this they were brought into functioning. If someone [on the LKMD or LMD] made a mistake or no longer agreed with the village head, he could be changed. The village head (*kepala desa*) now had more power—he could move things—whereas before things were restricted by what had occurred in the past . . . Before, the village head had difficulty implementing anything, but by the end [of his period of office], the *kepala desa* could determine things. Whereas before people would say this is how it has always been done, by this time you could say these are the regulations, and we have to go according to them.[77]

As the state normative system penetrated in this way, the symbolic power associated with state authority began to affect the village order. The process continued the centralization of village structures that had begun in the colonial period.[78] Moreover, it shifted the balance of power toward the village head, who now became an executive who acted with considerable autonomy from *adat* elders and the new village councils.[79] Henceforth, it seemed, decisions were to be justified more in terms of law and state administrative instructions and less in terms of custom.

While the former village head thought that this greater executive power enabled him to facilitate development projects and make decisions more efficiently, other villagers were less sanguine about the changes. With the increased executive power of the *kepala desa*, who was now more directly responsible to the *camat*, discussion (*musyawarah*) among the village elders had a diminished role in deciding village affairs. As one informant noted, before, all village affairs and disputes had to be dealt with by the *petuhapet* and only then could they be finished. This clearly has an effect on dispute resolution. Under the new structure many issues were not subject to *musyawarah* and were left unresolved,

Disputes should be taken to the LKMD and the LMD and then the *kepala desa*, and only then to the police if they can't be finished—at least in principle. However, in fact most issues are taken directly to the police, even though this should only be the case for big problems—for example war between *kampung* . . . However, the police need money [to solve a dispute].[80] While now there is [material] progress, issues can no longer be solved at the *petuhapet* level: those who are happy are even happier while those in difficulty face more difficulties (*yang senang tambah senang, yang susah tambah susah*) because those with money can fix things [with the police].

Yet the changes had effects outside the administration of justice, affecting the capacity of villagers to participate in decision making. While the *petuhapet* was able to resolve many village affairs through *musyawarah* within a village context, now decisions were often made without consulting the LKMD or LMD. He related that "Before, a lot of things were from below, whereas now [they proceed] from the *camat* down. Often problems are not even discussed by the LKMD—no consultation—so the basis [of the legitimacy of decisions] in the village is lost."[81] As a consequence, he concluded, the village was no longer "straight" (*lempang*).[82]

The Seuneubok

However, in addition to the *adat*/village government structures described so far, the Sama Dua people maintain institutional arrangements relating to cultivation in the territory known as the *seuneubok*. Although nutmeg farmers now utilize the *seuneubok* institution, originally a *seuneubok* was a pepper agriculture complex that consisted of ten to twenty pepper gardens. According to the pattern Ismail has described in the nineteenth century on the east coast of Aceh (where the pepper frontier moved after the South Aceh pepper boom), migrants from the Aceh hinterlands wishing to grow pepper formed groups of ten to twenty farmers. A head known as a *petua seuneubok* led each group. The followers of this head were known as *aneuk seuneubok*. The *seuneubok* itself constituted a form of territorial control over an area. As Ismail explains, when the farmers had located the area of land suitable for cultivation, they would establish the boundaries of what would then be known as a *seuneubok* (Ismail 1991: 67–69). The boundaries of the *seuneubok* were fixed on only three sides. Unopened forest remained on the frontier side of it as an area where, with the permission of the *petua seuneubok*, *seuneubok* members or newcomers could open new pepper gardens as required. In this way, the *seuneubok* included the intention of an expanding agricultural frontier.

In territorial terms, in contemporary Sama Dua a *seuneubok* consists of a specific area—all the forest gardens lying along a certain forest path. Once a village path leaves behind the last rice fields and village gardens of the coastal plain, the path winds eastward through valleys and begins to ascend and descend the spur lines of successive ranges of hills. Moving in the direction of the mountains behind the coast, each path forks in several directions. A *seuneubok* begins with the first forest gardens to the left and right of a main forest path. All the *kebun* on all the paths accessed by the main path of a *seuneubok* belong to it. A stream or a hill usually marks the boundaries with adjoining *seuneubok*, while the rear of the *seuneubok* consists of unbounded forest.[83] If a *seuneubok* continues to expand, eventually it will become too large and the head will no longer be able to manage the expanding frontier. He will then appoint a representative (*ketua kemplok*) to look after this area, and if farmers continue to open plots on the frontier end of the path, the area will eventually become a new *seuneubok* in its own right.

Accordingly, in addition to the territorial dimension of the *seuneubok*, a *seuneubok* also has a social dimension: besides belonging to a village, all forest farmers with *kebun* located off a certain main forest path belonged to a *seuneubok*. While the original farmers opening *kebun* in a *seuneubok* might have come from the same village, farmers from several villages might have *kebun* in a particular *seuneubok*. Consequently, the membership of a specific *seuneubok* did not correspond with a particular village: a *seuneubok* was not the territory of a village but rather a social and territorial entity in its own right. Indeed, as a frontier institution, a *seuneubok* was a nascent village and, in the days when villagers lived for extended periods in their *kebun*, oftentimes the *seuneubok* would eventually split from the parent villages and become a village in its own right.[84] In South Aceh today we find some contemporary villages bearing the name of former *seuneubok*.

The *seuneubok* had its own norms and rule-making and enforcing functions. As each *seuneubok* crafted its own regulations, the regulations tended to vary somewhat between *seuneubok*.[85] Moreover, with the support of wider institutions, each *seuneubok* was able to induce compliance amongst its members.

As my interviews and protracted discussions with *kepala seuneubok* and forest farmers in 1996–99 indicate, among other things the *seuneubok* were concerned with the control of property in the nutmeg gardens within the *seuneubok* territory (see Table 2.2 for a list of *seuneubok* regulations set down at a meeting of one of Sama Dua's *seuneubok*).[86] While the nutmeg gardens are theoretically subject to the state's legal regime, here the state legal apparatus is remote. Land titles are not formalized with the government land office, and property disputes seldom involve state courts. The personal property rights that a forest farmer enjoys over a garden are embedded in the collective arrangements that constitute the *seuneubok*. Within this *adat* regime, a farmer is considered to have permanent rights over a piece of land covered by perennial tree crops that he or she had cultivated, bought, or inherited.

Each farmer's property rights depend on the existence of an authority structure to enforce them. By taking responsibility for many of the everyday functions of the *seuneubok*, the head of the *seuneubok* (*ketua seuneubok*) provides the first level of the authority structure for the *seuneubok*'s everyday operations. For instance, a farmer wishing to gain access to land at the frontier end of a *seuneubok*'s forest path needs to ask permission from the *ketua seuneubok*. The *ketua seuneubok* then accompanies the farmer into the forest and allocates a piece of unused land. By clearing the land and planting trees there, a farmer becomes a member of the *seuneubok*. In similar fashion, someone wishing to buy or sell a *kebun* in a *seuneubok* also informs the *ketua seuneubok*.

While the *seuneubok* could craft rules to suit its own scale of organization, it was also embedded or nested within the next scale of organization. For instance, the wider village institutions supported the authority of the *ketua seuneubok* and had a role in the resolution of disputes. Usually, if a *seuneubok* member had a problem with another member of the *seuneubok*, he took it to the *ketua seuneubok*.[87] The *ketua seuneubok* listened to the issue, and if necessary he organized an *adat* session (*sidang adat*), calling in the village head and other *adat* functionaries (*para cerdik pandai*) for formal deliberation and decision making (*musyawarah*). The most common type of dispute arose when a member was found stealing nutmeg fruit from another member's forest garden. Usually, at the deliberation (*musyawarah*) with village elders and the village head the case was heard and sanctions were imposed on the offender according to *adat* principles.

Seuneubok rules, then, were "quasi-voluntary": *seuneubok* members chose to comply in situations in which they were not being directly coerced. However, this was *quasi*-compliance, in that non-compliance was subject to sanctions if the offender was caught (Ostrom 1990: 94). How-

TABLE 2.2
Regulations of a seuneubok

Meeting of Entire Management of *seuneubok*, Kec Sama Dua.
30 September 1987
Residence of . . . , under the chair of village head.

With the decisions as follows:

1. Whoever imposes a fee (*memajakkan*) on someone else [in exchange for the use of] a nutmeg *kebun* must give notice to the *ketua seuneubok*. If notice is not given, the *ketua seuneubok* will take steps in accordance with the valid regulations.
2. Whoever violates these rules, the *ketua seuneubok* will impose a fine, which will take the form of improving the *seuneubok* path for no less than 10 metres.
3. If a party is interested in selling a nutmeg *kebun* in this *seuneubok* to another party, they must give notice to the *ketua seuneubok*. This is because in the past *kebun* nutmeg have been sold without the knowledge of the *ketua seuneubok*, and the *ketua seuneubok* has been forced to become involved in this matter. If the *ketua seuneubok* is not given notice then the parties involved will have to take responsibility for all the problems involved.
4. Concerning empty land in this *seuneubok* or in the respective *kebun*, if another party is to work the land and notice is not given to the *ketua seuneubok*, and then a dispute ensures, the *ketua seuneubok* will not get involved in the problem concerned.
5. A party who has a nutmeg *kebun* in the *seuneubok* cannot take an outsider to carry produce except their own family. This is because already many times loss/theft of nutmeg has occurred – with the result that the owner of the *kebun* accuses the wrong person when in fact his friend who accompanied him was involved.
6. If the fruit of respective *kebun*, for example durian or other fruit, are stolen in the *kebun* of others, then the *ketua seuneubok* will take steps. This means a fine will be imposed such as constructing the path over not less than 10 metres according to point 2 above. The exception is when there is already permission.
7. Concerning the theft of nutmeg, if theft of nutmeg occurs in the *seuneubok* then whoever discovers it must report it to the *ketua seuneubok* together with the evidence. The *ketua seuneubok* will report this to the responsible authority.
8. These decisions are made with the serious intention that they are known by the members of the *seuneubok* and will be used wherever they are necessary.

With the ascent of the Village Head (signature)
ketua seuneubok (signature)
[with names and signatures of seventy four *seuneubok* members, and nine village elders (*para cerdik pandai*)]

Source: Results of Seuneubok Meeting provided by Seuneubok head, November 1997.

Figure 1. Tapaktoean port, Atjeh, circa 1900. Photo courtesy of Koninklijk Instituut voor Taal-, Land- en Volenkunde.

Figure 2. Three heads of the Soesoh principality, an *Aneuk Jamee* area to the north of Tapaktoean. IIde reis Oost-Indië, 33, circa 1894. Photo courtesy of Koninklijk Instituut voor Taal-, Land- en Volenkunde.

Figure 3. Heavily eroded hillsides directly behind Sama Dua. Farmers cultivate nutmeg gardens in the valleys. John McCarthy.

Figure 4. View eastward from the crest of a hill behind Sama Dua. Farmers cultivate nutmeg gardens in the valleys and hillsides up to three hours' walk behind Sama Dua. Patches of forest can still be seen on the peaks and steepest slopes. John McCarthy.

Figure 5. Nutmeg factory, South Aceh. Nutmeg oil is extracted by cooking nutmeg stones. John McCarthy.

Figure 6. The author with a *seuneubok* head, Sama Dua, November 1996. John McCarthy.

Figure 7. Mace (left) and nutmeg (right), drying in the sun. John McCarthy.

Figure 8. Nutmeg wholesaler inspecting nutmeg stones drying outside his house, Sama Dua. John McCarthy.

Figure 9. Nutmeg tree after the "trunk borer" infestation. John McCarthy.

ever, there were no formal policing or monitoring systems apart from other *seuneubok* members noticing infringements (such as theft) and reporting them to the head or to those whose property rights have been violated. However, as in other institutions of this type, besides the risk of sanctions, a thief also risked losing his or her good name in the village, or perhaps the attack of a vengeful tiger (see below).

Understanding Adat Institutional Arrangements in Sama Dua

Many theoretical models of institutions have used a logical framework derived from rational choice theory that is based on the assumption of a utility-maximizing, self-interested individual (Kato 1996: 554). However, in addition to individuals engaging in a rational calculus about consequences and preferences, in Sama Dua there were "institutionalised conceptions of action" that were connected with what March and Olsen (1996) have called "the demands of identity." According to this understanding, within an institutional setting individuals take on identities and roles that "are expressions of what is exemplary, natural, or acceptable behaviour according to the (internalised) purposes, codes or rights and duties, practices, methods, and techniques of the constituent group and of the self," argue March and Olsen (1996: 251). They also argue that "within an institutional framework, 'choice,' if it can be called that, is more based on a logic of appropriateness than on the logic of consequence that underlies conceptions of rational action. Institutionalised rules, duties, rights, and roles define acts as appropriate (normal, natural, right, good) or inappropriate (uncharacteristic, unnatural, wrong, bad)" (March and Olsen 1996: 252).

In Sama Dua, notions of identity—of what it was to be a part of the Sama Dua community—were clearly important guides to action. Yet it was unlikely that there was a simple binary opposition between situations where "identity-driven conceptions of appropriateness" drove the action of individuals and conditions where "conscious calculations of costs and benefits" dominated. It is possible for these orientations to co-exist: those who were acting in keeping with identity associated with a conception of proper behavior might also balance up the costs of behaving otherwise. Those who indulged in behavior that violated basic norms—particularly those connected with what it meant to be a member of the Sama Dua community—also faced social shame.[88] For instance, in Sama Dua *seuneubok* there was a precept that farmers should not pick fruit from the durian tree; rather, they should wait until the durian ripens and falls of its own accord.[89] This principle helped to ensure the quality of Sama Dua durians. While there were other ways of testing the ripeness of durian, this rule guaranteed that Sama Dua durians enjoyed a high reputation. As buyers could be assured that the Sama Dua durians were good, Sama Dua durians earned higher prices. However, the rule also caused farmers some inconvenience: in order to harvest the durian, they had to stay in their *kebun* during the durian season to ensure that forest animals or passing farmers did not consume the succulent fruit. Interestingly enough, this rule was enforced not by sanctions, but rather by the weight of shame attached to breaking such a strongly held norm of village life. If one earned a reputation for selling unripe or inedible durian, or were discovered picking durian from a tree, one would lose one's reputation in the *kampung*. As an informant explained, "they would be considered evil, because if one person does it, all can be affected. Because of this there are no sanctions. Therefore, durian from Sama Dua are highly appreciated in Banda Aceh."[90]

Another example of such a phenomenon involved the social norms inhibiting irresponsible use of the land. If a farmer keeps shifting his plot of land, "he will feel ashamed (*malu*)."[91] Other farmers did not want to be affected by the misuse of resources by their neighbors, by land degradation and erosion caused by careless use of steep land. Although the *seuneubok* rules (see Table 2.2) that one *seuneubok* formalized in a letter of agreement failed to mention this, there were strong feelings about this type of practice. If a farmer shifted plots without the *ketua seuneubok*'s permission, the head would be angry with the person concerned: because he was offended by a farmer's failure to ask permission before opening a new plot, the *ketua seuneubok* would no longer assist the offending farmer in the resolution of disputes. In this way a farmer would lose access to the dispute resolution process and other social functions provided by the *seuneubok*—he would be on his own. But what was worse, he would lose perhaps his most valuable possession—his good name in the *kampung*. According to one farmer, "if it is like this, for what reason would you live in the *kampung*."[92]

As discussed further below, the list of written rules in Table 2.2 was produced to meet a particular set of challenges, demonstrating that some aspects of the *seuneubok* arrangements could be drafted like a set of regulations. Yet, many of the norms that govern the behavior of *seuneubok* members (such as implicit notions of appropriateness) were not really analogous to a legal code. As studies of indigenous normative systems in other parts of the world have found, the rules and practices tend to be negotiable and flexible. While such systems tend to be referred to as "customary law," they will be misunderstood if they are taken to be straightforward sets of traditional rules (Moore 1986; Chanock 1998).[93] Drawing on a range of studies as well as their own research in northern Sumatra, the Dutch legal anthropologists Slaats and Portier (1992) found many of the principles guiding behavior did not depend upon a clearly recognized institutional apparatus for their operation, such as constables and courts. At least in northern Sumatra, unwritten traditional law was characteristically not organized in rules or even rule-like formulations: "Even if rules can be found in these traditional systems, it is questionable whether these have the same function and significance as rules in written law systems. More often than not what seems to be a rule turns out to have the character of a principle or even only a general guide-line for behaviour" (Slaats and Portier 1992: 6). Thus, rather than attempting to reduce the *seuneubok* to a set of law-like formulations, the approach here has been to consider the *seuneubok*'s institutional arrangements (and other *adat* institutions in this study) as a pattern of social ordering. These patterns of social ordering are associated with both implicit, deeply held social norms and more explicit rules.

The *adat* order worked on the assumption that the visible world existed in parallel and intertwined with an invisible and mysterious causal order. Consequently, the *seuneubok* incorporated religious, legal, and social functions, attempting to both influence and rely upon the invisible causal order to support it. At times the *seuneubok* combined its everyday business with a supra-mundane function, and the *seuneubok* carried out both a political and a ritual role.[94] This was seen when, in his role as the *seuneubok* leader, once a year the *ketua seuneubok* called what was known as a *kenduri seuneubok*, a ritual gathering that involved a common meal. At this time farmers and their families gathered in the *seuneubok* itself to discuss the functioning of the *seuneubok* and any conspicuous problems. Besides reviewing any outstanding *seuneubok* business, the *kenduri* was also a ritual feast. Members prayed for the continued prosperity of the nutmeg gardens in the *seuneubok*, and prayers were addressed to the *Aulia*, a spiritual forest being (*orang rimba*).[95] At the end of the prayers, each farmer took water from the *kenduri* and rice, placing an offering of rice under each nutmeg tree and blessing the trees with water. This ritual was said to help protect trees against illness and ensure that they bear lots of fruit.[96,97]

To ensure that the *seuneubok* runs well, farmers generally choose someone of standing as the *ketua seuneubok*; as an *adat* leader, he traditionally enjoyed high status within the village. Although the *ketua seuneubok* did not receive a salary, the person occupying the position did receive an honorarium: members of the *seuneubok* covered *seuneubok* transaction costs by paying the *ketua seuneubok* a contribution at the time of the nutmeg harvest as well as at the time of the *kenduri*. The party that was held to have offended *seuneubok* regulations usually met transaction costs for meetings to discuss *seuneubok* problems. The *ketua seuneubok* often also received a payment or contribution in kind on wood cut or hunting done in the forest behind the *seuneubok*.

Tigers and the Seuneubok

As the customary (*adat*) institution concerned with farming in the hills, in the past the *seuneubok* mediated relations with tigers. Over the course of this research, interviews concerning other issues would frequently turn toward the subject of tigers. Informants revealed that, both physically and metaphorically, the tiger in many ways preoccupied villagers. Older villagers remembered how earlier in the twentieth century tigers were quite common, and villagers walking down the village path at night would sometimes meet tigers sitting by the side of the path. Tiger attacks were reasonably rare, but they did occur from time to time, and many forest farmers were understandably afraid of tigers.[98] One villager admitted that he left his *kebun* because he was afraid of meeting a tiger. He said that people used to often see tigers in their *kebun*: "People were often disturbed by tigers, especially when they were cutting the grass around the nutmeg tree. They might look up and see a tiger. Not the *seuneubok* tiger [explained below]—that would leave us alone, only marking the ground on the path to show if an outside tiger was in the *seuneubok*."⁹⁹ Since many forest farmers were understandably afraid, villagers would usually go into the hills only in parties of three or more. If such a person was available, a *seuneubok* chose a man respected for his forest skills, known as a *pawang*, to act as *ketua seuneubok*, someone able to help mediate relations with resident tigers. A *pawang* had special esoteric knowledge: he could contact the *Aulia*, a guardian spirit of the forest, who would appear to him in dreams. In addition, *pawang* could call tigers.¹⁰⁰

By tradition, each *seuneubok* had an agreement with one or more tigers that were known as *harimau seuneubok*, tigers that resided part of each year in the *seuneubok* territory. A *seuneubok* head described the role of the resident *seuneubok* tiger as follows: "He is on duty (*dinas*) there. If there is someone who steals from the *kampung* and takes it to the mountain (*gunung*), he will be disturbed by the tiger."¹⁰¹ Villagers explained that there was something like a tacit agreement between *seuneubok* members and the tiger: the *harimau seuneubok* would hunt pigs and other pests in the *seuneubok* while leaving humans alone. The tiger would also warn farmers of the presence of dangerous tigers from outside the area by leaving distinctive claw marks on the main path of the *seuneubok* territory. When villagers saw these marks, they understood that there were wild tigers in the territory, and they would not proceed to their forest gardens that day. In return for the *seuneubok* tiger's benevolence, villagers also provided for the tiger. For instance, even to this day, an unwritten *seuneubok* norm holds that during the durian harvest, farmers leave five durian fruit from each tree for the *harimau seuneubok*. Once a year, at the time of the *kenduri seuneubok,* members would also provide a feast for the tiger. At this time, those *ketua seuneubok* able to act as *pawang* would call the tiger and provide rice, meat and vegetables. This meant that, in the course of their duties, a *ketua seuneubok* with these skills became familiar and might even befriend the *seuneubok* tiger, meeting it regularly in their forest gardens. At the time of the year when the *kenduri* was held, older villagers recall, resident tigers had been known to seek out the *pawang* to remind him of the feast, leaving signs in the dirt, calling out, or even sleeping under the *pawang*'s forest hut.

The *adat* rules relating to the *seuneubok* had a sacral/moral aspect to which both humans and tigers were subject. The *seuneubok* tigers were thought to enforce laws, and those subject to attack were held to be evil people who had broken *adat* precepts. At the same time, the tiger was subject to *adat* rules and would be caught if he violated them. If a tiger attacked and killed someone, the *pawang* would set out to trap it.

In South Aceh tiger attacks continue to cause fear in the villages. In January 1999 a tiger killed a schoolboy picking nutmeg in a forest garden in the nearby sub-district of Labuhan Haji. Newspaper reports on this tiger attack reveal the problems associated with tigers: "Entering the eighth day, yesterday (17/1), the tiger that pounced on an adolescent from the village of Hulu Pisang, Labuhan Haji, South Aceh, has not yet been apprehended. This makes local residents afraid because all signs indicated that this wild animal has begun to roam around the houses of residents." As the animal continued to cause fear, the report continued, villagers were unable to carry out their farming activities. If the forestry department did not deal with the problem, villagers threatened to poison the tiger. However, the *kecamatan* authorities and the village heads had already sought the assistance of a *pawang harimau* in Labuan Haji.[102]

As this report indicates, the Forestry Department regularly made use of the skills of the *pawang*. At one time the forestry department in South Aceh even had a *pawang* on staff. However, most *pawang* were now over fifty, and the next generation was uninterested in learning to become *pawang harimau:* like the tiger itself the *pawang harimau* were becoming increasingly rare. Afterwards the departmental *pawang harimau* subsequently died. Nonetheless, according to a forestry official I interviewed in early 1999, whenever a tiger attack occurred the forestry department hired *pawang harimau* to help track the tiger.

According to a forestry official, tigers tend to come down into the village areas during the western monsoon. Females bring their cubs down to avoid older males who could attack cubs, and older tired tigers also descend out of the hills at this time. "Before we had a *pawang* on our staff, but he died," he said, "they are usually fifty or over. Because of modernity, they are less common. A *pawang* is generally angry if people catch tigers, because he considers the tiger as part of his happiness (*kebahagiaan dari kehidupan dia*)." *Pawang* will catch tigers that have done wrong: "They say that a tiger won't want to enter a trap if he is not in the wrong." The tiger is an endangered species, and the forestry department generally wanted to avoid catching the tigers. If possible they would drive the tiger back into the forest, even though villagers wanted to catch it. "We have to be very sensitive handling these cases," he said. Poachers who catch tigers use poison or snares. Over 1998–99, the World Wide Fund for Nature found sixty-six Sumatran tigers ready for illegal sale in the markets of Sumatra. Traders sold tiger products such as skin, teeth, claws, and whiskers, mostly as ingredients in traditional Chinese medicines. The *Jakarta Post* reported that traders could earn between Rp 300,000 to Rp 500,000 per tooth.[103] Poachers who caught tigers used poison or snares, but these men were not *pawang harimau.* "We haven't met this," he said, "although there are stories."[104]

The Sumatran tiger is highly endangered: according to one estimate, in 2000 fewer than four hundred still survived in Sumatra's shrinking forests.[105] Nonetheless, in some villages in South Aceh villagers see them often. "People say the tigers are going extinct," an older villager said, "but those people haven't been here."[106] The forestry official confirmed this: "we don't know the number of tigers," he said, "but there are lots of tigers in some places, and here tigers often disturb the villages."[107]

When a villager was killed behind a Sama Dua village in the mid-1990s, a *pawang harimau* caught the errant tiger. Before the forestry department took it away, the tiger was put on exhibition for a week. When the forestry department released it near Tapaktuan, villagers were very disappointed: the tiger had killed someone and they felt it wasn't right to return its freedom.[108]

Institutional Change in the *Seuneubok*

When I first visited Sama Dua during 1996–97, several informants explained how the *seuneubok* had fallen into decline. For instance, the head of one *seuneubok* I interviewed during 1997 described how his *seuneubok*—located up a precipitous hillside behind the village where we sat—had fallen into disuse. He explained that villagers had once collected forest products, including *damar*, *getah*, and rattan in the forest there. In the colonial era, at various times, villagers had also planted *rambung (ficus* spp), rubber, and cloves as well as seasonal crops in forest gardens in a *seuneubok*. However, around the end of the Dutch period, farmers abandoned their plots, either due to the difficulty of managing the area or due to a fall in prices. Around 1950, villages shifted to planting *nilam* in small plots next to their houses. However, when nutmeg prices rose in 1979, the *seuneubok* was re-established under his leadership, and was now reached via a less steep pathway into the forest. Sixty-five villagers began as active members, but by 1983 many villagers had abandoned their *kebun*: nutmeg prices had collapsed and "it was easier to find building work or work cutting wood."[109] Subsequently, only a few farmers were still prepared to make the trip up the steep hillside path and actively work their *kebun.* By 1997 the *seuneubok* had all but ceased to function, and he no longer felt the need to hold *musyawarah* or *kenduri seuneubok.*

Other interviews across Sama Dua revealed that there were approximately twenty-three *seuneubok* in the sub-district. However, as one former village head estimated, "perhaps only one in four *seuneubok* are still strong, the other seventy-five percent are out."[110] This decline was the cumulative outcome of several changes affecting village life.

First, under the nutmeg regime that emerged in the 1960s, farmers invested less in their nutmeg gardens. As discussed earlier, due to price fluctuations, villagers could no longer depend on income derived from nutmeg cultivation. In response, many of them found other income earning opportunities. This contributed to a decline in the practice of farmers having permanent rights over some area and ephemeral use rights over temporary plots: farmers now only had enduring property rights over hillside plots.

As I noted earlier, in the past the *seuneubok* was about as important as the village, and the *ketua seuneubok* had about as much authority and respect as the *keucik*. Just as the *keucik* played an important role in the operation of village institutions, the *ketua seuneubok* was responsible for the operation of the *seuneubok* institution. Now, farmers invested much less time in their *kebun*. Rather than living in the mountains, as they formerly did, most farmers now lived down in the *kampung*. As the key institution governing what occurred in the nutmeg forest, the *seuneubok* became less important to local life. Farmers were now much less identified or involved with their *seuneubok*: the resource inputs in terms of labor fluctuated with the profits yielded by nutmeg farming. Accordingly, farmers were less willing to carry the costs of transacting—defining, protecting, and enforcing the property rights in the *seuneubok*. Farmers invested less time and energy maintaining the *seuneubok*'s institutional arrangements. As a consequence, *seuneubok* functions declined.

Second, cultural change also affected the *seuneubok*. According to a former village head, as the cash economy penetrated Sama Dua more thoroughly, "people now look for money and do not pay as much attention to *adat*." Previously, there was usually a program for a Friday or Saturday night, such as a prayer meeting at the mosque, and the village often held *kenduri*.[111] "But now there are satellite dishes and other entertainments. The village isn't so self-contained, and people are orientated to Medan or even Jakarta."[112] As people aspired more to the trappings of "development" (*pembangunan*), traditional arrangements came under challenge, and this included the *seuneubok*. As the former village head argued, the old role of the *ketua seuneubok* was no longer in accordance with the culture. Formerly, the *ketua seuneubok* was like a "commandant": "There were sanctions for many things. For instance, if you didn't attend a meeting you could be fined. But now people are much more individual."[113]

Third, many of the younger generation of *ketua seuneubok* were less capable of fulfilling the role. In the past, *ketua seuneubok* were chosen largely due to their considerable agricultural experience and other skills: the *ketua seuneubok* used to be an accomplished farmer and village member (*orang arif*). As a consequence, they were respected and considered wise: they could advise other farmers and make effective leadership decisions. However, as *ketua* grew old or died, the new heads that were elected were younger farmers who had spent less time in the *kebun* and did not possess the necessary expertise. As a former village head asserted, "while some *ketua seuneubok* were chosen by the *seuneubok* for their wisdom, others lobbied for the position, but maybe they are not competent."[114] The result was that "some *ketua seuneubok* don't weigh things up wisely; they don't have a head for this, or they misuse their position." For instance, in the past *ketua seuneubok* maintained their impartiality in disputes: they would even bring up cases involving their own extended family. But now some *ketua* treated family members differently than they treated other farmers. "And so the farmers are offended (*sakit hati*) and no longer respect the *ketua seuneubok*." Without a charismatic and respected *ketua seuneubok*, a *seuneubok* functioned less effectively.

Fourth, the increased activity of local government agencies has also affected the status of the *ketua seuneubok*. As I noted earlier, the state's legal and administrative structures were now reaching into the village. Yet *ketua seuneubok* tended to be uneducated, or even illiterate, and they lacked understanding of the wider legal/administrative system. One former village head noted: "As villagers began to know the law, they didn't feel that they had to follow the instructions of the *ketua seuneubok*. This was very different from before, when it was obligatory to follow *adat*: if you didn't, you would be fined."[115] As some farmers came to understand that the state failed to recognize the *ketua seuneubok*'s authority, some were less inclined to accept the word of the *ketua seuneubok*. Moreover, in the case of a dispute, rather than relying on the *ketua seuneubok*, if they saw some advantage in doing so, they could choose to take a case to the police.

This supports Ostrom's (1992) observation that when institutions governing resource systems are isolated, *de facto* organization by local resource users is sufficient. However, once an area falls more readily within the jurisdiction of wider state structures, the possibility arises that government agents will oppose "customary" arrangements and support those who refuse to follow the rules. Ostrom has argued that if the state does not support the authority of local organizations to make rules, effective local institutional arrangements may fall into rapid decline. Consequently, if such arrangements are to continue to function, it is important that state agencies do not undermine or otherwise oppose the rights of local users to devise local institutions for governing local resource use (Ostrom 1992: 75).[116]

Moreover, state agencies challenged *ketua seuneubok* in other ways: older *ketua seuneubok* lacked knowledge of the new agricultural techniques promoted by the government extension officers. As one informant said, "the *ketua seuneubok*'s knowledge is just ancestral practice . . . They are people of former times (*orang jaman dulu*) which were different: it is modern times now, and farmers go to school."[117]

When government extension officers came to the village with new agricultural techniques, the agricultural agency created their own farmer groups; extension officers did not work through the *ketua seuneubok*. As a result, many *ketua seuneubok* would see the new techniques as a threat. Moreover, "slowly the role of the *ketua seuneubok* is taken over by the extension officer (PPL) and the *ketua seuneubok* has no function."[118]

Fifth, as an institution suited to the frontier, the *seuneubok* became less important in long-established agricultural areas; therefore, the decline was more pronounced in older *seuneubok*, many of which no longer had a *ketua seuneubok*. As an informant noted, farmers carving new plots out of an area of remote, wild, and lonely jungle need the solidarity offered by the *seuneubok*. However, after a few generations, the *kebun* was already productive and the area had become benign. The solidarity of the first pioneer farmers was no longer even a living memory. Villagers no longer felt such a need for fellowship and support, and farmers were happy to work on their own. For this reason, the areas closest to the village were no longer organized into *seuneubok*.

In January 1999 my interviews with farmers and *seuneubok* heads revealed a complex range of situations.[119] In some places the *seuneubok* still had a name but no head; in others the *seuneubok* had a name and a leader, but it was unclear to what degree the *seuneubok* still functioned. In yet other areas the *seuneubok* retained their vitality. Moreover, as later discussion will indicate, some *seuneubok* began to revive after 1998. In two cases, the *seuneubok* head had died and villagers were unsure whether they had been replaced. However, in one village three *seuneubok* continued to function in many ways without a *seuneubok* head. In these cases, other village institutions had taken over *seuneubok* functions. The village head carried out the dispute resolution functions, while the religious head (*imam*) fulfilled the ritual function of leading the *kenduri seuneubok*. However, in other areas, farmers have to solve problems on their own.[120] An informant who bought a *kebun* in an area where the *seuneubok* no longer functioned complained that he missed the social and ritual functions offered by the *seuneubok*. For instance, to ensure the productivity and to protect his nutmeg garden, he had to organize his own small *kenduri* with neighbors.

There is usually a *kenduri* for the *kampung*, and a *kenduri seuneubok*. This hasn't happened because my *seuneubok* hasn't had a head these last twenty years or more . . . So when I bought the land, I held one myself—inviting orphans and having it in my house. Then I took some rice and placed a portion on each durian and on each nutmeg tree—and so after that the trees were very productive . . . The reason the pests are attacking the nutmeg is because there is no longer a *kenduri seuneubok*—if people did this correctly—[if] we prayed together and asked Allah for protection—then it would be safe . . . [121]

This pointed to the way that *seuneubok* fitted into a social, moral, and supernatural order.

In places where *seuneubok* no longer functioned, farmers lost the dispute resolution offered by this institution. While disputes could be taken directly to the *keucik*, many cases were also taken straight to the state legal system. This villager had taken a dispute over *kebun* boundaries with a neighbor directly to the law. The local court sent a legal team to check the issue before making a decision. Although he felt vindicated by the court's final decision, he found that compared to using *adat* arenas to solve a dispute, it turned out to be a long and costly process.

However, in those areas where *ketua seuneubok* still functioned, *ketua seuneubok* attempted to face these challenges. To reassert the authority of the *ketua seuneubok* and create a procedure for overcoming *seuneubok* problems, *ketua seuneubok* sought ways to embed their authority and the customary principles governing the *seuneubok* within the wider state authority system. As the former village head observed, "some *ketua seuneubok* found someone with knowledge of the law. By finding the overlap between the law and the *seuneubok* regulations, he could use the law to make the *seuneubok* stronger."[122] For instance, a *ketua seuneubok* explained how during the 1970s theft in the *seuneubok* began to become a problem that he had difficulty handling. Eventually, in 1987, he called a meeting to reach a consensus among *seuneubok* members regarding the rules that applied to theft and other most important problems in the *seuneubok* at this time. The meeting led to the writing of letter of agreement that all members of the *seuneubok* signed. This letter is reproduced in Table 2.2 (above). The letter of agreement did not present an exhaustive list of *seuneubok* regulations but rather specified the principal *seuneubok* regulations that were relevant to the most salient problems at this time. As this letter was legally binding, an infringement could then be taken to court. To further support the *ketua seuneubok*'s authority, he also obtained a letter from the police and army recognizing his authority. In this way, the *seuneubok* could continue to function, but now with official legal status and the support of the police.[123] Afterwards, if an offender continued to offend and disregarded a decision of the *ketua seuneubok* and a village deliberation (*sidang adat*), the person would then be taken to the police. Under a court's discretion, a letter could then be written stating that if the problem re-occurred, the culprit would be taken directly to jail, without further investigation, and held for a period of time proportional to the severity of the violation. He reported that this innovation had led to a large decrease in cases of theft.

The problem of theft in the *seuneubok* illustrates an instance where the *seuneubok* now nests in the wider state legal regime. Interviews with *adat* heads indicated that theft in a busy *kampung* was generally rare. However, up until the economic crisis, nutmeg prices were low and *seuneubok* were generally quiet places; farmers walking through deserted *kebun* were tempted to take nutmeg fruit as they passed. The ability of the *seuneubok* to protect the property of its members depended on other farmers reporting strangers in the *seuneubok* and instances of theft. However, even if they saw a theft, farmers were loath to report it because they were reluctant to make enemies in the village. I was told: "If there are two witnesses then they can report it. But one person alone is wary of reporting. At most he will report to the victim of the theft. Anyway he can't do anything without proof."[124]

When a case was reported to the *ketua seuneubok*, he was initially responsible for the handling of it.

Basically, after someone reports a theft case to me, and we look into it, and if need be, we hold a session (*sidang*), calling in the *keucik* a few other *adat* experts (*tokoh adat*) . . . The person is presented with a choice—we can handle it quietly (*secara damai*) in the *kampung*, or we can hand the case over to the police . . . In the latter case I write a letter to the police or go and report and the police go looking for the person.[125]

If a case was heard in a session of the village council (LKMD), in accordance with the principles of *adat* justice the session would make a decision and impose a penalty. However, sometimes the case could not be solved at this level, for instance, if one of the disputants refused to accept the decision and continued to offend. The next time the dispute arose, the case would be taken to the police and later perhaps subjected to the state legal system. However, taking a case to the police was considered a serious step that could lead to long-term enmities in the village.

If someone steals, the first time it is handled by a session of the LKMD. A person will be offended (*sakit hati*) if they are taken to the police the first time. If we use village (*kampung*) law, only this *kampung* will know, and so the culprit isn't so ashamed (*malu*). But after repeated offences, they will be reported to the police. The police will come and arrest them, and they will sit on the back of the police car bench in the open air before the whole *kecamatan*. After that, everyone will know and they will be very *malu* . . . [126]

Thus we see that at times villagers made strategic choices, "shopping" for the forum that best suited their needs.[127] If a disputant wanted to humiliate his opponent, he took the case directly to the police.

However, a *ketua seuneubok* observed, cases were often not brought to a conclusion if they ended up with the police. To illustrate this conclusion, he cited a recent case.

Two people recently had a dispute over ownership of two nutmeg trees. Deliberation (*musyawarah*) occurred in phases. The first time just with the *ketua seuneubok*, the second also with the village head, and the third time with the police. Yet, one side didn't want a settlement (*damai*). But the police couldn't solve it and in the end they also called in the *ketua seuneubok* who solved it in the police station in the name of the police. The trees are divided into two, but in the end one side still takes revenge, stealing nutmeg, so the case re-emerges. The person is a stubborn (*keras kepala*) egoist. In October he was contented, but now the agreement has been broken again. *Petuhapet, keucik,* and the *ketua seuneubok* will gather for another session . . . Now twenty people will join in the session: members of LMD, LKMD [village councils], elders (*cerdik pandai, tokoh adat*), and religious leaders.[128]

These examples suggest that individuals could opportunistically make the most of a context of institutional pluralism. If a villager saw some strategic advantage in doing so, he could shop for the forum that suited his needs, choosing to involve the police and ultimately the state court system rather than *adat* dispute mechanisms. However, the involvement of the police or other state officials was clearly contingent. For example, if the police could not solve a village dispute, the dispute would then be referred back to mediation by village leaders. In many cases village disputes involving matters of honor tended to be protracted, and state officials might gain only extra troubles from becoming involved. In such cases they would be happy to support village *adat* solutions to intractable village feuds. In this sense, village disputants were not the only actors who accepted and tried to make use of the coexistence of the *adat* institutional arrangements and the parallel state legal system.

Over time the *seuneubok* authority system has necessarily become enmeshed in a complex way with the wider authority system of the state. When the *seuneubok* were more isolated, *de facto* organization by members was sufficient. As the area fell more readily within the jurisdiction of wider state structures, a problem increasingly arose: would government agents support those who refused to follow the *seuneubok* rules? In this context some *ketua seuneubok* were able to reassert their authority: by making use of congruencies between the *adat* order and the parallel state legal framework in this way, *seuneubok* heads could reinforce the *seuneubok* institutional order. As we will see, this is not the only way that *seuneubok* heads made use of a state institutional order with a seemingly contradictory value orientation to shape the direction of social change.[129]

Revival

As I noted above, during several interviews in 1996–97, villagers observed that the *seuneubok* were in serious decline. Yet, one *ketua seuneubok* interviewed at the onset of the economic crisis believed that there was a strong connection between the viability of gardens in the *seuneubok* and the strength of the *seuneubok* as an institutional arrangement. "If prices of *kebun* products are high," he said, "many people go to their *kebun*. People support the *seuneubok* and are enthusiastic: the *seuneubok* is strong."[130] When I returned to Sama Dua in early 1998, it was clear that villagers were responding to the economic crisis by turning back to agriculture. At this time, a *ketua seuneubok* noted that, due to the strong connection between the viability of hillside gardens and the *seuneubok*, a consequence of the economic crisis would be the revival of *seuneubok* institutional arrangements.[131]

With a boom in the price of patchouli oil, farmers began planting *nilam* plants in new forest plots. To facilitate land pioneering, farmers formed groups of ten to fifteen and selected their own head (*ketua ladang*) to coordinate their activities. Before heading off into the forest, it was long-standing practice to report first to the *ketua seuneubok*. If a party failed to return, the *ketua seuneubok* would take responsibility for organizing a search party. Moreover, the *ketua seuneubok* carried the institutional memory of the *seuneubok*. If farmers opening new plots wanted to avoid possible conflicts over property, they needed to approach *ketua seuneubok* to ask information regarding land suitable for farming as well as to seek leave to open new areas. Moreover, by seeking permission first, the new groups ensured that the *ketua seuneubok* would provide advice and guidance as well as helping to resolve problems.

When I returned to Sama Dua in February 1998, the price fetched by agricultural export crops (such as nutmeg and *nilam*) had risen roughly in proportion to the rapid decline in the value of the Indonesian rupiah. As the economic crisis hit Sama Dua, villagers began to approach the *ketua*, seeking out unused plots in the *seuneubok*. "There are maybe five or seven active in the *seuneubok*," the head of a *seuneubok* that had ceased to function (discussed earlier) explained at the time, "but now about six people a day come asking about opening a *kebun*."[132]

A year later, in January 1999, another *ketua seuneubok* described how his *seuneubok* had become very busy. He was now the *seuneubok* head for thirty farmers working *kebun* located along the *seuneubok* forest path. At a certain point, the path forked to the left and right. During the *nilam* boom, some farmers had opened *kebun* two hours further up the left-hand fork and three hours up the right-hand fork. Feeling unable to handle these areas, he appointed heads to look after the left- and right-hand pathways. "These are representatives," he said, "if they need my help they can call on it. Otherwise they handle problems on their own."[133] These nascent *seuneubok* had their own names; thirty farmers worked plots on the left-hand pathway, while ten farmers worked the right-hand pathway.

In a similar fashion, in January 1999, a farmer with wide experience in the logging and agribusiness sectors across South Aceh, described how he joined a group of ten farmers opening *nilam* plots four hours' walk into the mountains. The group chose to open a flat area some ninety minutes' walk past the last cultivated area—beyond the mountain that marked the furthest boundary of that particular *seuneubok*. As the new *kebun* were so far out, the *ketua seuneubok* said he was unable to take responsibility for affairs there. In response, the group asked him to act as a head of what now constituted a *seuneubok* in its own right. To facilitate this process, the *seuneubok* head helped him to learn the functions of the *ketua seuneubok*, inviting him to follow mediation sessions (*sidang*) to learn how to settle disputes. The group of pioneers worked together to construct a large camp with a large hut (*pondok*), where they stayed up to one and a half months at a time. Working together, they opened an area sufficient to give each farmer a plot of around one hectare. At first they planted *nilam*, later supplementing it with nutmeg and durian.[134]

The crisis demonstrated that *adat* arrangements remained important: in time of need farmers would fall back on *adat* property arrangements that offered them some degree of social security. An expected impact of the crisis, then, was the renewal of these arrangements. Given the tenacity of these arrangements, this outcome is hardly surprising.

Overlapping Territorialities:
Right of Avail in the State Forest

As I discussed earlier, although the colonial state had embarked on a process of state territorialization, colonial foresters did not zone the forest in this area for timber exploitation or for plantation use. Later, after a long hiatus in state territorialization, a series of state-sponsored mapping exercises during the New Order period renewed this process.[135] In Aceh the governor coordinated a team involving several provincial-level sectorial agencies, which produced a set of forest consensus maps (TGHK) setting out the boundaries and respective classifications for Aceh's state forestry estate. After the governor of Aceh and the minister of agriculture signed the TGHK map for Aceh in December 1981, it served as the framework for allocating forestry concessions (*Surya Karya* 1990). State planners calculated an area that villagers would need for agricultural expansion. They classified an area of forest near villages as "unrestricted state forest" (*Hutan Negara Bebas*) and left it outside the permanent forest zone (*kawasan hutan*).[136] Accordingly, the revised RePPProt map (1988) for the Sama Dua area shows a thin band of forest behind Sama Dua that is classified as "unrestricted state forest" (see HNB in Map 5).[137] This area did not extend further than five kilometers from the coast. Although these classifications did not correspond to local notions of the extent of Sama Dua territory and although the process was hardly consultative, they do point to the fact that government planners had some knowledge of *adat* assumptions: it amounted to a *de facto* allowance for a village "right of avail."

Beyond that area lay a band of "limited production forest" (*hutan produksi terbatas*). This strip of land expanded from a width of around five kilometers at the northern end of Sama Dua to around ten kilometers in width further south to include the Kluet River (RePPProt 1988). Forest legislation held that this forest forms a part of the state forest estate (*kawasan hutan*), and as such should not be converted to other uses but be maintained under permanent forest cover. However, as the name suggested, state forest policy allowed that "limited production forest" could be exploited for logging.

Since the TGHK maps were associated with many problems and conflicts, subsequent legislation required that the system be revised (Dephut 1992; Steering Committee 1998).[138] In accordance with the 1992 Spatial Planning Act, each level of government needed to prepare a spatial use management plan. As most of the hilly forest areas surrounding Sama Dua had a slope of over 40 percent, the South Aceh district government's

TABLE 2.3
*Protected and Cultivation Areas in Sama Dua District
according to district Spatial Plan*

Status	Total
Protected Area	
Nature Reserve (Hutan Suaka Alam)	0
Protection Forest (Hutan Lindung)	9,671
Cultivation Area	1,129
Total Area	10,800

Source: Pemerintah Kabupaten Daerah Tingkat II Aceh Selatan (1991/1992).

district spatial plan reclassified it as a "protected area" (Pemerintah Kabupaten Daerah Tingkat II Aceh Selatan 1991/92). Subsequently, district and provincial government regulations specified that most of the Sama Dua territory previously listed as "limited production forest" should be excluded from logging and agricultural use (see Table 2.3) (Serambi Indonesia 13/11/95).[139]

Moreover, the Leuser Development Programme (LDP) carried out another territorialization and mapping exercise. Preliminary Leuser Management Unit (LMU) maps of the ecosystem reveal that in the Sama Dua area the Leuser Ecosystem included the Limited Production Forest and some limited areas of the Unrestricted State Forest. While LMU had developed guidelines for marking the ecosystem's boundaries and had begun to designate these in some areas, in Sama Dua at the time of this research, these boundaries remained unmarked (see Leuser Ecosystem boundaries in Map 2). However, in most respects in Sama Dua the territorialization process carried out by the state and later by LMU has created forestry boundaries—for the state forest zone and the Leuser Ecosystem—that remain so many ink markings on maps. As noted earlier, local residents consider this area to be the territory of Sama Dua.

During 1989–90, in response to two attempts to exploit local forests in accordance with the TGHK territorialization process, villagers of Sama Dua acted collectively to defend their territory. In 1977 the Ministry of Forestry had issued a timber concession to PT Dina Maju south of Sama Dua in the steep mountainous forest of the Kluet watershed behind the coast. Just to the north, the timber company PT Remaja Timber had obtained a concession to log over 40,000 hectares of forest in the limited production forest behind Sama Dua and neighboring coastal sub-districts. In a series of colorful and emotive articles in July 1990, the Medan-based paper *Waspada* reported that the concession included thousands of hectares of local people's plantations, graveyards, *adat* land, and other community areas.[140] As the concession encompassed very steep mountains containing the headwaters of rivers subject to flood, *Waspada* reported that seven sub-districts (including North Kluet and Sama Dua) were "threatened with sinking" (*terancam tenggelam*) (Waspada 1990). The district head (*bupati*) vehemently opposed this and other concessions (Kompas 1991). When interviewed in late 1997, he pointed out that government regulations did not allow logging on slopes greater than 40 degrees. However, the forestry department had issued concessions on steeper land. These permits were based on the TGHK classifications and made without accurate information concerning local conditions. The *bupati* saw the consequences for local people if the logging of steep mountainous areas behind village settlements occurred: "It is very rich forest here, with *Kruing* and *Damar* trees, but is also very steep. The rivers are very swift and so if the upstream areas are cut, the areas of settlement will be damaged."[141] In his view these areas should never have been set aside for logging. "The TGHK was completed before more sophisticated knowledge was available," he said, "and so I said to the people in Jakarta if there is already a mistake, we shouldn't continue it."

When the forestry department opposed him, the *bupati* organized local meetings to mobilize villages against logging companies. After deliberations (*musyawarah*), villagers in Sama Dua corporately decided that they did not want to lose arable land to outsiders. They would only allow logging to occur in their territory on certain conditions: the company could take the wood on condition that the 1,500 hectares of arable land at Alur Rimbia would be given to Sama Dua. Moreover, the company would have to build a road to enable local farmers ("local transmigrants") to settle there. In June 1990, the dispute came to a head. The *bupati* threatened to resign if logging of the mountainous area behind the coast went ahead. He told reporters:

I cannot imagine if all the HPH cut the forest, causing West and South Aceh to change function and become ocean. Therefore it is better if I stand down from the position of *bupati*—because I cannot take responsibility to the people, to nature and to God. For what purpose do I become *bupati*, if my people and region are sacrificed to enrich oneself only on the basis of a permit from the central government. (Waspada 1990)

Articles in the regional and national press brought his stand to national attention. The governor of Aceh, Minister of Home Affairs Rudini, and Minister of Forestry Hasjrul Harahap all made statements on the issue (Peristiwa 1990; *Serambi Indonesia* 1990). Meetings with the governor in Banda Aceh and the minister of forestry in Jakarta followed, after which the minister agreed to review the two timber concessions. Eventually, the company decided not to go ahead.[142] According to the former *bupati*, PT Remaja Timber was not able to continue "because they were not brave enough."[143]

Around this time, another challenge to the Sama Dua territory occurred. The Department of Transmigration announced a plan to settle 200 transmigrants in arable area behind Sama Dua. To prepare for a transmigration settlement, the land would have to be cleared. As in other areas, this would involve granting timber interests a Timber Harvest Permit (*Ijin Pemanfaatan Kehutanan*, or IPK) to remove and process valuable logs. In 1989 officials had scoped out the area by helicopter to consider the feasibility of opening a transmigration site behind Sama Dua in a 2,000-hectare area (Sinar Indonesia Baru 1989). In response, the Sama Dua people acted collectively, attempting to defend their property rights by establishing a prior claim to the area. The villages of Sama Dua pooled their resources and built a road with village development funds (*bangdes*) and voluntary labor (*gotong-royong*). The idea was to open land, plant crops, and establish a village before the government could give the area away for transmigration or logging. "The area is ideal for a HPH," a villager said; "it is flat land and the trees are very big. But really this is our ancestral land—not for them. So we should establish twenty huts (*pondok*) there. But after we built the road around 1990, nothing happened and the road turned back into forest or was subject to landslide."[144]

These incidents revealed that even though the Sama Dua *adat* territory had no official status, local residents, supported on this occasion by the district head and local journalists, were prepared to defend their territory. As long as Sama Dua villagers could successfully contest development plans based on the territorialization process carried out by state agencies, these forestry boundaries remained so many ink markings on maps.

This case also demonstrated that, contrary to many accounts that portray local communities as mere victims of development projects imposed from the center during Suharto's regime (1966–98), at times local communities have defended property rights in their own territory.[145] Other examples of conflicts across South Aceh indicated that, particularly since the onset of the crisis in 1997, it was not unusual for a community to defend territories that were held to be subject to local "right of avail."[146] Two exceptional circumstances during the New Order period may account for Sama Dua's unusual success in defending its turf. First, the personal qualities of the district head at the time when the logging concession and transmigration proposals arose had a key role in this outcome. The *bupati* at that time, Sayed Mudhahar, refused to accept the payments offered by logging interests attempting to buy favor from his administration.[147] Moreover, he showed unusual courage in standing up to the Ministry of Forestry. As the national daily *Kompas* reported, "district heads usually always agree with projects from the centre," yet, "for the sake of the environment" Sayed threatened to resign if the Ministry of Forestry agreed to the new logging concessions. Sayed's behavior was extraordinary, and he became somewhat of a hero of the environmental movement (Kompas 26/2/91). Second, compared to other districts in South Aceh, the Sama Dua territory is mostly mountainous and relatively inaccessible. Because of this, outside interests have tended to concentrate their efforts elsewhere.

Official and De Facto Control of Forest Territory

A reading of Dutch colonial reports concerning this area exposes the early history of the dynamic relationship between these competing regimes.[148] In maritime Southeast Asia, pre-colonial states had depended on controlling the bounty of the forest resources in their hinterlands, and kingdoms rose and fell based on their ability to maintain this control (Peluso 1992: 52). In this context, the sovereignty of the pre-eminent local chief was tied to control of the stream of benefits derived from forest and other resources within the territory over which he claimed control. In the Aceh region, the local heads sustained their resource base and personal positions by levying taxes on forest products (known as the *pantjang alas*) and levying a fee on pepper and pinang trade (the *wase uleebalang*).[149]

A 1931 report by a Dutch official (*controleur*) named van de Velde noted that in pre-colonial times, the local chiefs had a right known as *hak Allah* ("right of God") over the unopened forest surrounding their territories.[150] In accordance with this right, outsiders wishing to collect forest products had to first ask permission from the local head. The head then levied a fee (commonly known as the *pantjang alas*) of around ten percent on the collection of forest products in surrounding territory. (Elderly informants in Sama Dua told me that the *Datuk* of Sama Dua "enjoyed" the *pantjang-alas*.) However, there were also partial payouts made to local villages heads, *sjahbandar* (port chiefs), and almost certainly the *ketua seuneubok* who collected the forest fee (as well as the tax on pepper) for the area on behalf of the pre-eminent local head (Adatrechtbundels 1938: 137). The forest products on which this tax was levied tended to vary from area to area. However, the major products included timber, bark, cinnamon, camphor, benzoin (incense wood), *getah* (latex tapped from native trees), *damar* resin, thatch (*atap* and *nipah*), rattan, wax, honey, rhino horn, deer horn, and elephant tusks (Adatrechtbundels 1938: 137). Since forest products were a large part of the exports from the area, these fees constituted a key source of income for local chiefs. In 1901 the first Dutch administrator stationed in Tapaktuan, a man named Colijn, reported that large groups of up to three hundred people from outside the area were making expeditions into the forest to collect forest rubber (*getah*). Colijn noted that these expeditions were only possible with the permission of the local head, who levied taxes of five dollars per *pikul* (60 kg) on the forest rubber (Adatrechtbundels 1938: 136–37).

Because the Dutch colonial authorities in Aceh first encountered the fee in the ports, they initially believed that the *pantjang alas* was an export duty (Kreemer 1923: 142).[151] Colijn, the first Dutch administrator in the Tapaktuan area, held that this fee was a regular tax that should be taken over by the colonial authorities. Only later did they begin to realize that it was a fee paid to local heads for access to forests under local "right of avail."[152] Although the fee was paid to the head who exercised control over the resources, interestingly enough, the fee applied to members of the *adat* corporate group only under certain conditions: members would pay this fee only if the products that they collected were for sale or trade; outsiders had to pay *pantjang alas* whether the products were for their own use or for export. In other words, the "community" distinguished between resources extracted for use by villagers and resources extracted for exchange for individual profit. From this it was clear that what Dutch *adat* scholars termed a "right of avail" was at work here: the *pantjang alas* was an access fee paid for use of what was considered a common pool resource of a "community."

After the conquest of Aceh, according to colonial texts, the Dutch gradually created a state forestry regime. In its thirst for revenue, over time the colonial government attempted to take over control of the *pantjang alas*. At first, the Dutch took over the fees levied on the export of forest products at the ports: these fees now became a colonial government excise. However, later the Dutch discovered that local heads continued to levy a *pantjang alas* in addition to the export duty. In 1912, when Swart, then governor of Aceh and its dependencies, was looking for extra state revenues, he decided to take over this fee as well. Henceforth, according to colonial regulations, the self-governing territorial heads (*Zelfbestuurders van de Landschappen*) such as the *raja* of Sama Dua were empowered to levy a 10 percent *pantjang alas* fee in addition to the export tax on forest products "for the benefit of the *Landschap* coffers" (Kreemer 1923: 23; Adatrechtbundels 1938: 138). For the years 1916–18 the *pantjang alas* collected from the port of Tapaktuan were quite significant.[153] However, it is unclear how successful the colonial authorities were in implementing these ordinances or collecting this tax. Van de Velde's 1931 entry in the *Adatrechtbundels* noted that the *Zelfbestuurders* gave their consent to the changes that now gave over the *pantjang alas* to the territorial coffers. Van de Velde describes the colonial regulations that pertained to the forest thus: those wishing to take forest products for commercial reasons now needed to obtain permits for harvest and transporting the wood from the forest service (Adatrechtbundels 1938: 143). However, as far as implementation went, he noted that there were no changes in the ordinances implemented by the *Zelfbestuurders*. This implied that local heads might not have implemented the regulations or might have continued to collect the fee outside the colonial tax regime.[154]

At the same time as the colonial regime attempted to ensure that the fees entered the coffers of the local heads, it enacted forestry ordinances that also expanded the concept of "right of avail" to the point that (in legal terms) it almost lost its original meaning. Thus, in keeping with long-standing practice that allowed that village members had free access to forest resources for non-commercial use, the colonial forestry ordinances for Aceh allowed that subjects of a territorial head in Aceh and dependencies were free of the *pantjang alas* if they were collecting forest products for their own use (or "self-use"). In a note published in the *Adatrechtbundels*, Van de Velde observed that the definition of "self-use" had been extended to include forest exploitation by the regional government for uses such as road and bridge building. Under colonial ordinances, the concept of self-use expanded further: in 1927 the right to collect forest products for local use was opened to non-residents in all forest areas not subject to a legal concession. Consequently, while the colonial statutes began by recognizing this residual "right of avail," in the course of broadening its tax base the colonial government gradually attempted to alienate access fees flowing from forests adjacent to communities from the local communities as well as ensure that the local government did not have to pay such fees to local communities.

These developments can be understood as a process of adjustment, in which colonial administrators attempted to understand and then co-opt *adat* concepts to serve the interests of the parallel state order. Nevertheless, as in other cases, the dominant colonial legal order failed to penetrate fully, encountered pockets of resistance, and was absorbed and even co-opted (Merry 1988). In practice local heads have continued to levy these fees up until the present day, demonstrating the long-standing reality of overlapping, competing *adat* and state regimes. Moreover, *seuneubok* heads have also embarked on a parallel process of co-opting elements of the state regime to serve the interests of the parallel *adat* order.

As various informants noted, nowadays in a official legal sense the un-opened forest behind Sama Dua belongs to the state. The law has failed to recognize *adat* claims over surrounding forest territory. Land became subject to the *adat* regime once an individual had opened it and planted it with trees. However, in practice the residents of Sama Dua maintained that the *adat* territory extends to Tanah Hitam (see Map 6). As one informant noted, "Up to *tanah hitam* this is land for all Sama Dua people to open, and if someone else wants it, Sama Dua will not give it up."[155]

Certainly, the rights enjoyed by local villagers were embedded in village practices. Interviews with villages and officials alike led me to the conclusion that the local forestry regime allowed for a local "right of avail" (*hak ulayat*): local villagers opened *kebun* in the Unrestricted State Forest (*hutan negara bebas*) immediately behind Sama Dua without asking for permission from the state. Although villagers did not use the term, this "right of avail" also allows local villagers to cut wood for non-commercial uses. Villagers have always obtained timber from the forest to build a house, a hillside hut (*pondok*), or a fishing boat or to meet some other timber need. Villagers built their houses, shops and coffeehouses (*warung*) from forest timber.

In 1996 I interviewed a villager who had cut logs to build a coffee shop (*warung*). He indicated that he was aware that the forest was considered state property, explaining that villagers were forbidden to take wood for commercial purposes—without an official logging permit under the state licensing system.[156] If villagers cut large quantities of timber for sale, this constituted "illegal logging" (*penebangan liar*), and they were liable to arrest.[157] Because a major provincial highway passes through Sama Dua, villagers transporting timber need to be careful: large-scale logging might draw the attention of local officials or the police. In this case, according to one informant, to avoid legal sanctions, loggers might be able to negotiate an unofficial payment to the official concerned. However, since the Sama Dua area was not directly adjacent to an active logging concession and did not directly face on to national park land, forestry officials were rarely in the immediate vicinity.

Although the law prohibited logging without a permit, unlike Menggamat and other areas in South Aceh, in Sama Dua logging was taking place only on a small scale. Given the sensitive nature of the issue, generally *ketua seuneubok* avoided discussing *seuneubok* rules relating to timber extraction. During 1996–97, villagers noted that when farmers were opening new plots of land, *seuneubok* norms allowed that, rather than leaving valuable timber to rot, they could cut it into planks and carry it back to the village for use or sale. In addition to cutting wood on their own plots, farmers wishing to obtain wood could negotiate with others who had more trees on their land. Alternatively, they would need to travel to the *seuneubok* frontier, where all *seuneubok* members enjoyed timber rights to what constitutes a common pool resource. However, with the forest frontier moving farther and farther out, and given the distances and the type of terrain involved, carrying larger quantities of wood back to the village had become increasingly arduous.

In an interview, one *ketua seuneubok* revealed that, at least in some *seuneubok*, there were rules relating to timber within its territory. In this *seuneubok*, local villagers could take large trees from abandoned *kebun*, which they would cut and drag down the stream back to the road. During periods when there was little agricultural work, occasionally someone with sufficient capital would fund a team of up to fifteen villagers to make an expedition into the *seuneubok*.[158] The organizer and the team members would take a chainsaw up, fell large trees, cut them neatly into planks, and carry them back down. According to this informant, outsiders wishing to cut wood or villagers wishing to take timber for sale needed to inform the *ketua seuneubok*. If they were not members of the *seuneubok*, they would need to negotiate with the *ketua seuneubok*; they would usually have to make a contribution to the village development funds or to the local mosque. Alternatively, they might make payment in kind by donating timber.[159] When the *seuneubok* had been operating effectively, the *ketua seuneubok* had held a meeting at which—in a similar fashion to that described earlier in Table 2.2—members of the *seuneubok* signed a letter of agreement specifying the *seuneubok*'s principal regulations. Members signed under a provision that recognized what he called the "village development rights" (*hak bangunan kampung*). According to this provision, outsiders taking timber for any reason at all or *seuneubok* members extracting timber for sale were obliged to pay this fee (*uang pembangunan*) to the village.[160] To some extent, by giving what used to be known as the *pantjang alas* the appearance of a state legal form, the *seuneubok* head was trying to make use of the status associated with the law to support this old *adat* property concept. Ironically, doing so involved using the state law and development discourse to give legitimacy to an *adat* concept that actually operated in contradiction to the assumptions of the state forestry regime.

He also revealed that as *ketua seuneubok* he was prepared to defend local property rights. Ironically, on at least one occasion he had done so by mobilizing the state law enforcement agencies to protect local *adat* claims over what was officially "state forest." When someone took timber without first reporting to the *ketua seuneubok*, the *ketua seuneubok* in turn reported him to the police. In another case, the head recalled how during the 1980s police from the neighboring district of Sawang (*Polsek Sawang*) went up into the *seuneubok* in disregard of the *seuneubok* rules governing access:

They said they were looking for deer, but they had a chainsaw and were looking for wood. When they came down it was clear that they had gone after wood. So we called the head of the Sama Dua police station (*Kapolsek*). In the end the case was taken to district headquarters (*Kapolres*) in Tapaktuan, who oversaw a consultation (*musyawarah*), wrote a letter, and the problem was solved. Anyone who seeks wood or rattan must report and pay a fee (*uang pembangunan*) which is used for the mosque. If they [*seuneubok* members] use the wood themselves then they don't have to report. Outside people have to [pay], whether for sale or for use. People from here pay this if they sell the wood on—[in which case they] also pay a fee to the village. Payment depends on the amount [of wood]. 1 cubic metre is 10,000 to 15,000 Rp.[161]

These examples demonstrated that a large body of rules associated with the state legal system—including those relating to the forest—enveloped the *seuneubok*. The *seuneubok* operated on premises regarding property rights and legitimate authority that not only existed outside the state rules but even contradicted these rules. Yet, where a *seuneubok* head could find some concurrences with corresponding priorities within the parallel state order, at strategic moments a *seuneubok* head might be able to mobilize the state rules, or threaten to do so, to support *seuneubok* tenets. By opportunistically making use of the competing order, a *seuneubok* head might attempt to control social action by drawing on the symbolic capital of the state law (Merry 1988). In the course of this process, *seuneubok* heads adjusted to the parallel state order, ensuring that, in indirect or unexpected ways, state legal norms found a place in the many binding obligations that were not legally enforceable but affected the choices of those wishing to extract timber from the forest.

The economic crisis that struck Indonesia during 1998 once again revealed how, by providing resources for villagers at a time of crisis, *adat* territory served as the villagers' source of social security of last resort. Even at this time, to some extent the *seuneubok*'s controls over logging continued. Whilst reporting to the *seuneubok* head and paying *uang pembangunan* was voluntary, a villager who logged the *seuneubok* forest without obtaining permission risked alienating the *seuneubok* head. As one villager noted, the *seuneubok* head had to be told or otherwise he would not take responsibility for any disputes.[162] As it was generally accepted that *adat* forests should provide a form of social security of last resort, there was a degree of tolerance on the part of the corporate group and the *ketua seuneubok* for poor villagers' logging the forest. There were also clearly limits to exploitation. Village norms precluded taking more than one's fair share: "If they take one or two logs, this is OK, but it is not acceptable to continuously take. Some ask for permission first, and there is some acceptance if the person doesn't have anything to eat. It is just a temporary need. Once they have enough they can find another source of livelihood."[163] Yet, in the face of this crisis, at times loggers also cut *seuneubok* forest surreptitiously. As one villager noted, "there are also those who just go and take trees without permission, but [these are] only one-off acts . . . Once or twice, taking a group up there and logging. But this can depend on the existence of buyers."[164]

Following the fall of Suharto and at a time of increased tension in Aceh just prior to the outbreak of violent conflict, it was clear that the state forest authorities were even less able to control logging of what was nominally state forest. According to one informant, if the forestry office was notified, the loggers might be arrested and the chainsaws confiscated:

According to the authorities, this is destroying the forest. I had a friend in Bakengon who was arrested, hit, and his chainsaw was taken away. But now there is no one who is arrested here. The authorities are afraid because of *reformasi*—they are afraid of people. Under Suharto officials, people in responsibility, took the profits for themselves and so trust in the army and the government apparatus is lost. For example, they even shoot children in Aceh. It will take a long time to go back to basics and rebuild.[165]

As in other areas, during this crisis the state forestry agencies no longer had the capacity to implement these laws in Sama Dua.

Conclusion

In the specific context of Sama Dua a particular set of *adat* arrangements operated with respect to forest and agricultural areas. Here *adat* arrangements pertaining to the forest (the *seuneubok*) had evolved during a long history of cultivation focused on producing cash crops for export. As an institutional arrangement concerned with farming, *seuneubok* were primarily concerned with protecting the property rights and socio-religious relations of farmers in their permanent hillside gardens.

An intimate connection existed between this set of arrangements and agricultural activities in a number of respects: this *adat* regime was also tied to the livelihood needs of farmers. Over a long period of time the *seuneubok* had provided a framework for the transformative use of the forest: the clearing and conversion of native forest into gardens, as well as the collection of forest products and some timber extraction by local residents. Therefore, during the 1980s and 1990s, when villagers were less reliant on hillside gardens, the *seuneubok* became less important. Later, when villagers turned back to their agricultural activities during the economic crisis, they necessarily returned to the *adat* arrangements (the *seuneubok*) that sustained them.

Sama Dua's villagers have largely depended on natural resources and agricultural products whose scarcity and value fluctuate wildly in response to markets located elsewhere. Consequently, *adat* institutional arrangements in Sama Dua have had to govern access to and use of land and forest resources under varying conditions. In doing so, they have needed to adjust to the shifting character of local livelihoods. Therefore, as *adat* arrangements were constantly renegotiated, they proved to be both resilient and dynamic. As the customary *adat* order also incorporated legal, social, and supra-mundane religious functions, it is clear that local institutions are not built merely on the logic of rational choice. Rather, this customary *adat* system encompasses a village socio-ritual order, local notions of identity, and associated notions of appropriateness.

In Sama Dua, *adat* was not merely organized in rules or even rule-like formulations: rather than attempting to reduce *adat* to a set of law-like formulations, *adat* institutional order(s) are more accurately conceptualized as ideas and patterns of social ordering. These patterns of ideas regarding ordering are associated with both implicit, deeply held social norms and more explicit rules. This suggests that attempts to reduce the *seuneubok* or other *adat* systems here to "customary law" understood as a set of legal formulations might not only be rather ineffectual, but might also risk reifying something that was by its nature constantly subject to mediation, compromise, and change. Although the term "customary law" "sounds as if it designates a straightforward set of traditional rules," the entity to which it refers is a pattern of social ordering or a set of ideas that is embedded in social relations that are historically shifting (cf. Moore 1986: xv)

In Sama Dua, *adat* has adapted to successive political situations. Previous work has shown how colonial scholarly and administrative practices constructed the notion of the "customary law community" for colonial purposes and according to Dutch concepts of bounded territorial, indigenous polities (Holleman 1981; Burns 1989).[166] Just as colonial scholars and administrators adapted concepts and practices used in Indonesian communities to construct the terms used in the edifice of *adatrecht* literature, so did colonial administrators in the field adapt and co-opt indigenous institutions to create the "jural communities" or "adat law communities" through which social control was maintained.[167] As this process occurred in Sama Dua, to some degree members of the local elite collaborated: as in other cases, in some measure colonial administrators and an indigenous elite co-produced the customary order (cf. Chanock 1998). The *adat* structure in Sama Dua was not simply an artifact of colonial policy; the colonial intervention seems to have helped to solidify local identities and fix local notions of territoriality on district maps.[168]

After the Indonesian revolution, the Nationalists embarked on a more ambitious nation-building project that entailed establishing the primacy of the state system (Sonius 1981: vxiv). In keeping with this strategy, in Sama Dua the state gradually dismantled the indigenous institutional structure that had defined Sama Dua as an "*adat* law community." In appointing officials down to the sub-district level, during the Sukarno period the state removed the positions of the *raja/landschapshoofd*. Then during the New Order the *datuk/kepala mukim* lost their positions at the interface between local communities and the state. In effect the state cut off the head of what had been the Sama Dua "jural community." Yet, as we have seen, even today *adat* regime(s) have continued to operate in the villages and *seuneubok* of Sama Dua.

In the lived experience of Sama Dua there continue to be referents to the concept of an "*adat* jural community" found in colonial *adatrecht* and post-colonial NGO discourse. As in other cases where indigenous polities were dismantled during the post-colonial period, what might be understood as a customary order continues to affect social relations in hamlets and villages, continuing to legitimate patterns of resource access and use in surrounding areas. Clearly, at times this customary order does affect the state's attempt to implement its own decisions. For instance, although local people do not use the term, the *adatrecht* notion of a territorial "right of avail" was not merely a contingent concept: when it counted, Sama Dua constituted a "community" that was capable of collective action to defend its territory. This became apparent on more than one occasion during the 1980s, when the territory of Sama Dua—the area subject to Sama Dua's "right of avail"—came under threat. Consequently, while the colonial state may have developed the conceptual edifice of "*adat* law" and subsequently shaped indigenous arrangements for their own purposes, at least in Sama Dua, the concepts developed within "*adat* law" continues to have referents in aspects of local life and identity.

A reading of colonial reports reveals the origins of the disjuncture between *adat* regulatory order(s) regarding access to and use of local common pool resources and state regulations governing access to what was to become (officially) state forest. Under Suharto's regime the Basic Forestry Law (Act. No. /1967) placed all "forest estate" under the management of state forest agency, and the TGHK forest agreement specified and even expanded the area subject to this regime. Accordingly, this

disjunction not only continued; with the granting of concessions and the
setting up of protected areas, it even widened. Irrespective of this, at the
same time local *adat* practices regarding access and use rights over the
forest continued to operate. Despite unsympathetic national laws, the
communities maintained *adat* territoriality. In particular, the practice of
levying village taxes on commercial uses within *adat* territory proved the
continued vitality of long-standing *adat* rules that contradicted state law,
attesting to the unremitting history of local resistance to the state forestry
order. Just as local people during the 1990s attempted to thwart outside
claims over local resources, in colonial times local heads eluded the long,
gradual process whereby the state laws attempted to alienate property
rights over local forests. Despite increased state intervention, *adat* concepts and practices had proved surprisingly enduring.

While state and *adat* institutional arrangements often stood in opposition to each other, since the state started to show an interest in this area
there have been mutual adjustments between *adat* and state law.[169] Over
recent decades, the *seuneubok* came to depend upon external sources of
authority and legitimacy. Where *adat* leaders could locate corresponding
priorities in the state order, at particular moments they found ways to
mobilize the state regime to support customary *adat* arrangements. This
suggests that the binary approach of contrasting state law and a local
customary institutional order, of dominance and counter-resistance, can
fail to see the real connections between local customary arrangements and
state law. As a close examination of this situation reveals, as well as being
at odds, over time these two orders have constantly made mutual adjustments and accommodations. This suggests that, rather than focusing on
how a discrete normative field (*adat* or the state) regulates resource use, it
is a necessity to understand how ecological outcomes emerge amidst the
complexity of "shifting patterns of dominance, resistance, and acquiescence, which occur simultaneously" (Wilson 2000: 16).

Thus, analysts might do well to avoid simplifying, reifying, or roman-
ticizing *adat*. Campbell has called for a "nuanced understanding of *adat*
as a dynamic and evolving process of village level decision-making interacting and interlocking with external legal, political, social and religious
influences." *Adat*, he has suggested, is not necessarily "a glorious living
tradition of harmony with nature that is fully operative in forest dependent communities" (Campbell 1999: 4).[170] The Sama Dua case bears out
the good sense of this counsel. In Sama Dua at least, as farmers have
converted native forest into cash crop producing gardens, village livli-
hoods and the property relations that sustain these rather than ecological
sustainability remain a central preoccupation. In Sama Dua, as *adat* has
developed over time, it has been extremely adaptable to new economic

situations. This *adat* order proved neither to be principally opposed to commercial and economic development, nor has it necessarily supported sustainable forest use. Rather, ecological change can more readily be understood in terms of the economic and political dynamics driving farmers to open land on the frontier and to log the frontier forests.

3

Menggamat: Turning in Circles

Introduction

In South Aceh district, some twenty-eight kilometers south of the district capital of Tapaktuan, lies the township of Kota Fajar, the center of North Kluet sub-district. From Kota Fajar, a road turns north and follows the Kluet River (*Krung Kluet*) upstream as it meanders along a river valley. The road narrows as it winds higher up the escarpment, offering views of a wide river plain of verdant rice paddy, set among steep hills covered in the deep green of the tropical rainforest. Some forty-five kilometers from Kota Fajar, the road terminates among a complex of villages, the village league (*kemukiman*) of Menggamat (see Map 7).

The Kluet ethnic group is indigenous to the upper end of the Kluet River valley, a small pocket between the mountains stretching inland from the coast behind Sama Dua (see Maps 4 and 5). This small population is distinguished by their use of the Kluet language (*bahasa Kluet*). Although the Kluet use their language for everyday communication, in primary schools, and in *adat* ceremonies, it has never been written down. Written sources regarding the Kluet people—either in Dutch or in Indonesian—are very limited, and, according to Ismail (1990: 1–2), the exact number of native speakers of *bahasa Kluet* is unknown. Based on the population of the twenty-eight villages where the Kluet people are said to predominate, and using statistics from the North Kluet Office of Statistics, in 1992 Khairuddin estimated that there were 20,279 people of Kluet ethnicity (Khairuddin 1992: 15).[1]

Despite its proximity to Sama Dua, Menggamat, in the upper reaches of the Kluet River valley, presents a sharp contrast. Until recently the people of Menggamat lived in comparative isolation. Although the area has exported non-timber forest products in the past, Menggamat lacks Sama Dua's long history of productive agriculture for the world market, and there are also comparatively few historical records of the area. Yet, Menggamat also adjoins a substantial area of forest-protection forest and the Gunung Leuser National Park to the east.[2] After the district government constructed a new road into the area during the 1990s, Menggamat attracted the notice of district logging networks. To stabilize this forest frontier, during 1994–96 the World Wide Fund for Nature's Leuser Program (WWF-LP) chose Menggamat for community-based conservation activities. As these events occurred so recently and in such a dramatic fashion, Menggamat presents a convenient opportunity to study the emergence of logging networks and an attempt by a community-based conservation intervention to contain them.

This chapter first explores the history of Menggamat and the development of *adat* arrangements pertaining to agriculture and forestry in Menggamat. I will then consider the institutional dynamics surrounding the emergence of illegal logging and subsequent community-based conservation efforts in the area. Finally, I will discuss the impact of economic and political fluctuations over the crisis period of 1997–99.

Menggamat

Oral history and the limited written information concerning the area indicate that the first settlers came from the neighboring Alas peoples (Khairuddin 1992: 13). Informants explain that *bahasa Kluet* is very similar to the language of the Alas people, and is closely related to the languages of the Singkil and Pak-Pak ethnic groups.[3] However, as the Kluet valley is close to the coast, waves of migration from Aceh and West Sumatra down the coast also affected the valley's demography. Although Aneuk Jamee and Aceh speaking communities now reside on the coast, historically people from these ethnic groups were absorbed into the Kluet speaking people of the upper Kluet River and they subsequently adopted the Kluet language and customs. The term Kluet is said to originate from the Acehnese word for "wild" (*liar*). Other accounts derive it from the word *kalut*, which means to undertake ascetic practices (*bertapa*) in the forest (Hidayah 1997: 136). Both terms indicate how neighboring coastal people regarded this formerly isolated group. During the course of my research, villagers in Sama Dua often told stories of mysticism and fearsome black magic practices in Menggamat. Villagers from the coast

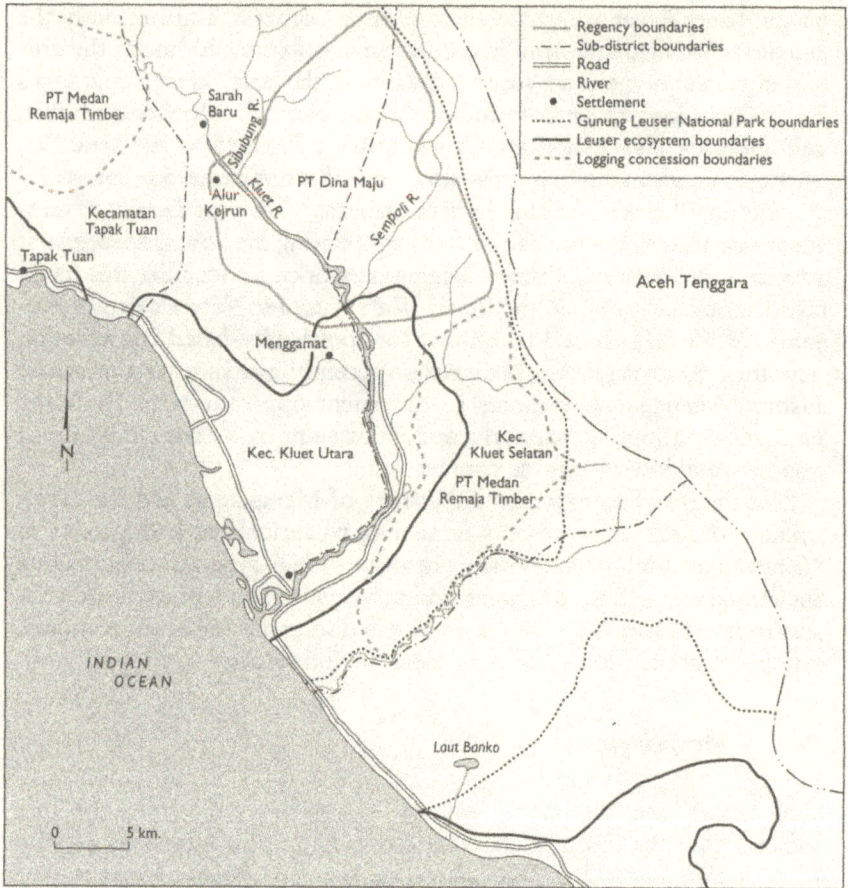

Legend:
- ———— Regency boundaries
- —·—·— Sub-district boundaries
- ═══ Road
- ——— River
- ● Settlement
- ········ Gunung Leuser National Park boundaries
- ——— Leuser ecosystem boundaries
- ---- Logging concession boundaries

PT Medan Remaja Timber — Sarah Baru — Sibubung R. — Kluet R. — Alur Kejrun — PT Dina Maju — Sempoli R. — Kecamatan Tapak Tuan — Tapak Tuan — Aceh Tenggara — Menggamat — Kec. Kluet Utara — Kec. Kluet Selatan — PT Medan Remaja Timber — INDIAN OCEAN — Laut Banko

N

0   5 km.

MAP 7. North Kluet Sub-district: forest concession and National Park boundaries.

were also afraid to visit Menggamat because of allegations of widespread use of poison.[4] Given the remoteness of the area and its proximity to inaccessible forest, since colonial times the upper Kluet River has reputedly served as an ideal hiding place for guerrillas and others out of favor with the authorities.[5] These stories reflect a pattern found elsewhere in Indonesia in which lowland people harbor disparaging representations of interior and marginal upland peoples.[6]

According to stories handed down from the past, around nine generations ago the ancestors of the present residents of Menggamat opened land around the present-day village complex.[7] People of Alas, Acehnese, and Padang descent divided the land, and each group took a section of the immediate Menggamat area, establishing three distinct settlements.[8] They placed boundaries between the different areas made of *gamat* wood. Subsequently, as an older informant noted, the settlement was known as Menggamat, which means "to use the signposts."[9] Kinship formed the basis for organizing the life of these settlements, and each group came to form an exogamous clan group (*marga*). The group of Acehnese identity, known as *Pinim*, settled in the south of the current settlement and the Alas *Selian* clan settled in the central area,[10] while the settlers of Padang descent, known as Candiago, settled to the north. At that time, the *marga* system was also connected to how newcomers gained access to land:

> In the past someone coming to open land would be adopted into a *marga*. They would be taken as family—as younger brother—even if they came from another ethnic group (*suku*). From then onwards they would be considered to be part of that *marga*. In this way, the three original kampungs gradually grew. After finding land they would then tell the *keucik* (village head) that they were working so-much land in such a location.[11]

Eventually, Acehnese political influence on the coast dominated what became known as the Kluet kingdom (see tables 3.1 and 3.2): Teuku Kilat Fajar, a descendant of the Sultan Marhum Kahar of Banda Aceh, assumed the title of lord (*kejuren*) (Ahmad 1992: 278–79; World Wildlife Fund and ID 0106 Gunung Leuser National Park Conservation Project n.d.: 1).[12] In setting up their system of indirect rule, the Dutch recognized this *kejuren* as the territorial head (*landschapshoofd*) of the Kluet subdivision (*landschap*), which included the wider area of Bakongan and Kandang (Kreemer 1923: 234) (see Table 3.1).[13] Descendants of Teuku Kilat Fajar continued to hold the *kejuren* position up until the Japanese occupation in 1942 (Ahmad 1992: 279).

Nowadays, the Kluet area consists of two sub-districts (*kecamatan*) separated by the Kluet River, each incorporating several traditional community leagues (*kemukiman*). The sub-district of South Kluet, with its center in the town of Kandang, lies south of the river. Here, people of Kluet ethnicity predominate in two village leagues on the northern end of the valley. In North Kluet sub-district, the Kluet also predominate in the twenty-one villages that make up the two northernmost *kemukiman* (Khairuddin 1992: 4). The 5,500 residents of Kemukiman Menggamat live in thirteen villages located north-south along the Menggamat and Kluet rivers (Nababan 1996: 2).

TABLE 3.1
Administrative Structures in Menggamat

Colonial period—Menggamat (1926)[i]	Contemporary (1999)
Keresidenan Aceh	Governor of Aceh Province
Asisten Residen stationed at Meu-laboh administered the Division of West Coast of Aceh (*Afdeeling West-kust Van Aceh*).	
Controleur administered Tapaktuan Subdivision (*Onderafdeeling Tapaktuan*)	*Bupati* of Kabupaten of South Aceh (South Aceh District) administered from Tapaktuan
Kejuren (*Zelfbestuurder*—self-governing head)	*Camat* administers sub-district (*kecamatan*)
Uleebalang[ii]	*Kepala mukim* (ceremonial *adat* role only), head of village league (*mukim*).
Keucik advised by village elders (*petuhapet*)	Village head (*kepala desa*) with LKMD/LMD village councils (under leadership of village head)

[i] This was the administrative structure in 1926. There were several administrative reorganizations between the inception of colonial rule and World War II.
[ii] As explained in the following footnotes, the positions of the *uleebalang* in Menggamat differed from those found in historical Aceh.

Agriculture and Forest Products

The villagers of Menggamat have traditionally cultivated wet rice (*sawah*) in the easily irrigated areas surrounding the Kluet and Menggamat rivers. To supplement rice cultivation, farmers also opened dry-land gardens, where they cultivated legumes (*kacang-kacangan*), chili peppers (*cabe*), and other vegetables as well as banana, coconut, coffee, and fruit for their own use. Due to the historical isolation of the area, the farmers of Menggamat were poorly integrated into the cash economy. According to farmers, nutmeg and cloves, the tree crops that dominated the permanent forest gardens (*kebun*) of South Aceh's coastal villages and have histori-cally sustained their cash economy, are not well suited to Kluet.[14] In the past, *nilam* (patchouli), a crop introduced by the Dutch, was the sole cash crop. A plant long associated with shifting agriculture, it grows well in Menggamat's hilly soils.

In addition to *nilam* cultivation, following the opening of the road to Menggamat in the 1980s, farmers increasingly cultivated other crops such as areca nut (*pinang*), rubber, coffee, peanuts, and most recently chocolate in mixed-crop gardens (*kebun*). Since it was introduced around 1985, as in the Alas valley, candlenut (*kemiri*) has also become a common tree-crop. In the 1990s, farmers would cultivate *sawah* and permanent gardens (*kebun*) close to the village. When *nilam* prices were high, some farmers would also open temporary plots (*ladang*) under shifting cultivation further away from the village.

As the market value of the various agricultural products fluctuates widely, farmers shift between crops and agricultural strategies, and the villagers also supplement agriculture with other activities. Living on the edge of a large forest area, local people have hunted animals for protein and collected firewood, medicinal herbs, fruit, and building materials from the surrounding rainforest. The object of forest product collection has tended to vary over time. The collection of forest rubbers (such as *getah balem*) was significant before the colonial power effectively controlled Menggamat after the Bakengon war, a guerilla conflict that occurred in the 1920s. During the colonial period *damar* resin (*damar mata kucing*) had a high price, but the market for *damar* in Menggamat disappeared during World War II and never really recovered.[15] Villagers had also collected rattan during the colonial time, but never in large amounts. After independence, rattan had more of a market and significant amounts were collected.[16] Villagers have also harvested *gaharu* (the diseased heartwood of *Aquilaria malaccensis*), *kemenyan* (benzoin), and other forest products for cash income. In recent years, as we will see, many have also been drawn into logging.

Institutional Arrangements in Menggamat

In Menggamat, a customary authority structure evolved to supervise the governance of cultivated village lands and forest resources within the Menggamat territory. Before discussing the complex interactions surrounding resource uses, I will briefly consider the nature of the *adat* authority structure and associated concepts of local tenure and territoriality over which it has exerted varying degrees of control.

Due to the greater degree of Acehnese cultural influence in Kluet compared with Sama Dua, the *adat* authority structure operating in Kluet was closer to that found historically in the heartland of northern Aceh. In Menggamat, the *kemukiman* consisted of a league of villages with a unitary *adat* regime that governed village life. *Adat* regulations and

TABLE 3.2

Forest Uses and Associated Institutional Arrangements in Menggamat

Forest Use	Village/customary regime	State regime
1. Collection of forest products by villages in forest	• Jurisdiction of customary authorities over Menggamat territory limited by Forestry Department division of territory into production and protection forest; De jure local tenure limited to area classified as *Hutan Negara Bebas*—the cultivation area (village lands) directly behind the core villages. • Following the creation of a Community Forest (CCF), *adat* leaders were able to participate in decision-making regarding Menggamat territories classified as State forest lands; however local State authorities had a veto on all decisions regarding forest use. • Historically these resources were so extensive that generally there were no rules on extraction by local villagers. • During 1920s–1930s exclusive property rights over blocks of damar trees were given to local figures. • Outside collectors or traders paid *uleebalang* or other *adat* head "taxes" on those exporting forest products from the area.	• Operational rules of forestry department set out in State regulations, ministerial decrees and laws regarding State forest lands. According to TGHK, *adat* tenure regime operates on land outside specific forest boundaries; large area of Menggamat territory allocated to HPH and other areas set aside as protection forest. • State rules prohibiting local use of protection and production forests not enforced. • Following creation of Community Forest in degraded production and protection forests, collection of non-timber forest products was to be allowed under conditions of comanagement agreement with local community organization (YPPAMAM).
2. Opening land for agriculture	• Historically, free access to residents of Menggamat village communities. • After the 1970s need to ask permission from *adat*/village head to open plots of land in village territories close to Menggamat core villages. • In remote villages of Alur Kejuren and Sarah Baru free access continues.	• On State forest land (especially in "Production Forest" and "Protection Forest") transformative uses are prohibited by forest authorities. However, these rules are often not enforced.
3. Logging	• Historically, no commercial logging. Plenty of wood for villagers' own use close to villages. • Last decades: (in theory) need to ask permission from village head before taking wood in village territory of core villages. • With commercial logging, limited ability to control logging of areas outside core village lands; village heads collude in illegal logging.	• Official operational rules prohibit logging in State forest land without a valid permit. • Collusion by *oknum* allows for logging operations with implicit backing of local State agents. Decisions regarding enforcement made surreptitiously by local State agents or depend on informal understandings prevalent among officials and loggers.
4. Hydrological functions	• Due to awareness of the limiting conditions under which agriculture takes place – that degradation to the Kluet watershed will affect the productivity of agricultural land – the adat regime has developed some rules to prevent irresponsible farming and environmental damage, such as cutting trees in an inappropriate place or at the wrong time of the year.	• Menggamat area classified as part of the Leuser Ecosystem; State laws classify area as protection forests and surrounding GLNP areas.
5. Non-extractive use[i]		• Menggamat area classified as part of the Leuser Ecosystem; State laws nominally protect areas classified by the district spatial plan as protection forests and or areas included within the National Park (TNGL).

[i] As noted in the Introduction, "non-extractive uses" include: biodiversity conservation, recreation, research, carbon sequestration.

decisions were "carried out and maintained in an institutional structure under the leadership of an *uleebalang*" for the whole of the *kemukiman* (Nababan 1996: 2).[17]

Tenure

Up until the 1960s, Menggamat had abundant accessible land. As rice land was under constant cultivation, rights over these areas naturally tended to be permanent. Menggamat lacked the extensive permanent tree gardens found in Sama Dua. This meant that in dry land areas under the swidden system then operating, villagers had use-rights (*hak pakai*) over temporary plots (*ladang*) rather than permanent private property rights (*hak milik*). Villagers would open new plots of land or shift plots without reporting to the village head first. Rights in land tended to be less abiding, and once a farmer abandoned a *ladang*, the property rights of a former user lapsed and another villager could use it. Rights in trees were more abiding: "If a durian tree was on my former plot, another person could then use it as long as they didn't cut down the tree. In the past people didn't have to even ask—rights were lost over the land if a farmer wasn't using it."[18] As one village head noted, in the past *adat* was less concerned with rights over land, and more focused on village affairs.[19]

This situation began to change during the 1970s, when *adat* rules were extended to tenurial issues and the new practices were enshrined under a new *adat* agreement.[20] From then on, villagers had to report to the village head before taking over a new piece of land. The head of one village explained how, under this regime, farmers were required to ask permission from the village head before shifting plots. If a farmer opened a new plot without permission from the village head, the village head would call an *adat* session in which the errant farmer would be called to account. This type of offense would usually fall into the category of *gempar malu*, for which the main sanction is primarily social rather than material. As an offender of *adat*, the farmer loses his standing and good name in the eyes of the village. If a farmer commits a further offense, this is more serious because on-going breaches of the authority of the *adat* head will endanger the whole *adat* regime. As such offenses dishonor the authority of the *adat* head, they are known as *malu rajo* ("shaming the head"). The village head can then call the family head (*ahli waris*) of the offender, who will give the person a warning. Continual infringements can lead to a combination of material and social sanctions: from loss of access to village territory and finally to disinheritance or even expulsion from the village.[21,22] As in other customary systems, ultimately this fear of loss of access would be sufficient to induce a reasonable level of compliance (Moore 1986). Under this *adat* system, if a farmer used a plot of land and subsequently abandoned it without leaving perennial trees, after three years the land would revert to the village and the village head could allocate it to another farmer. If there were any planted trees on the land, whoever wanted to make use of the land would also need to ask permission from the previous user.[23]

This new regulation brought Menggamat's tenurial arrangements more into line with those operating in other areas, helping to facilitate the collection of government land tax in keeping with government policies that disapproved of shifting agriculture. Moreover, it reflected wider changes as, following the opening of the road, Menggamat increasingly integrated with the wider cash economy. With greater access to market, this change helped stimulate farmers to cultivate perennial tree crops. By planting newly introduced tree crops (such as candlenut), farmers now secured abiding property rights over land. Alternatively, if they abandoned a swidden area without planting permanent trees, the village head could reallocate the land to another farmer, meaning that the farmer would lose the labor time invested in clearing the land of large trees. If the land was located strategically close to the village, the farmer would also lose the convenience of cultivating land where the task of carrying farm produce back to the village was less onerous. Accordingly, farmers choose to preserve their rights over plots close to the villages. For instance, after harvesting a *nilam* crop, they would plant the land with some perennial trees. In the meantime, they would move to another plot, planting younger plants there, only returning to the first plot when the tree crops established there became mature.[24] In this fashion, as in Sama Dua and elsewhere in Indonesia, rather than existing for staple food production, in the core area of Menggamat the swidden practice became a part of the establishment of an income-generating agro-ecosystem (Michon et al. 2000; Dove 1983).

Yet, the tenurial regime that has developed since the 1970s still allows for shifting cultivation, particularly in the frontier areas further out from the core village complex. Successive plantings of short-term crops, such as chili peppers (*cabe*) or legumes (*kacang-kacangan*) on the same plot of land without fertilizer will lead to a gradual fall in production. Moreover, after a single crop of *nilam*, without fertilizer the land will be infertile for up to two years. Partly to overcome this problem, swidden systems exploit the biomass of living forest by using fire to transform forest into its base elements, which then become available as nutrients for annual food crops planted on the site (Dove 1985). To make use of the forest biomass in this fashion, from time to time villagers need to shift plots.[25]

This system resembles what, in a study of the Maine Lobster Industry, Acheson has termed "nucleated territoriality" (Acheson 1989).[26] That is, farmers working land closer to the core settlement of Menggamat tend to invest in permanent tenurial rights, whereas those working further out have a weaker sense of ownership. Accordingly, in the remoter villages, the regime described above does not operate. Alur Kejuren is located some six hours on foot from the Menggamat core villages, or two hours by motorboat up the Kluet River, and the furthest hamlet is even more remote. In the past villagers would move here temporarily to plant cash crops such as *nilam* in the fresh, fertile soils. A few years later they would return with the capital they accumulated to a more comfortable village life in the core villages of Menggamat. The land here, referred to euphemistically as "temporary land" (*lahan lari*), was not subject to the *adat* regime imposed closer to the core villages. Even today, because the village head is not resident, the area is not subject to the *adat* regime operating in the core villages of Menggamat. Villagers can still readily shift plots, just as they used to in the core villages of Menggamat.

The nucleated nature of Menggamat's tenurial regimes was illustrated by the *nilam* boom of 1997–98.[27] At this time many farmers who owned *sawah* were unable to combine *nilam* farming with wet rice cultivation: they entrusted the task of cultivating wet rice in the irrigated flat lands close to the village to other family members resident in the village. Many also maintained permanent dryland mixed plots of candlenut, coffee, fruit trees, and other crops in the village territories subject to the *adat* regime described earlier. In addition, after negotiations within the village, many villagers opened *nilam* gardens in abandoned plots of land within the village territories, particularly on the fertile banks of the Kluet River. However, as in Sama Dua, farmers preferred to plant *nilam* in the extensive forest areas further out, on areas previously logged or on fertile virgin forest land. A farmer I interviewed in early 1998 described how, like many farmers, he had a plot in the forest a day's walk from the Menggamat village complex. Besides the increased fertility provided by burning a forest area and thereby using nutrients in the ashes to sustain newly planted crops, the distant forest areas provided him with wood for cooking the *nilam* leaves. He said that one drum of *nilam* leaves required a cubic meter of wood. "If the *kebun* is further out," he said, "the wood is close by. It is much harder in the kampung: it is difficult to find wood. As more time passes, the further away people have to go."[28] As these tasks are so time consuming, villagers would build huts in their *nilam* gardens and take up residence there. Some villagers with distant plots take their wives with them, returning only once a month.[29]

According to a village head interviewed in Menggamat, ideally within the village territory the functioning of the *adat* system would largely depend on the discretion of the *keucik*.[30] For instance, with the consent of the village head, villagers could obtain timber in the village lands. In granting his consent, the village head aimed to prevent villagers from taking trees on steep lands in a way that would contribute to erosion, landslides, or floods that would affect other farmers. Moreover, those wishing to cut wood also needed to explain to the head how they intended to use the wood; the head needed to be satisfied that it would be used within the village and not for sale outside. When considering a villager's request for use rights over a piece of forest for opening a *ladang*, the village head also took into account the farmer's need for land and the availability of suitable plots of the village land. In judging the authenticity of an individual's requirement, the *keucik* has to consider the request alongside the village's overall needs.[31] Therefore, the *keucik* have articulated a generally accepted sense of correct practice, and have gained respect for their ability to make good decisions on behalf of the whole village.

In Menggamat, there were several *adat* rules directly related to maintaining ecological conditions conducive to farming. The village head explained how the *adat* regime attempts to prevent irresponsible logging in the village territory. If someone cut timber in abandoned agricultural land without the permission of the village head, he noted, penalties would be light. However, if he cut a tree in virgin forest, or on hillside subject to erosion, or in the watershed of the river, where cutting was forbidden altogether, sanctions would tend to be heavier. Depending on the severity of the offense, sanctions could range from a fine of one goat to losing the right to use the *adat* forest, or in extreme cases even to exile from the village. Accordingly, sanctions relating to infringements of *adat* rules regarding use of village territory were graduated according to the seriousness and context of the offence.

Furthermore, under the rules administered by the *kejuren blang*, an *adat* official in charge of irrigation, villagers were not permitted to cut trees close to the river or a water source to prevent erosion. Those who do take trees from the village land were obliged to replant a specified number of trees on village land to replace those cut down. There were also sanctions prohibiting the cutting of the forest during specific times of the year. Long experience had shown that the rice crop would be vulnerable to infestation by pests disturbed from the forest. According to the *keucik* of one village,

There is a ritual feast [known as the *kenduri ulelung*] that is held three months before the harvest, when the *padi* (rice plant) flowers. During this *kenduri*, the *keucik* will make a speech. He will say that, if you want to cut wood in the forest, the sound of a falling log cannot be heard for three months—because it is dangerous for the *padi* to hear this [i.e., no trees are to be felled in hearing distance of the rice fields]. After this time villagers are forbidden to take forest products or logs past the *padi* without first leaving it for a night in the forest. Otherwise *orang halus* (spirits) come and wreck the *padi*. [Otherwise] logging in this area is forbidden after this speech because it will disturb pests, who will then destroy the harvest. Whoever infringes these sanctions will be fined one goat, four containers of rice (*bambu beras*), and four coconuts.[32]

Territoriality

Beyond the village lands under tenure, a large area of steep country covered by rainforest has long been considered Menggamat territory. According to the former *uleebalang* of *kemukiman* Menggamat, the *adat* territory once included the whole watershed area of the Kluet River.[33] To the northwest, the Kluet watershed is cut off from the coast by a forbidding range covered by thick rainforest. Logically enough, this natural feature marked the boundaries of their area with the communities living on the coastal side of this mountainous country. During colonial times these mountainous boundaries were inscribed on maps, and they now constitute the boundaries with Tapaktuan and Sama Dua administrative areas (*kecamatan*) to the west. Maps show how a river that descends from the wilderness of the Leuser Ecosystem to join the Kluet River forms the eastern boundary of Menggamat.[34] To the south, the mountain peaks set the boundaries with the other *kemukiman* within the Kluet area.[35] To the north, along the Kluet River the people of Menggamat tell of a history of earlier land use here and the Kluet territory is said to extend up to where a stream flows into the Kluet River.[36,37]

As in Sama Dua, the physical characteristics of the wider Menggamat area limit the ability of local communities to control forest resources found within Menggamat territory. The geographical boundaries of this territory may be well known; however, many forest products are widely dispersed in the forest, and their regeneration may be erratic and unpredictable (Michon et al. 2000). Although at times the collection of forest products has been an important source of income, the price of forest products often fluctuates considerably. Consequently, villages have been unable to depend on the collection of forest products for their livelihood. Even so, when prices are high, there are large commercial incentives for unsustainable extraction.[38] This low "salience" means that local communities have less incentive to invest in creating institutions that ensure sustainable harvesting of forest resources within their territories (Ostrom

1997). Accordingly, in the extensive forest behind the villages, access to forest products for villagers has tended to be "free."[39] As in Sama Dua, historically there was one simple rule regarding the collection of forest products: the collector has to be a member of the village to use these resources. Otherwise, during the colonial period, a trader buying forest products from the area or a collector gathering products directly would need to ask permission from the *adat* head concerned. The collector or trader would then be subject to a "tax" (known as the *pantjang alas*) paid to this *adat* head (Adatrechtbundels 1938).

The exception to the rule emerged during the colonial period. In this period, as elsewhere in the archipelago, when one particular resource became increasingly salient upriver, aristocracies developed territorial strategies to extract benefits from the resource by attempting to control access.[40] According to local accounts, a significant market for *damar mata kucing* (*Shorea javanica*) resin developed during 1925–28, at the time of the guerrilla war between colonial forces and local rebel forces. At this time the guerrillas had taken to the jungle. Local informants hold that the Dutch increased the price of *damar* to stimulate collectors to range across the forest, harvesting *damar* while looking for those who were hiding. During this period local notables began to employ collectors who went to the forest to gather *damar* on the thickly forested, steep hillsides of Menggamat. Eventually, these people, usually influential figures of the locally powerful *pinim* group (the exogamous clan group of Acehnese identity), "set out boundaries and reported these to the *uleebalang*."[41] The *uleebalang* issued property rights over these blocks of forest for the collection of *damar* resin. From this time on, "other people could no longer enter, only the owner's people."[42] Although these figures are said to have promised to report any *Muslimin* activity in their block, many were in the trade for financial reasons and failed to inform on guerrillas. After paying tax to the *uleebalang*, the owner of the block would then sell *damar* on to outside traders. The price of *damar* fell during the Japanese occupation, never to fully recover, and the block system fell into disuse.

The preceding discussion discloses how *adat* institutional arrangements operated prior to the arrival of logging networks in Menggamat. As we will now see, this system has come under considerable stress during the last decade. To understand how this occurred, it is necessary to recount the history of state forestry planning in the Menggamat area.

The State Forestry Regime and Menggamat Territory

As we discussed in the last chapter, the colonial state and its successor, the Indonesian state, have engaged in a strategy of territorialization whereby the forest is divided into specific zones. Within each zone, to varying degrees the state has attempted to regulate the activities of people and the extraction of natural resources. Contemporary forest maps of South Aceh reveal a mosaic of forest classifications that reflects the local history of state territorialization (see Map 7). This history can be broken down into three phases.

First, when the Netherland-Indies colonial government created the Leuser Nature Reserve in 1934, this included a mountainous area to the east. Contemporary maps show that the National Park now contains the upper headwaters of the Sampali River, some fifteen kilometers east of Menggamat. However, the advocates of the Leuser reserve were disappointed: the new reserve failed to protect a full range of ecosystems stretching from the coast up to Mount Leuser, as was originally envisaged. Accordingly, in 1936 the colonial government added 20,000 hectares of lowland rainforest and coastal swamp, an area considered to be of special importance as a corridor for elephant herds (Wind 1996: 7). Originally known as the Kluet Nature Reserve, this area borders South Kluet. It lies some fifteen kilometers from Menggamat, and extends down to the coast between Kandang and Bakongan.

Second, as we saw in Chapter 2, during the 1970s the state embarked on a territorial strategy that classified most of "outer island" Indonesia as state "forest zone" (*kawasan hutan*). The "forest use agreement" (*tata guna hutan kesepakatan* or TGHK) included 88 percent of the mountainous and heavily forested district of South Aceh as "forest zone" (Bappeda 1992). In 1978 Strien noted that "in Aceh it is the policy of Forestry to hand out concessions without regard to the topography and the potential for logging. The whole forest area has been split up among the various candidates and only later the feasibility of logging is investigated. The result of this policy is that concessions have been given out in areas that are totally unsuitable for logging" (Strien 1978: 50). In the 1980s a mapping exercise (the RePPProT process) refined the TGHK maps, leaving some 39 percent of a total area of 8,910 square kilometers of South Aceh classified as "production forest" earmarked for logging concessions and other "productive uses" under the direction of the Ministry of Forestry and its line agencies. The RePPProT map for the Kluet region showed the steep upper reaches of the Kluet area as limited production forest, an area that was granted to the concessionaire PT Dina Maju.[43] A corridor of land some five kilometers to the east was classified as protection forest. The cultivated area around the Menggamat settlements was left unclassified (*Hutan Negara Bebas*), while virtually all of the South Kluet area was classified as limited production forest.

Third, in the early 1990s, in accordance with the 1992 Spatial Planning Act, the district government prepared a Spatial Use Management Plan for South Aceh. This plan reported that the hilly forest areas surrounding Kluet had slopes of over 40 percent (Pemerintah Kabupaten Daerah Tingkat II Aceh Selatan 1991/1992). This meant that, following the requirements of the new legislation, the Menggamat territory that was previously classified as limited production forest in the RePPProT maps now needed to be reclassified as "protection forest" and to be exempt from logging. In 1994 a district government regulation determined that the legal status of the forest area should be changed to "protection area," a decision ratified by the provincial governor (Serambi Indonesia 1995b).[44] The Ministry of Forestry had already given out long-term leases over forest lands. This meant that here, as elsewhere, the new Spatial Use Management Plan for South Aceh awaited integration (*paduserasi*) with Forestry Department procedures and the expiration of previous leases. In early 1998, the twenty-year logging concessions in South Aceh, such as the one in North Kluet, remained technically valid.

In accordance with the earlier process, in the 1970s the Forestry Department had granted a logging concession to the company PT Dina Maju in the upper Kluet area. As Strien described it, the area is "very steep, deeply intersected mountains" (Strien 1978). A map prepared by WWF-LP shows that the concession began just beyond the village lands of the core Menggamat villages (see Map 7). To the east, the concession extended virtually to the boundary of the national park, while to the north it included both sides of the Kluet River to just beyond the Sibubung stream. The concession also encompassed two peaks of over a thousand meters in height, most of the watershed of the Sampali River, and the lands of Alur Kejuren village.

As local informants remember, PT Dina Maju faced considerable difficulties logging here. Village leaders did not give the company permission to open a road through the village territory. While carrying out logging operations during 1975–77, the company could not drive their heavy machinery into the forest and were forced to extract timber via the river. The giant logs extracted from the Sampali and Sarah Baru areas were up to 2.5 meters in girth. PT Dina Maju floated them down to Kandang, where they were loaded onto boats for export. During exploitation Dina Maju faced significant problems.[45] The current was not strong enough to carry such giant logs, and at times it took months to reach Kandang. In the meantime the logs would often rot in the water. At other times the Kluet River flooded, and logs were carried out to sea by the current or marooned high on the river bank, where they would also rot. Local informants said that because PT Dina Maju's activities in the forest disturbed local spirits (*setan*), logging operations were subject to all kind of accidents and mishaps.[46] In any case, after two years of struggling, the company gave up and moved to a more accessible concession in West Aceh.

Despite the official status of surrounding areas as "forest estate," in Menggamat this official status remained virtual. The local forestry agency (*Dinas Kehutanan*) and the national park staff had never established a presence in Menggamat. As in other areas of Indonesia, the forestry agency suffered from limited resources, low wages, and insufficient personnel.[47] In early 1999 a Dinas official noted that the forestry region where Menggamat is located (*cabang 7*) had forty-three forestry staff, ten of whom were desk-bound administrators. The remaining thirty-three staff members had responsibility for 463,775 hectares of state forest. The Menggamat area is not a site of active operations by a HPH, nor is it directly adjacent to the national park (TNGL). The closest offices of the Dinas Kehutanan and TNGL are in Kota Fajar and the TNGL office for Kluet Selatan (close to Bakengon), respectively (see Map 7). This has meant that, unlike the situation described in Jambur Lak-Lak in the Alas valley on the other side of the forested ranges (see next chapter), apart from the brief period of logging concession operations, there had never been an attempt to assert the state's territorial claim over the forest here in accordance with the forestry agency's TGHK maps. This was the case until WWF-LP attempted to support the conservation of protected and National Park buffer zone areas in South Aceh.

Logging Networks in South Aceh

The widespread logging of Menggamat's forests that began during the 1990s can be viewed from two perspectives. Observed from the outside, this phenomenon can be explained in terms of district-level interests extending their activities into Menggamat. From the villagers' perspective, people explained how many took up logging in response to economic and institutional changes within Menggamat itself. I will begin this narrative from the district-level perspective, and later shift to the village viewpoint.

Logging Networks and District-Level Networks of Power and Interest

Tensions between the state and local interests over the control of forest resources in "outer island" Indonesia have a history that extends back to the colonial period; however, the policies of the New Order that gave logging concessions in *adat* territory sharpened and intensified these tensions. As villagers watched surrounding forests being logged by outside interests irrespective of long-standing local notions of territoriality, resentment arose among these communities, and it often led to various forms of overt and covert conflict between logging concessionaires and local actors.[48]

At the same time, district entrepreneurs (known as *cukong*) also sought ways to exploit the lucrative opportunities offered by logging operations. Lacking capital or connections, district entrepreneurs could not obtain logging concessions from the Department of Forestry in Jakarta. At the district level, however, there were other avenues to gain access to the forest. Those wishing to open timber plantations or agricultural plantations could obtain smaller leases in the state forest zone. If the land to be cleared still contained productive stands of timber, they could acquire Timber Harvest Permits (*Ijin Pemanfaatan Kehutanan,* or IPK) to remove and process valuable logs. Consequently, during the 1990s, those wishing to open timber operations applied for an IPK, which was allocated by district and provincial authorities.

According to an informant who attempted to open such an operation, doing so entailed obtaining three types of permits, each requiring a long process of application to various government agencies. Given the complex administrative system of the time, with parallel, deconcentrated agencies of the central government and "autonomous" provincial and district government agencies, the process for acquiring a permit to manage wood legally was long and involved: it entailed numerous steps, each facilitated by payments.[49]

1. An entrepreneur (*cukong*) with capital and aspiring to open a logging operation needed to begin by obtaining a valid "stock area" for obtaining wood. This process began with the Forest Management Unit (*Resort Pemungutan Hutan*) and culminated with the Land Office (*Kantor Pertanahan*). Various approvals had to be obtained in a twelve-step sequence also involving the Forest Management Agency (*Badan Kesatuan Pemangkaun Hutan*), the District Forestry Office (*Dinas Kehutanan*) at the district and provincial level, the Governor's Office, and the Provincial Office of the Forestry Department (*Kanwil Kehutanan*), who would finally issue an IPK permit for one year.

2. Once aspiring operators obtained an IPK for a valid area, they still could not operate unless they acquired a permit to manage the wood. The *Kanwil Kehutanan* office in the provincial capital issued this permit (*Rencana Pengurus Bahan Baku Industri,* or RPBI). Having an RPBI allowed a sawmill operator to obtain permits to transport timber out of the district. Permits for transporting logs were of two varieties: those for unprocessed logs (*Surat Angkutan Kayu Bulat,* or SAKB) and those used for transporting sawn timber (*Surat Angkutan Kayu Olahan,* or SAKO).

This step was relatively straightforward: the entrepreneur needed to approach the district forestry office (*Dinas Kehutanan Tingkat II*), which would then issue a recommendation to the provincial forestry office (*Dinas Kehutanan Tingkat I*) to issue the permit.

3. The would-be operators then needed to obtain a permit for a sawmill, a process that bypassed the Forestry Offices (*Dinas* and *Kanwil Kehutanan*). This involved seven steps that began with the village head, proceeded to the *camat*, two departments within the *bupati*'s offices and culminated with the approval of the *bupati* himself, who issued a location permit for the sawmill (*Surat Izin Tempat Usaha*). Armed with this permit from the district government, the entrepreneur could finally approach the sub-district and provincial Industry Offices (*Dinas Perindustrian*), which would then issue a license for a sawmill (*Industri Pengolahan Kayu Hulu,* or IPKH).

In a study of regional government finances, Devas (1989) allowed that the regulation of certain activities was essential. However, at the district level excessive regulation, such as the requirement for numerous permits, created obstacles for economic development and scope for corruption, "thereby ensuring that the intended purpose of regulation is not achieved" (Devas 1989: 71).[50] Accordingly, the long, convoluted process for obtaining these permits has led to widespread abuses. The informant estimated that (in 1998) a *cukong* would need to spend more than Rp 40 million and could not expect to complete the business within one year. "So nobody is going to do it this way," he said, "I tried myself and gave up." He concluded that it was better to find a shortcut, paying local officials to turn a blind eye:

It is better to steal than to process a permit, making constant payments to officials. Fifty thousand rupiah is the smallest payment to the district police (*Kapolsek*) . . . Everyone at the local level obtains payments . . . The police or army set up posts and every truck must pay, or people are held until payments are made. Payments are then made up and down the chain of command. All levels take the opportunity offered by the wood trade: district and sub-district officials get a *setoran*—a monthly payment.[51]

For *cukong* who managed to complete the process, the cost would be so high that they would seek to recoup the investment by disregarding regulations. According to a report in *Serambi Indonesia*, "as well as cutting in their own area, they also cut in the area around them"—including areas inside the Gunung Leuser National Park (Serambi Indonesia 1994b).[52]

While most operations did not obtain all the required permits, given the practical utility of the symbolic capital provided by state law, *cukong* preferred to draw on the symbols and meanings of the state legal system and operate with a semblance of legality. The nature of the sawmill licenses (IPKH) granted by the local department of industry office also facilitated this. According to the regional industry office (*kanwil*), a sawmill permit would be issued after the raw materials were available. It should be based on the IPK issued by the forestry office, which was only valid for a single year. However, the industry office issued IPKH sawmill licenses for the life of the sawmill—based on an IPK for just one year.[53] Thus, the sawmill could continue to operate when the IPK had expired, generating a demand for illegally obtained logs. Moreover, *cukong* could obtain an IPKH license from a sawmill whose logging area was exhausted, or move a sawmill with a valid permit into a new location. Many sawmills operated hundreds of kilometers from their registered "stock area." When the police checked the stock area, rather than finding tropical rainforest, they found themselves visiting developed parcels of land complete with washing and bathing facilities (Serambi Indonesia 1995a). In 1994 the *Dinas Perindustrian* office estimated that of eighty-five sawmills operating in South Aceh, thirty-seven were operating illegally; other observers put the number at fifty (Serambi Indonesia 1994b). Once the sawmill processed the timber, to carry it to the market in Medan, as we noted above, required a transport permit (SAKO). Sawmills with valid RPBI permits could issue their own SAKOs, and other sawmills bought SAKOs from legal sawmills or directly from forestry officials (*Dinas Tingkat I*) operating beyond their responsibilities.[54]

The term *cukong* referred to sawmill operators and other entrepreneurial patrons. These could be businessmen operating in their own right. Alternatively, they could be *oknum* such as army (ABRI), police, or even government officials who—working beyond the law—headed logging networks.[55] Local state agents could back the boss of a logging team (*tauke*), who would then carry out logging operations at arm's length from his patron. These *tauke* depended on the patronage of *cukong*, borrowing money and chainsaws to organize operations. To operate effectively, *cukong* needed to seek the patronage of key officials. In return for support, these officials would offer protection to the *cukong*'s operations.

As many articles in the Acehnese newspapers during the 1990s confirm, logging networks spread across South Aceh and beyond. For example, in August 1995, the head of the ruling Golkar faction in the provincial parliament, retired Major General HT Djohan, told *Serambi Indonesia* that many officials were involved in illegal logging in many areas of Aceh, including South Aceh, either as support (*membeking*) or directly acting as bosses (*tauke*):

Every week there are community figures that report the involvement of government figures in wood theft . . . Many officials (*aparat*) have chainsaws. Observe
the scores of illegal timber trucks on the roads each day. See also the scores of
illegal fee collection posts (*pos pungutan*) along the whole length of the highway
. . . The wood business also involves many central government and provincial
government agencies. Regional forestry offices, the army, and even TKPH [spe-
cial anti-logging teams] are involved. Observing the potential and power of the
agencies involved, he said, it is hard not to believe that—if the wood and forest
problem is not handled with justice—this won't offend the community, who see
this taking place before their eyes. (Serambi Indonesia 1995c.)

In the context of South Aceh, forestry officials and the police were unable to enforce the law. On the one hand, the *cukong* maintained logging
operations with the protection of patrons occupying strategic positions
within the local administration. An official from the district forestry service (*Dinas Kehutanan*) interviewed in early 1999 described the result:

I have been involved in forest surveillance since 1986, and I have not yet seen a
cukong caught . . . It is like a vicious circle (*lingkaran setan*) . . . The Police (*polri*)
have people in the field, but when we surrender someone to them, they are just let
go. Or when we take someone to *polri*, they have a personal contact there. Or the
forestry person has a contact, or is involved. So there is a network, and between
friends someone who breaks the law cannot be brought to court. So we tend to
let our friends go.[56]

On the other hand, forestry officials were reluctant or afraid to arrest
poor villagers with few economic options other than logging the forest. In
January 1997, *Serambi Indonesia* reported that a combined security operation (*tim gabungan*) had caught eight hundred illegal loggers in a single
day. However, they were let off with a warning. They were "community
members who took wages from *tauke* who stood in the background."
The real culprits, the ones who should be arrested, it was suggested, were
the *tauke* and *cukong* (Serambi Indonesia 1997b). A forestry official I interviewed in January 1999 explained that the reluctance of law enforcers
to deal with villagers carrying out logging has a basis in experience:

But I don't want to just see the little people getting caught—people who are just
trying to eat go to jail. If they just obtain money, these people don't pay attention to who owns the forest . . . There is only one alternative: increase welfare so
people don't want to go to the forest. If you close the sawmills, there won't even
be wood for coffins . . . During the economic crisis (*krismon*) I couldn't do anything. When the price of *nilam* was high, no one went to the forest—but now it is
busy in the forest. They are busy cutting, and we cannot forbid it or (he draws a
finger across his throat) . . . [In other offices, *jagawana*] have caught people, and
then a crowd of people have descended on the office and wrecked it.[57]

Although *jagawana* forest police carry pistols for self-protection, they are often intimidated by the large number of illegal loggers they meet in the field. When outnumbered, they cannot take action other than to write a report to their superiors. Moreover, as one *jagawana* told me, after the onset of the political and economic crisis (*krismon*), the authorities placed higher priority on avoiding conflict than on defending state forest. Thus, particularly in Aceh, they were more reluctant than ever to take action against villagers carrying out illegal logging to meet pressing economic needs.[58]

Kompas reporters visited the district of South Kluet (adjacent to Menggamat) in February 1997 to investigate illegal logging around GLNP, finding further evidence that those opposing illegal logging placed themselves in danger. The reporters observed seven logging trucks driving towards a sawmill laden with freshly cut timber. The trucks drove past TNGL forestry staff, "who were in the village at the time, but did not make the smallest move to apprehend the illegal wood." When *Kompas* interviewed Nataruddin, the head of the TNGL resort for Kluet Selatan, he said that "we are unable to prohibit the illegal logging because all are connected with a 'mafia' and we don't know who is the boss. If we carry out actions without the support of security officials, perhaps the only thing that will return home will be our names" (Kompas 1997a). According to a witness interviewed by *Kompas*, in October 1996 an official involved in an anti-logging operation became caught up in a stand-off, presumably with police or military officers involved in logging.[59] The official was "almost involved in a shootout with an individual because he was accused of securing confiscated wood." A source told *Kompas* that the difficulty of regulating TNGL wood was a "vicious circle." "Rather than becoming a former hero (*pahlawan kesiangan*)," he said, "better to become involved." "This is the origin of every informal payment," *Kompas* concluded. "The Pucuk Lembang village head imposes his own tariff: every truck that carries wood from his village must pay 3,000 Rp per ton. Besides this, there are also payments to every party" (Kompas 1997a).

This suggests that the social field around logging encompassed extralegal exchanges between *cukong* and *oknum*. The pattern of exchange and accommodation between *oknum* and timber interests has generated its own norms and means of enforcing compliance. To operate, *cukong* need to seek the patronage of an *oknum* who would offer protection to the *cukong*'s operations. Moreover, this context accommodated the need for *oknum* to engage in exchanges with clients, peers, and patrons while it also served to secure district government priorities.

Logging and Revenue

Under the New Order, the final decision regarding the appointment of a *bupati* rested with the Ministry of Internal Affairs in Jakarta. Candidates for *bupati* needed to cultivate support from key officials within this ministry. Currying political favor in the ministry reportedly also involved financial generosity on the side of the candidate. However, as well as gaining the assent of the center, an aspiring local politician needed to be nominated by the district assembly (DPRD). A large entourage of supporters sustained via extensive patronage would help a figure to secure office. At the same time, it would also assist a serving *bupati* to obtain this office for a second time. For these purposes, the *bupati* and other district politicians might enter into exchanges with their bureaucratic peers and colleagues, entrepreneurs, and powerful figures in the district and beyond. They could also use the promise of future discretion over budgetary allocations and the awarding of contracts, permits, and other assets at their disposal to win financial and political support from a wide range of actors.

Once in office, a district head who was unable to raise the district's own funds would have to depend on funding allocated by the central government. Most of these funds were already earmarked for paying the incomes of officials, or allotted to specific budget lines and development priorities. Without significant locally generated revenue, the district government would not be able to service the debts of the district. A district government lacking a sufficient tax base would only be able to support high-profile projects or otherwise distribute patronage by finding revenue outside the official district accounts.

Revenue generated directly by the district—known as *Pendapatan Asli Daerah*, or PAD—constitutes a small component of the total district government budget (*Anggaran Pembangunan dan Belanja Daerah*, or APBD). For instance in the 1990/91 fiscal year PAD contributed only Rp 469 million to the APBD of Rp 12.2 billion, or only approximately 3.8 percent of district revenues (Badan Pusat Statistik Kabupaten Aceh Selatan 1992). However, it is hardly surprising that PAD has played a significant role in the considerations of district government. The district assembly tends to evaluate the performance of the district administration in terms of its ability to raise revenue and initiate high profile projects. In this context, for a *bupati* and his administration, the ability to generate funds has been one measure of the effectiveness of his period in office.[60]

In a study of district government taxation across Indonesia, Devas (1989) observed that district government has always desperately sought

TABLE 3.3
Self-generated District Government Incomes (PAD), South Aceh

Fiscal Year	PAD (Rp)
1987/88	85 million
1989/90	560 million
1990/91	469 million
1997/98	2,881 million

to augment district revenue. *Bupati* have embarked upon ambitious revenue-raising drives to build up the district's self-generated income (PAD). Setting ambitious annual targets, each district typically levied a number of district taxes, the vast majority exacted under regulations (*perda*) set by the district administration itself (Devas 1989). These include charges known as *retribusi*, fees for a wide variety of permits and official forms as well as charges for government services that contribute to district government income (Devas 1989: 81).

In South Aceh each *bupati* has faced the problem that, unlike many areas of Aceh that produce oil and natural gas (*migas*), this district has few industries on which to levy taxes for the district treasury. While South Aceh has had a timber industry, most of the taxes levied on logging concessions have accrued to the central government, with only a small proportion returned to Aceh's districts.

A district forestry office (*Dinas Kehutanan*) official I interviewed in the course of this study said that the government has attempted to prevent the development of large differences between provinces. Accordingly, the government in Jakarta took the timber royalties (*Iuran Hasil Hutan*, or IHH) to the center and split them up according to a complex formula for redistribution across Indonesia's regions. On average, South Aceh received Rp 500 million each year from the timber concessions (HPH) operating in the district. In addition, on average each HPH paid a tax on land and building (*Pajak Bumi dan Bangunan* or PBB) of around Rp 40 million per year, some of which was reallocated to the district. While the seven HPH in South Aceh created revenue for the district, it amounted to only a small percentage of the total timber revenue generated in the region.[61] On May 15, 1998, *Analisa*, a Banda Aceh broadsheet, reported that fifteen HPH concessionaires across Aceh had contributed $US 4.5 million towards the Aceh provincial government budget for the 1997/98 fiscal year. As *Waspada* reported in 1998,

According to Sayed Mudhahar, actually the receipts from timber royalties (*Iuran Hasil Hutan*, or IHH), the Reforestation Fund (*Dana Reboisasi*, or DR), and the

TABLE 3.4
Total District Government Budget, South Aceh[i]

Fiscal Year	APBD (Rp)
1987/88	1.5 billion
1990/91	12.2 billion
1995/96	24.4 billion

[i] A complete set of local government reports were not readily available at the time of research. Nevertheless, overall trends are clear from the data available.

Tax on Land and Building (PBB) are an immense sum, as large as Rp 2 to 2.5 trillion. Yet, he said, for the region [i.e., the province], as indicated by the budget (APBD) for the forestry sector, it [i.e., the revenue returned] is only as large as Rp 15 to 20 billion each year. (Waspada 1998a)[62]

Only a percentage of this amount was then distributed to the district government.

Forest outcomes also depended heavily on the personal attitude and aptitude of successive district heads and the policies they chose to pursue. The *bupati* over the period 1988–93, Sayed Mudhahar, proved a successful lobbyist and fundraiser. As Table 3.3 illustrates, he oversaw a six-fold increase in district incomes over his period of office, while the district government budget also grew significantly. In the late 1980s, the main provincial road through the district was unsealed over many sections, and the trip to Medan was an arduous journey. In the absence of bridges, buses and cars could only cross rivers by raft, and this was particularly difficult when wet season rains swelled rivers. As an article in *Kompas* described the situation, South Aceh was considered one of the least developed parts of Aceh, itself a province at best considered remote from Jakarta: "South Aceh is a name almost identical with backwardness. In Java, perhaps people have never known the name Tapaktuan, the center of South Aceh government. Even in the policy framework for the development of Aceh Province itself, this region is still categorized as left behind by other districts" (Kompas 1991). In this context, the national planning board (*Bappenas*) helped the district to obtain significant development funds from the central government as well as loans from the Asian Development Bank and the World Bank. This enabled his administration to facilitate a range of development projects, and most of the district roads, bridges, and bus terminals were built during this time.[63] Yet, as I discussed in the last chapter, this *bupati* gained a reputation for fighting logging companies from outside the area that were held responsible for widespread environmental damage. At one stage, the *bupati* even threatened to resign if the Department of Forestry agreed to new HPH in the district (Kompas 1991). He also issued decrees supporting conservation, endorsed law enforcement against illegal logging, and initiated efforts to have the status of the Menggamat forest area changed from production forest to protection area (Kelompok Kerja WWF ID 0106 n.d.: 2). According to Sayed Mudhahar himself, he also refused overtures from logging interests attempting to bribe him.[64]

However, the district policy changed under the man who held the position of *bupati* from 1993 to 1998. He also proved to be adroit at raising district income. Over his period of office, PAD increased sixfold, while total district budgets (APBD) approximately doubled (see tables 3.3 and 3.4). Under this *bupati*, the strategy of levying district taxes (known as *retribusi*) expanded to include taxes on logging operations often operating with only a fig leaf of legality. An informant observed that this *bupati* had to service the debts accumulated by his predecessor. However, as he lacked the ability to lobby in Jakarta, he increasingly turned to *retribusi* levied on logging. "This *bupati* set targets of generating of say Rp 1 billion per month from logging," he said. "But, he received Rp 2–3 billion, but the forest is now wrecked."[65] As a WWF-LP document politely put it, the *bupati* "was more interested in increasing regional government income than efforts to protect the environment" (Kelompok Kerja WWF ID 0106 n.d.: 3).

Over this period, those operating sawmills enjoyed close connections with district government. According to informants, the *bupati* gave permission to officials to gain revenue from their operations. As one source put it, "in the name of the region, Pak [*bupati*] made his own regulations to use the forests for funds to develop the region."[66] According to a journalist interviewed in the course of this study, in South Aceh, during this period, the district gained an estimated 60 percent of its income (PAD) from official and informal fees levied on the timber industry, mostly on illegal logging.[67] District government fees levied on timber included:

The "industry and natural resource tax" (*Retribusi Hasil Bumi dan Industri,* or RHBI), a charge imposed on timber at the point of export;

The "third party tax" (*Sumbungan Pihak Ketiga*), levied on products exported from the district. On the provincial border with North Sumatra, trucks would pay Rp 500,000 to leave the district;[68]

The "wood collection location charge" (*Retribusi Tempat Penumpukan Kayu*), levied on all companies engaged in logging operations (*Analisa* 1996).

The police or army also set up "scores of illegal fee collection posts (*pos pungutan*) along the whole length of the highway" (Serambi Indonesia 1995c).

Practices of generating extra-legal revenue could claim some local legitimacy. As I noted earlier, outside interests and central government gained the bulk of economic rents derived from legal logging concessions. Meanwhile, local villagers had to bear the negative externalities of the logging. This gave populist local leaders scope to criticize the way the official state forestry regime facilitated access for outside corporate interests and to condemn the lack of revenue accruing to the district budget from these logging operations. Local loggers encroaching on the state forest estate could then claim that they were only local people taking their share. In other words, resentment against what were considered unjust laws that benefited outside interests—either logging concessions seen to be controlled by ethnic Chinese entrepreneurs or foreign conservationists who (in this view) had privileged access to local resources—was used to justify resource extraction by local elites and villagers alike.

As an informant familiar with district politics told me, however, the impact of the district "policy" that permitted this logging on the local environment did worry some observers:

> It was an open secret that the local government (*Pemda*) was obtaining the bulk of its finance—increasing *pendapatan asli daerah* (PAD)—from wood . . . At the time we wrote articles warning against the dangers of this . . . Many saw the need not to become dependent on this sector alone. Lots of people suggested this, that the *Pemda* seek out other sources of revenue that didn't destroy the environment. But it is not easy to move away from this dependence on wood revenue.[69]

Several informants observed that a large portion of the funds collected this way remained in the personal pockets of the tax collectors and their superiors. This was confirmed by a 1996 report in *Analisa*, an Aceh-based broadsheet. The report carried a statement from a member of the district assembly (DPRD) who complained about the operations of the district government office responsible for collecting the PAD taxes in South Aceh (*Dispenda*). It was estimated that, if the region produced 4 million cubic meters of wood each year with an Industry and Natural Resource Tax (RHBI) of Rp 1,000 per cubic meter, and if there was no "leakage" or "corruption," RHBI receipts from timber alone would amount to Rp 4 billion. This figure did not include other charges placed on timber. However, in 1996 the district government had set a target for PAD of only Rp 1.7 billion. As a Tapaktuan journalist estimated, less than one-sixth of the money collected entered the official accounts.[70] One of the problems, the *Analisa* article had noted, was that, before collected funds entered district government accounts, *Dispenda* placed the money in the personal account of a certain district government figure (Analisa 1996).

Logging in Menggamat

How do these dynamics affect Menggamat? At the outset, it is important
to note that a series of technological, social, and economic changes set
the context for the emergence of logging networks in Menggamat. The
local situation had changed drastically since the opening of the road to
Menggamat during the 1980s. Before logging began in the Menggamat
watershed, the upper Kluet area was inaccessible. However, the advent
of new roads, chainsaws, and sawmills made it possible to extract trees
that now had a high economic value. Nonetheless, newspaper reports
from this time link the genesis of logging networks with the arrival of entrepreneurs, financial backers, or *cukong*, or financial backers. Initially,
these operators concentrated their activities on other areas of South
Aceh, including Kluet Selatan.

In this neighboring sub-district, just south of the Kluet River, PT
Medan Remaja Timber (MRT) operated a timber concession. MRT had
obtained its concession here in 1975. The concession encompassed virtually all of the 6,859 hectares of Pucuk Lembang's village territory, leaving aside only 500 hectares (4.6 percent) of the area—the village rice
fields and a small amount of dry farming land (Barber 1997: 15; Kompas
1997a). Over time the presence of MRT brought considerable change.
Due to severe damage to the forest, wild boars increasingly attacked cultivated fields, and farming activities in the village were thwarted.[71] In the
steep surroundings, logging also affected the hydrological functions of
the forest, and it became difficult for villagers to obtain water for their
daily needs during the dry season (Kompas 1997a). As logging-generated
erosion clogged up local irrigation systems, rice fields in the neighboring
village of Durian Kawan dried up. When MRT began operations, bull-
dozers had also opened a road to the village, connecting Pucuk Lembang
to the outside world. Subsequently, local values began to change. In 1997
Barber reported: "The presence of the concession has introduced new
materialistic values which have caused internal conflicts and diminished
social cohesion within the community, resulting in revolts against traditional precepts that are perceived as contrary to individual interests"
(Barber 1997: 19). With agriculture beset by problems caused by logging,
local people discovered the lucrative opportunities for obtaining ready
cash: by working in logging teams, they could earn up to Rp 200,000 per
week. "Rather than die of hunger," the village head said, "better to encroach on the forest" (Serambi Indonesia 1997a). Many villagers became
involved in carting logs out of the forest for chainsaw operators from
outside the area. However, a few locals purchased chainsaws with capital

borrowed from entrepreneurs, first working locally, but later throughout
the district. Operating on a large scale, the loggers sold their logs to the
sawmills that now operated in nearby Kota Fajar. As Barber reported
after visiting the area in mid-1996,

illegal loggers . . . operate in areas already logged by MRT, as well as in the adjacent national park . . . Illegal logging in this area is on an extremely large scale.
MRT has done nothing to stop illegal loggers who operate openly, piling cut logs
on the company's main road. As in other parts of the Leuser Ecosystem, it is an
"open secret" that this illegal felling is backed by local government and military
officials. Indeed, it is rumoured in the village that one operation active in early
1996 involved National Park officials themselves (Barber 1997: 19).

In other words, as the logging of state forest by concessionaires proceeded, the illegal logging of state forest in Pucuk Lembang gained legitimacy. Because logging concessions controlled by outside business interests harvested the timber, there was very little benefit to local villages.
The HPH failed to recognize that, from a village perspective, adjacent
forest areas are considered the territory (or common pool resource) of
the nearest village, or that historically local *adat* heads levied fees on
behalf of their communities on outsiders exploiting local forests. Since
logging damaged local agriculture or otherwise caused economic hardship, local people increasingly resented it, together with the violation of
local understandings concerning the local property rights over surrounding forest territory. Thus, the scene was set for the emergence of illegal
logging operations employing large numbers of local people.

In November 1996, the forestry police, the army, local police, and a
special provincial Integrated Forestry Security Team (*Tim Pengamanan
Hutan Terpadu,* or TPHT) carried out a "collective operation" (*operasi
gabungan*). The operation caught eight hundred loggers "red handed" in
South Kluet. By the middle of 1996, widespread logging had already begun in neighboring Menggamat. As their "room to maneuver" in South
Kluet was now limited, according to a report in *Serambi Indonesia*, loggers increasingly shifted their attention to neighboring North Kluet (Serambi Indonesia 1997c).

Adat, Leadership, and Logging

To gain access to *adat* forests, outside logging interests gradually found
ways of accommodating key Menggamat leaders. To understand how
these networks gained entry into Menggamat, it is necessary to examine
the changing nature of local institutions.

As I discussed earlier, under the Dutch system of indirect rule, the lowest colonial official was the *Controleur* of what was known as the Tapaktuan sub-division (*Onderafdeeling Tapaktuan*), occupying a position somewhat analogous to the *bupati* in the present-day governmental structure.[72] Under this system, the indigenous heads, especially in remote pockets which had no commercial interest to the Dutch (such as Menggamat), retained a great deal of autonomy in ordering local affairs according to local *adat*.[73] The Indonesian Republic inherited this system from the Dutch, and for many years the "traditional" structure of village government associated with the *adat* of a specific cultural group continued to function with a great deal of autonomy—especially in inaccessible places like Menggamat.

Writing of the changes in Aceh following the passing of the 1979 Basic Village Government Law (UUPD 1979), Mattugengkeng (1987) has contrasted the traditional Acehnese form of "village" ("*gampong* government") with that of the Indonesian "village" (*desa*) arranged according to the new law ("*desa* government"). As I noted in the last chapter, in the new structure there was no recognition of the village leagues (*mukim*) so characteristic of Aceh. The villages were now to be arranged into sub-districts, and the village head (*kepala desa*) was to be directly responsible to the *camat* (sub-district head). Thus, there was no longer a official position for the *imam mukim* as head of the village league, despite the key role the *imam mukim* had played both in the structure of "*gampong* government" and in the implementation of *adat* (Mattugengkeng 1987). However, invoking Aceh's status as a special region within the Republic (*Daerah Istemewa*), the provincial government attempted to retain this position by passing a provincial government regulation (Perda No. 2 1990) recognizing it (Nababan 1996: 4). After these changes, the *imam mukim* was retained as an *adat* leader, but this was now largely a symbolic role; in administrative terms the village head was formally responsible to the *camat*.

The changes had a significant effect on Menggamat. As Mattugengkeng has argued, whereas "*gampong* government" had integrated government, religious, and *adat* principles within one institutional structure oriented to the village, now religion and *adat* were more or less disconnected from the principles of "village government" (Mattugengkeng 1987). As in Sama Dua, the council of elders (*petuhapet*) that had previously overseen village affairs was now replaced with two councils—the village security council (*Lembaga Ketahanan Masyarakat Desa*, or LKMD) and the village consultation council (*Lembaga Musyawarah Desa*, or LMD)—both under the leadership of the village head.[74] The councils did not have the status of the *petuhapet*, and members could ultimately be appointed and dismissed by the village head. Therefore, the village elders were less able to effectively supervise and counsel the *keucik*. As the lowest rung of government responsible for implementing government policy and priorities, the *keucik* now became less responsive to the village "below" (Nababan 1996: 5).[75]

As an informant in Menggamat explained, "Now the *keucik* is more involved with the *camat*. Before he couldn't meet with the *camat*. Previously there were only village regulations (*undang-undang desa*). Things were sorted out by the village on its own—by the *petuhapet* and *keucik*. The *camat* was a long way away and so the *kepala mukim* was more available. Very little went to the *camat* . . . The road [now] makes [this] communication possible."[76] In Menggamat, the dependence of the *imam mukim* on district officials was exacerbated because the *imam mukim* was a newcomer of Aneuk Jamee descent who had married into the area. He took over from the previous head, who had been the son of the former *uleebalang* of Menggamat. The new *imam mukim* was "first picked (*dicalonkan*) by the *camat* based on an instruction of the *bupati*, then elected by the 13 village heads."

This intervention certainly influences the side . . . the *imam mukim* takes in carrying out his duties. As a logical consequence he will be inclined to secure the priorities of the *camat* and other superior officials rather than pay attention to the protest of the community towards the uncontrolled logging, which almost every wet season causes floods and water shortages for irrigation in the dry season. Another reason for this is the dependence of the *imam mukim* on the *camat,* because the *camat* had the power to directly contact the *kepala desa* without proceeding through the *imam mukim,* with the result that his official position is very weak. (Nababan 1996: 5)

In Menggamat, the *cukong* had made arrangements with local government officials, including the *camat*, the head of the sub-district police section (*kapolsek*), and the head of the local military command (*danramil*), thereby gaining access to the Menggamat forests (Nababan 1996: 3–4). By 1995, there were five sawmills operating along the road to Menggamat. Though these sawmills depended on wood extracted from Menggamat, not one of them had a valid IPK permit for the area. One of them even operated with a permit that established it as a business of the district government (*Badan Usaha Milik Daerah*) (WWF-LP 1995b). With the support of local officials, the *cukong* accommodated the *imam mukim*. According to Nababan, the *imam mukim* "operated as an agent for sub-district officials":

The *imam mukim*, as well as operating four chainsaws, also has a role as the surrogate of officials in the sub-district. Several community leaders say that for his role the *imam mukim* receives a "share" from the *camat* for securing the interests of civil and military officials who are his superiors. The *imam mukim* recognizes his involvement in logging as an effort to secure the economic needs of his family that could not possibly be met with his salary of only Rp 40,000 per month (Nababan 1996: 3–4).

After negotiating at this level, the *tauke* then approached the village heads. Because the *tauke* had the support of local army officials (*kora-mil*), civil officials (*camat*), and police (*kapolsek*), village heads were unable to enforce *adat* rules that previously regulated where and when the cutting of trees could occur. As a WWF-LP worker later explained, "the *keucik* were afraid of coming into conflict with the *camat* and wouldn't really act."[77] Moreover, the functioning of another *adat* official, the *kejuren blang*, was also compromised. Previously, the *kejuren blang* had been responsible for protecting forest in the headwaters and regulating the flow of irrigation water. However, due to the political power of *ok-num* involved in logging, the *kejuren blang* could not control the felling of trees in the watershed.

Over time village heads, in collaboration with other village decision-makers, attempted to reassert some control over access to the local forest territory as well as gain some material benefits for the village from the logging of what was still considered *adat* property. In several cases, particular village heads and their village councils came to an agreement to impose fees on those logging Menggamat territory. They charged loggers a "development fee" (*uang pembangunan*) for access to the forest under their authority. As the name suggests, the revenue raised in this way was to be used for developing village facilities such as mosques and schools.

As one village head later recalled, "At first there was no *uang pemban-gunan* because we were still stupid (*bodoh*). We could still be provoked by outsiders who said that the state owned the forest, not the village . . . We didn't know the regulations at the time, and still could be tricked . . . and officials (*aparat*) had given permission."[78] Yet, by April 1995, according to a WWF report from this time, village heads were imposing fees on loggers, requiring that they report first before carrying out logging: "A chainsaw can operate in the forest surrounding a village by paying a registration fee (*uang pendaftaran*) of 5,000 Rp per chain saw and *uang pembangunan* of 25,000 or 50,000 per chainsaw, depending on the village" (WWF-LP 1995b).

As I noted in the last chapter, historical sources for Aceh and West Sumatra reveal that the practice of outsiders paying a fee to *adat* heads for access to village territory had precedents. In historical times, this fee was known as the *pantjang alas* or *bunga kayu*; at one time, the fee had been incorporated into colonial legislation governing forest use (Adatrechtbundels 1938). An informant in Menggamat argued that *pantjang alas* of the colonial period should be distinguished from the *uang pembangunan*. In colonial times the *pantjang alas* had been a fee paid directly to the *adat* head concerned and, he argued, was a component of personal income; in contrast, the *uang pembangunan* was a fee paid to the village. Rather than just "contributing to the *keucik*'s own pocket," people from outside the village needed to pay the village a "fee" (*retribusi*) to gain access to the forest. Yet the price was not fixed: someone wishing to log in the forest would need to discuss (*musyawarah*) the issue first with the village head and establish a price. The price would depend on who wanted to log and for how long.[79] Of course, this informal system gave village heads ample opportunity for extracting rents for themselves.

The position of village head involved a conflict of interest that was at least partly related to the nature of village finances. In Java, each village head is allocated a certain piece of village-owned land (*tanah bengkok*) to cultivate, and this land then provides him with an income in return for the work he performs on behalf of the community. In contrast, in South Aceh there are no village-owned rice fields. Village heads receive only a small honorarium from the government, and at least one village head complained that it was in no way commensurate with the work involved in carrying out their duties. Therefore, to support themselves, it has been common practice for village heads to levy fees for certain services and permits, and these sorts of fees have been a common practice in the Outer Islands of Indonesia (Devas 1989: 37). These include administrative charges for letters of recommendation that are required for different purposes. Usually the scale of these fees has not been set down in any formal way but has depended either on the generosity of the person seeking the patronage of the village head or on negotiation between the parties. The *adat* fee on access to forests (*pantjang alas*) fits into a wider pattern. As well as acting on behalf of the collective to maintain some control over village territory and levy a "development fee" on behalf of the community, by charging fees on their own behalf, village heads themselves directly benefited from logging. As the distinction between public and private domains remained rather unclear, and in the absence of clear historical information, it is difficult to distinguish *uang pembanguan* from *pantjang alas*. Moreover, clearly the same *adat* assumptions were at play in both historical periods.[80]

However, despite attempts to assert control, with more than two hundred chainsaws operating around Menggamat by 1995, Menggamat could not maintain its *adat* regulations in the face of the loggers. In theory, it should not have been difficult to control access to the Menggamat forests. There is only one road out of the area, and one river for floating logs down stream. Indeed, over time villages set up posts on the road to enforce the collection of village taxes (*uang pembangunan*). But, while the boss of a logging team (*tauke*) may at first have reported to the village head, he would invite his friends, who would proceed to the forest without requesting permission. Thus, many logging teams were operating without reporting to the village head. As certain members of ABRI, the police, and district government were involved, community control towards the logging became increasingly weak: "The community with its *adat* regulations cannot withstand the tide of loggers. For example, in the village of Simpang Tiga, there was an ABRI official who carried six or seven chainsaws, then logging in areas protected by *adat* (watersheds and hills). However, members of the community cannot forbid this. This logger enters without permission and without paying *uang pembangan* to the village head" (WWF-LP 1995b).

In this situation, if village leaders did not reap the rewards of logging for themselves, like other villagers who did not join in the logging of Menggamat's forests, they would end up sitting by while outside parties enriched themselves. Just as Peluso found in a similar case from West Kalimantan, in this context, village leaders—along with villagers involved in logging—were practical about the implications of not participating: "a total loss of benefits as opposed to the enjoyment of benefits in the short term" (Peluso 1992a: 217). Therefore, village heads were opportunistic, allowing logging, and obtaining benefits from their position.

Gradually, village heads also began to become involved in extracting rents directly from the forests for themselves. In 1995, WWF-LP reported that the *tauke* were cooperating with the village heads. Soon, in addition to levying the "development fee" for the village, many village heads operated as *tauke* themselves, buying chainsaws and providing capital to those carrying out the logging. One village head even operated sixteen chainsaws on behalf of another *tauke*. As a WWF-LP project worker remembered, "of the village heads there, only two were not involved, and I am only certain that one of them was not involved." The village heads built houses and bought satellite television and new motor bikes from the profits of logging.[81] As village heads followed the logic of the situation, many were absorbed into the webs of power and interest that involved *oknum*, *tauke*, and *cukong*.

Nababan (1996) concluded that the onset of "wild logging" in Menggamat's forests was linked to institutional change. The rearrangement of village government in accordance with the village government law, together with the injection of capital by outside entrepreneurs, weakened the strength of the *adat* community facing outside intervention. In effect, because the locus of power had shifted upwards, *adat* leaders lost their control over access to and use of the resources found in the Menggamat territory. The *imam mukim* became an instrument of outside interests, on whose patronage he depended, and the village heads were no longer under the tutelage of the *petuhapet*, but were directly responsible to the *camat*, who now informally sponsored the logging. *Adat* sanctions could not be brought to bear on loggers violating *adat* principles or on local leaders who used their office corruptly. Menggamat's villages had lost their ability to control logging, which damaged forests and endangered surrounding farming lands.[82] In the process, the short-term interests of those mining the forest eclipsed the long-term interests of Menggamat's people. Nababan concluded that "observing the complexity of the problems connected with the uncontrolled logging shows that the destruction of forest resources is a consequence of the weak bargaining position of the community towards various outside interventions into the area, both as a result of formal government policy implementation as well as the injection of capital" (Nababan 1996: 5). However, despite the impact of outside interventions, it is clear that *adat* leaders did not sit back idly and watch. On behalf of their villages, they reasserted *adat* property rights by imposing a tax on timber extracted from what was considered to be the *adat* territory of Menggamat's residents.

Logging: The Village Perspective

In counterpoint to this institutionalist account, local informants describe how villagers took up logging in response to the economic situation. Outsiders had first begun introducing chainsaws into Menggamat in 1992, and the first uncontrolled logging began at that time. Yet before 1995, an informant explained, villagers primarily worked their own gardens.[83] It was only in 1995, when the price of *nilam* fell sharply, that a significant shift occurred. As one informant explained, "When one *bambu* (approximately one liter) of unhusked rice has the same value as one *bambu* of nutmeg, then we can say that nutmeg farmers will prosper. When one kg of *nilam* is equivalent to one *mayam* (3.33 gm) of gold, then *nilam* farmers will prosper."[84] In 1995 *nilam* fell to 25,000 Rp/kg, at a time when gold was 80,000 Rp/*mayam*. Consequently, people moved to other occupations, primarily the lucrative business of working on logging teams. When logging on a wide scale began in 1995, it happened largely because *tauke* offered villagers a strategy of survival at a critical moment.

The *tauke* employed local people to carry out the logging, organizing teams of six to fifteen loggers, including a sawyer skilled in operating a

chainsaw, usually from outside the area, and teams of bearers from surrounding villages. By joining a logging team, villagers became the clients
of the *tauke* who provided the resources for logging operations: "Before
the logging team goes to the forest, they take money from the *tauke* to
pay for their expenses in the forest and for their family needs. This money
forms a debt that must be paid according to the amount of wood produced . . . the income of the sawyer, bearers and others depends on the
quantity of timber" (WWF-LP 1995b). Logging teams gradually moved
into the forest, beginning close to villages and gradually penetrating further from the village settlements. The loggers concentrated their efforts
in the forest upstream from the Menggamat village complex, an area
classified as limited production forest on forestry maps. As Menggamat
villagers interviewed in 1998 reported, "the village forest (*hutan desa*)
is left, because people have their *kebun* here . . . They can log state forest (*hutan negara*)."[85] In other words, they did not allow loggers to take
wood from areas adjacent to their gardens (*kebun*), where tree felling on
steep land would damage crops or create landslides. Moreover, as a former WWF project worker reported, "There was not much logging in the
village forest because they sought *meranti* and *semantuk*—the most valuable trees, and there weren't any in the steep [secondary forest] areas behind people's *kebun* . . . Or very few with a commercial value compared
with further up. So the *hutan adat* doesn't look damaged compared with
the area further up which had excellent wood and was heavily logged."[86]
In other words, visitors to Menggamat could see intact secondary forest
(known as *hutan adat*) in the village lands close by. In contrast, loggers
had significantly deforested the tract further out, in an area that the forestry department had classified as limited production forest.

According to villagers, felling trees and sliding them down the steep
slopes was a dangerous operation. "People often die," one said, "they
can be cut with a chainsaw, or hit by a falling tree."[87] The logs were usually six meters long, and chainsaw operators cut them into planks on location. After a tree was felled and turned into planks, the bearers carried
the wood down to the river. In places close to the river, they made a raft
and floated it down the river. Otherwise, bearers would cut a path with a
machete (*parang*) and drag fifteen planks at a time down the steep slopes,
one person on each end. "If the person behind slipped," said a villager,
"you could be killed, but it isn't hard once you get used to it." A former
logger described watching his friend be decapitated by some wood that
had gained speed on a steep slope. "It doesn't matter how much they
pay," he concluded, "I will never do it again."[88] For this reason, villagers
preferred to use buffaloes to carry wood down steep slopes. The buf-

faloes were harnessed to the logs, and on slippery slopes the buffaloes could pull up to slow down the logs.

As a Menggamat resident noted, a large web of people benefited from logging operations. First, logging operations directly employed villagers and migrant workers from surrounding areas. These included the chain-saw operator, the logging team, timber carriers, those floating the wood down the river, the workers loading and unloading trucks at the sawmill, truck drivers, sawmill operators, and administrative staff. In addition, a wide range of others were indirectly employed, from the boat operators taking logging teams up-river, to *warung* (kiosk) attendants, to mechanics and motorcycle vendors. The villager estimated that one chainsaw led to the employment of up to two hundred people.[89] In addition, beyond Menggamat itself, there were the *cukong* and *tauke*, and a wide network of district officials collecting taxes and receiving payments from these operations.

In 1998 loggers described the division of the benefits. In general, bearers were paid Rp 3,000 per plank, and for carrying wood they could earn at least Rp 20,000 per day. "But as you are paid per ton," a logger said, "if you are diligent (*rajin*) you can earn 30,000 per day."[90] At this time, participation in logging networks could change a person's status in the village economy: according to the *camat* of North Kluet, this was more than a *camat*'s official salary.[91]

Once the wood reached the village, the bearers would load it onto a truck and take it down to the mill at Kota Fajar. On the road to Kota Fajar, the truck would pass two police posts (*pos pungutan*), and the *tauke* would pay the police Rp 20,000 at each post. Furthermore, the *tauke* would pay the *jagawana* (forest police) Rp 60,000 per truckload. At the mill, the *tauke* paid lumberjacks to unload the wood. The wood was then sold to the sawmill. In 1998, villagers reported that the price for local wood was Rp 250,000 per ton; export quality timber earned Rp 400,000 per ton. Export quality timber was shipped to Medan. As each truck could hold four tons, the profits earned by the *tauke* were extensive: between Rp 200,000 and 400,000 per ton.[92]

In 1996, Barber had investigated the economics of logging networks in South Aceh. Barber reported that in 1996 Meranti (*Dipterocarpus* spp.) sold in Medan at U.S.\$ 267 per ton for domestic grade and US\$ 356 for export grade. In comparison, Damar (*Shorea* spp.) sold at US\$ 311 and U.S.\$ 500 per ton (Barber 1997: 36).[93] Although villagers—financed by local moneylenders—operated a few chainsaws, many more worked transporting illegally cut logs from the forest to the outskirts of the village, earning between U.S.\$ 2.25 and U.S.\$ 4.50 per day. The few local people involved in actually cutting received a slightly higher income (Barber 1997: 19). Furthermore, he added, "The gross sale price realized by the operators selling to local sawmills in the area is between $89–$111/ton, less the cost of felling, splitting and transport from the forest, which comes to about $57/ton, or 65 percent of the sale price, providing the owner of the chainsaw with net profits of $31 per ton" (Barber 1997: 40). The operators selling to sawmills, sawmill operators, and the *cukong* selling timber on to Medan made most of the profits. This demonstrates the significant degree to which actors with control over strategic points at which the timber resource leaped in value could extract a disproportionate share of the value timber earned in the wider market (Ribot 2000).

Consequently, to the extent that the outside *cukong* and *tauke* recruited local villagers to work in logging teams and paid "taxes" to local leaders, they ensured that, at a time of economic need, villagers would share in the stream of benefits derived from mining local forests. In contrast to logging concessionaires (HPH), these operations accommodated the property rights asserted by local communities over surrounding forests. Therefore, the *oknum* and the timber entrepreneurs were able to gain the cooperation of local communities in logging those forests. However, this change meant that local *adat* authorities now allowed rapid exploitation, sacrificing the long-term values of the forest for short-term gain.

Reports from the colonial period reveal that this phenomenon had historical precedents: getting local *adat* heads on side by seeking their authorization and presenting them with payments and royalties, delegating responsibility for timber extraction to local middlemen (including *adat* heads), and/or subcontracting local villagers to carry out the logging amount to a widespread and very successful system of extraction.[94] Dutch timber companies had operated concessions in some areas of Aceh, particularly on the island of Simeulue off Aceh's southwest coast. Given the difficult terrain, logging operations in these areas had often proved arduous, and sometimes there were also problems with local populations (Kreemer 1923; Goor 1982). However, in 1911, the head of the directly administered territory of Singkil (in southern Aceh south of the Kluet area) allowed that a Dutchman interested in exporting timber from the area could operate a sawmill on the condition that local people carried out the logging. The sawmill owner discovered that, compared with the effort involved in operating a timber concession directly, it was efficient and profitable to buy timber from the local population. After other traders noticed how lucrative this was, they also placed large orders with the local population: "As a consequence almost the entire male population of upper Singkil went into the forest to log valuable timber. As can be imagined, this resulted in the reckless destruction of considerable timber" (Kreemer 1923: 125). In a similar fashion, during the 1990s sawmill operators discovered that this was a most effective method for extracting timber from local territory.

In this context, as we will see, outside agencies—including NGOs and large donor projects—that intervene to change the complex local rules of the game faced substantial obstacles.

Community-Based Natural Resource Management in Menggamat

In keeping with the community-based natural resource management (CBNRM) focus, which tends to see communities as "the locus of conservationist thinking" (Agrawal and Gibson 1999: 631), in a number of cases across Indonesia, conservationists have sought to reconstruct *adat* institutions as conservation and environmental management institutions (Zerner 1994; Eghenter 2000). Doing so has typically involved legitimizing conservation in terms of *adat*, the foundational discourse that legitimizes and structures political and ceremonial life at the village level (Blackwood 2001). When WWF's Leuser Project (WWF-LP) chose Menggamat as a site for community-based conservation activities in 1994, WWF-LP began investigating pre-existing *adat* rules for managing the forest, looking for a local conservation ethic in *adat* that might provide the rationale and the foundation for community conservation in the area. As the WWF-LP project leader noted, WWF-LP had chosen Menggamat as a site for community conservation because the *adat* system here was judged to fit best with the idea of *adat* as an ecologically sustainable indigenous management system and it was hoped that it would provide a model for environmentally sustainable resource management.[95]

WWF-LP first embarked on wide-ranging consultations with Menggamat leaders regarding WWF-LP project activities and local problems, including the likely impact of the uncontrolled logging in the Menggamat watershed on villages and agricultural areas. At this time, virtually all the village leaders were involved in logging. Even so, WWF-LP found that the *kepala mukim* and other key *adat* leaders supported the proposed community conservation activities. WWF-LP workers felt that these *adat* institutions provided an institutional framework with the legal and moral authority to bring people together and make decisions. As one project worker later reported:

If the *adat* leadership agrees, the *keucik* agrees. In principle, the *kepala mukim* carries the thirteen village heads, so it is the *kepala mukim* that decides. We didn't look to see if he was involved in logging or not, what was clear was that he

backed it [the project], which meant that he could change, this was our principle
... We didn't attempt to prohibit logging; we didn't have power to prohibit them
... Also we understood that the community needed to meet their everyday needs
... We offered an understanding to them that if this [logging] continued, what
the impact on the community would be like ... We tried to offer an appreciation
like this to the *adat* heads.[96]

Following experience gained elsewhere, WWF-LP recognized that if the
Menggamat forest were to be managed according to revitalized "customary resource management rules and practices," a community management organization would be needed (Barber 1997: 16). Eventually
the discussion with village leaders led to the establishment of the Menggamat Community Representative Body for Conservation Forest (YP-
PAMAM).[97] YPPAMAM would be under the leadership of key *adat* figures, who would act as advisers and decision-makers. While community
conservation would be based on revitalized *adat* principles, YPPAMAM
would also seek to obtain a more valid legal status for the proposed
community forest. By making written regulations for the community forest that nested within state regulations, YPPAMAM wanted to use the
state regulations to legitimize the status of the CBNRM initiative. To do
so involved straddling the divide between official law and *adat* rules. In
this way, it was hoped, YPPAMAM could find a legal validity for applying "revitalized" customary rules to the management of the surrounding state forest, now to be known as Community Conservation Forest
(CCF).

    According to the CBNRM rationale, gaining greater recognition for
"traditional community-based rights" and increasing the dependence of
villagers on the surrounding resource base would provide "official incentives for sustainable use" (Lynch and Talbott 1995: 6). Advocates of
CBNRM have argued that, if villagers' very livelihood depended upon
maintaining the local resource base, "they would do so unless prevented
by ineluctable forces" (Lynch and Talbott 1995: 24). According to the
CBNRM model this would involve developing community-based institutional arrangements empowered "to police the forest and prevent outsiders and members of the own communities from overexploiting forest resources" (Lynch and Talbott 1995: 6). To support the argument
in favor of the CCF, WWF-LP and YPPAMAM sought to demonstrate
that the area directly adjacent to the village that was subject to *adat*
regulations (*hutan adat*) had escaped the brunt of the logging, whereas
the forest most heavily logged was the unprotected state forest (*hutan
negara*) further out from the core villages.[98] They argued that, if *adat*
regulations were extended to the *hutan negara*, forest management could
be improved.

By this time, the notion of a community conservation forest (CCF), or as they are known elsewhere, an "extractive reserve," had gained currency in Indonesia and elsewhere (Peluso 1992b; Salafsky et al. 1992). An extractive reserve is an area of existing rainforest "set aside for low-impact use by certain residents of the rainforest or its environs" (Peluso 1992b). The most famous extractive reserves were those of the rubber tappers of the Amazon. In a similar way, WWF-LP envisaged that this CCF would safeguard the protection forest behind Menggamat, securing the local water and forest resources. Because local villagers would collect forest products in the CCF on a sustainable basis, the CCF would also provide income while securing local water and forest resources.

In order to develop the idea further, WWF-LP surveyed the forests behind Menggamat. These surveys estimated that there were approximately 10,660 trees (*Shorea* spp.) in the Menggamat area that could yield *damar* resins. WWF-LP calculated that each tree could produce four kilograms of *damar mata kucing* every three months. Based on a market value of Rp 600 per kg, in 1995 the forest could produce Rp 10.6 million per quarter. With higher prices, the *damar* forest could yield up to Rp 170.6 million per year. In addition, the Menggamat forests could also yield *keruing* oil, rattan, and other forest products. In comparison to the long-term income from sustainable harvest of the forest and the protection the watershed forests offered to local rice farming, a one-off logging operation could yield only an estimated Rp 26.7 billion. In addition in that case, villagers would have to live with the agricultural damage caused by lost watershed functions, without the income from forest resources, and they would also have to wait another 150 to 400 years for forest regrowth (WWF-LP 1995c; Effendi 1998). If Menggamat was to take the first option, Menggamat collectively needed to support WWF-LP's argument that the long-term welfare of Menggamat needed to be placed above the short-term interest of village loggers in obtaining immediate cash.

Nesting Within the Official Regime

For WWF-LP to develop a Menggamat Community Forest (*Hutan Konservasi Kemukiman Menggamat*) within the legal framework provided by the Ministry of Forestry, WWF-LP and its local partners would need to operate within the constraints of current legal formulations. By this time the Ministry of Forestry had developed a "community forestry initiative" that aimed at addressing community and environmental priorities simultaneously (Suryohadikoesumo 1997). In 1995 the minister had specified "Guidelines for Community Forestry" that allowed for "community forestry activity" on designated "critical forest lands"—including degraded

Figure 10.   The Kluet River valley, Menggamat. John McCarthy.

Figure 11.   Menggamat village land. According to *adat*, farmers post a flag some two weeks before opening a new *kebun* to ensure that no other villagers have prior rights over the land. John McCarthy.

Figure 12. Secondary forest in village lands, Menggamat. John McCarthy.

Figure 13. A "limited production forest" after the logging networks, Menggamat. John McCarthy.

Figure 14. Sawn timber extracted via the river, Menggamat. John McCarthy.

Figure 15. Timber merchant, Tapaktuan. John McCarthy.

protection and production forests "where rehabilitation is necessary."[99] The guidelines excluded community forestry from active timber concession areas, national parks, and other reserves. By participating in community forestry on what was officially state forest land, villagers would be entitled to plant trees and harvest non-timber forest products (Menteri Kehutanan 1995). In return, the government hoped to rehabilitate degraded forests, increasing their productivity and quality while protecting the surrounding environment (Suryohadikoesumo 1997). By allowing local users to make use of protection forest, this initiative represented a significant policy shift. Previously, all such uses were illegal, irrespective of their ecological impacts or the concept of territoriality maintained by surrounding communities. Now, the ministry would allow "that communities with viable traditional forest management institutions should be granted rights and responsibilities over forest areas up to 10,000 ha or so" (Barber 1997: 17).

In effect, under this form of co-management, the forestry department would maintain or—insofar as state control in remote sites such as Menggamat was only virtual—extend state control and villagers would obtain use rights subject to a lease agreement entered into with the head of the Provincial Forest Service (*Kanwil Kehutanan*). This type of agreement also made legal entitlement conditional upon narrow, potentially discriminative, environmental prerequisites (Li 2002). Moreover, it implied that villagers would collectively relinquish more pervasive *adat* rights over Menggamat territory for limited usufruct rights. By doing so, villagers would unwittingly acknowledge the status of Menggamat territory as state forest land and admit that they have only specified rights of use, rather than traditional tenure (Haverfield 1999: 63).

Despite these constraints, the initiative allowed WWF-LP some space to move forward with a CCF initiative, and after consultations with regional government and the forestry department, a group of Menggamat leaders drafted community regulations concerning the proposed community forest. On 17 May 1995, the thirteen village heads, the *imam mukim, kejuren blang,* and religious leaders (*ulama*) signed an official village government agreement establishing the CCF (Pemerintahan Desa di Kemukiman Manggamat 1995).[100] The development of the CCF encompassed several objectives:

Rehabilitating the logged forest and its ecological functions—managing the watershed and protecting the fertility of local soils;

Increasing the role of the *adat* community in efforts to protect the environment while directly sustaining community livelihood over the long term;

Providing a bufferzone for Gunung Leuser National Park by supporting the function of this protection forest area;

Helping to confer legal recognition to the *adat* village forests by the regional government, thereby protecting the *adat* traditions still operating there.[101]

The main section of the regulations clarified the rights and duties of those wishing to use the CCF. With permission from the specified authorities, villagers could collect forest products in allocated areas for their own use and for community industries.[102] However, there was neither an allowance for commercial logging nor any recognition of the *uang pembangunan*, the informal taxes village leaders had levied. The regulation simply prohibited the cutting of wood for sale to logging concessionaries or sawmills.[103] As these activities were not permitted by the Community Forestry Guidelines, according to the head of YPPAMAM, in lieu of the notion of a community tax (*uang pembangunan*), the regulations required that whoever cut trees in the CCF would have to plant twenty-five trees.[104] Where *adat* and state legal formulations disagreed, to obtain legal status the CCF guidelines needed to comply with state legal formulations rather than Menggamat *adat* practices.

By citing twenty Basic Laws (*Undang-undang*), regulations, and other legal instruments in a preamble, the regulation attempted to nest the CCF in the embrace of positive law. However, the decision simultaneously drew its authority from long-standing *adat* principles operating in Menggamat. Those violating the regulations would be dealt with according to "*adat* sanctions that are in operation" (Pemerintahan Desa di Kemukiman Manggamat 1995: 18). In accordance with the flexible, unwritten nature of local *adat*, the sanctions were graduated according to the severity of the offense, and the regulations did not establish specific sanctions for specific violations. The articles concerning sanctions simply stated that, when an offense is committed, an *adat* sanction will be decided upon "according to an assessment from the *adat* consensus institutions" (*lembaga musyawarah kemukiman*) after "community members give appraisement and admonition" (Pemerintahan Desa di Kemukiman Manggamat 1995: 18). If offenders did not change their behavior and continued to violate the regulations, they would then be reported to state authorities and be subject to much heavier sanctions laid out in the state penal codes and other state laws mentioned in the preamble. By making use of both state legal discourse and *adat* discourse in this way, this process tended to erase the previously clear line between the *adat* and the state legal orders (Agrawal 2001).

In April 1996, WF-LP and YPPAMAM made an official approach to the Regional Forestry Office (*Kanwil*). In requesting recognition for the CCF in accordance with the Minister of Forestry's "Guidelines for community forestry," they argued that the heavily logged state forest (*hutan negara*) further out from the core villages would be better managed by revitalizing the *adat* regulations (*hutan adat*) that previously pertained to the *hutan adat* areas directly adjacent to the village, and managing the area as a CCF.[105] Finally, on 5 December 1996, the governor signed an agreement between the Leuser Management Unit (LMU) and YP-PAMAM, recognizing the CCF and setting out the terms for its management. The agreement granted a five-year, extendable management and use contract to a 13,810–hectare area (Barber 1997: 18). The exact borders and zoning of uses remained unresolved until November 1998, when the head of the Regional Forestry Office (*Kanwil Kehutanan*) issued an official decision establishing YPPAMAM's rights over the CCF (Dephutbun 1988). By this time WWF-LP's project activities had ended, and LMU had assumed responsibility for developing the CCF.[106]

Community Forestry and Logging Networks

Although the creation of YPPAMAM and the community forest signified progress towards project goals, the initiative faced problems that would prove intractable. The first problem was that, although village leaders had signed the document setting out the regulations to be applied in the community forest in May 1995, "wild logging" (*penebangan liar*) continued. In response, YPPAMAM appealed to the villagers directly through village discussions. In addition, YPPAMAM posted banners around Manggamat requesting a halt to logging and pointing out the threat of state sanctions:

> Help, don't cut the forest, we want to be safe from flood and poverty.

> Possession and use of chainsaws without permission of the Forestry Department can be penalised with 10 years jail or fines of 100 million (*Keppres No. 21 1995*).

> Those felling trees in forest without permission from appropriate authorities can face the penalty of 10 years jail or fines of 100 million (*Peraturan Pemerintah No. 28 1985*) (Serambi Indonesia 1995b).

When these rhetorical appeals to state law largely failed, YPPAMAM resorted to more drastic means. In October 1995, with a great deal of coverage in the local press, YPPAMAM began pounding nails into trees in the watershed forests of the Koto Indarung village territory extending to the border with South Kluet (Waspada 1995). Under the heading "Overcoming Wood Theft: Nail Booby Traps Gain Support," a regional newspaper reported that Bintaro Yakob, the head of YPPAMAM, "reminded people not to carry out illegal logging from now on because the embedded nails in the trees were dangerous for those using chainsaws" (Serambi Indonesia 1995b). A WWF-LP staff member, Abdul Hamid, stated that

> this action is born from a feeling of disquiet in the community about the recent conditions resulting from increasingly violent and arbitrary acts by illegal loggers. This expresses the concerns of the community regarding the protection of forest resources that bear directly on the life around the forest—for example, the supply of water for sawah, preventing the danger of flood and plant pests such as pigs and rice blight. (Serambi Indonesia 1995b)

Touching on the vexed issue of law enforcement, Abdul Hamid suggested that, to maintain the Manggamat area, government agencies and decision-makers should consistently implement the law. When the CBNRM failed to gain a consensus amongst local stakeholders within Menggamat fora, its advocates ended up turning to state law. However, WWF-LP was a small player with no law enforcement mandate to fight (illegal) logging.

These developments demonstrated that, despite attempting to elicit participation, WWF-LP tended to include those interests supporting its CBNRM initiative. On the one hand, some key *adat* and village leaders supported the presence of WWF-LP in Menggamat because WWF-LP activities, and the subsequent creation of a new community organization (YPPAMAM), "gave anti-logging traditional leaders a forum to express their views." As Barber pointed out, "Teuku Titah Aman, the most popular traditional leader in the area, noted that 'up to now, whenever anyone spoke about traditional law, they were always branded as voices of the past, people out of tune with current reality'" (Barber 1997: 16). The WWF-LP claimed that they did not try to prohibit local people from logging. But by threatening legal sanctions and driving nails into trees, rather than appealing to those with other points of view, YPPAMAN activities generated resentment and resistance to the CCF approach.[107] The CCF proposal struggled to gain support from some of the village heads (*keucik*) who were heavily involved with logging. As I noted earlier, the village heads' attitude was: "if I withdraw from logging, I would just have to sit by and watch outsiders collect the spoils." If these village heads became involved in logging and collected "taxes" levied on loggers and logging trucks or even functioned as *tauke* themselves, it was not in their interests to support the implementation of the proposed CCF. In 1996, this conflict hindered the reaching of agreements in two annual meetings of the thirteen village heads comprising Kemukiman Menggamat. Every time supporters of the conservation initiative brought up the illegal logging issue, village heads owning chainsaws left the meeting (Barber 1997). Clearly, some local actors resisted the conservation agenda. In a similar fashion to elsewhere, the CCF rules involved "the imposition of an alien and judgmental consciousness which privileged conservation" (Zerner 1994: 1093). Those involved in YPPAMAM might claim to speak on behalf of the "community," but the CBNRM process failed to fully represent the whole "community." Clearly, the underlying questions remained difficult to resolve: Who could speak for Menggamat as a whole? And who could define what Menggamat's "custom" was? Indeed, could "community stakeholders" with such different interests be brought to a consensus?[108]

The second problem was that, in attempting to build a conservation program on *adat* principles, community-based conservation in Menggamat had raised a particular set of problems. WWF-LP/YPPAMAM consultations with older *adat* leaders revealed that elements of *adat* had been associated with maintaining the local environmental balance. WWF-LP proceeded on the assumption that these elements could be "revitalized" or reshaped to support community-based forest management. Taking it a step further, in the name of *adat* WWF-LP/YPPAMAM developed a CCF based on the idea of generating income by extracting non-timber forest products rather than logging. However, doing so involved a rather one-sided and ahistorical understanding of *adat*. *Adat* regimes in South Aceh had long included *adat* procedures (e.g., the *pantjang alas*) for levying fees on those wishing to access local forests for commercial use of non-timber forest products. In the past these *adat* procedures were generally implemented, but at that time timber had not been harvested here for commercial gain. After commercial logging began, however, *adat* leaders could easily begin charging such "*adat* fees" on logging operations. As we saw earlier, in the colonial period Dutch timber interests had been able to gain the collaboration of *adat* heads and local populations in neighboring areas by paying fees to them and involving them in logging *adat* areas (Kreemer 1923; Goor 1982). In other words, drawing on past practices, *adat* arrangements could allow for logging, as long as *adat* property rights were respected. A "traditional" cultural form (*adat*) could be extended to encompass new forms of resource exploitation such as logging. Indeed, in a sense Menggamat's *adat* had shifted to allow for logging: by the mid-1990s concessionaires and outside illegal logging interests were engaged in logging South Aceh's forests, and villagers had grown accustomed to gaining income from extracting timber resources (Perbatakusuma et al. 1997: 4).[109]

Perhaps, over a period of time, WWF-LP might have been able to come up with project strategies to face this problem. To this end, according to a former WWF-LP field worker, WWF-LP had recognized the need to begin with small activities and then gradually build their program in Menggamat: "To get community participation is difficult; we must begin with something small and develop from there." Project objectives would only be achievable over a long period of sustained activity.[110] Through education, community-level discussion, and advocacy, WWF-LP hoped to foster a community learning process. When project workers explained to villagers what the long-term effects of logging would be, villagers were encouraged to consider the impact of uncontrolled logging over the long term on local livelihoods. This strategy would not bring results immediately, but because the villages experienced the environmental impact of the logging in terms of floods and other environmental damage, WWF-LP was confident that many villagers would support conservation activities. However, this strategy required a long commitment to the area, something that circumstances failed to allow.

In this case, the CCF regulations differed from the practices favored by many of Menggamat's *adat* leaders, and some village leaders opposed the proposed CCF. As a possible solution to this problem, in 1996 some villagers had advocated that WWF-LP "regularise the cutting of timber in a sustainable manner as a source of income and support for conservation" (Barber 1997: 18). By granting the villagers limited logging rights within Menggamat forests, such an initiative might attempt to regulate *de facto* practices in the surrounding forests. The CCF agreement for Menggamat could not allow for commercial logging, however, because it was against the law. Therefore, while WWF-LP allied itself with some *adat* leaders, the constraints of state law at this time hamstrung the CCF intervention. Because the Community Based Conservation intervention only involved devolved authority to implement state rules, WWF-LP and its local partners had limited scope within the law to craft rules that would suit the Menggamat situation. An effective conservation intervention would need a policy framework that more fully allowed for local-level management, enforcement, and dispute resolution (Agrawal and Gibson 1999).

The outcome of this battle over "the soul of *adat*" seems to be at least partly determined by how effective either the WWF intervention or the logging could be in offering villagers a viable strategy of survival.[111] To succeed, the CCF strategy needed to offer the Menggamat communities a viable economic alternative to logging. Yet, in 1996, few believed that non-timber forest products by themselves could supply "the growing demand for cash income needed, for example, to send children to school in the cities" (Barber 1997: 18). Although the idea of harvesting non-timber forest products made sense on paper, villagers doubted whether it would work. In 1999, as a former WWF facilitator observed, a well managed *damar* tree could yield one kilogram of *damar* per day. By way of evidence he pointed to the situation in Krui, South Sumatra. After *damar* prices rose from Rp 2,000 to 2,500 per kilogram before the economic crisis to 7000-8000 Rp/kg during the crisis, Krui farmers tapping *damar* trees had become comparatively well-off. He added: "We are sure [it will work]. But the community is not sure, because the community only appreciates facts that already exist. Only when they see it as a reality, then they will try."[112] As well as requiring the provision of a reliable market and better techniques for gathering the *damar*, developing *damar* collection as a sustainable economic basis for the conservation of the CCF would require several years of acculturation as well as demonstrated success.[113] To this end, following the end of WWF-LP activities, former WWF-LP staff formed another organization (*Yayasan Bina Alam Indonesia* or YBAI), and with the support of LMU, continued to nurture the development of the CCF. There was subsequently little progress in securing the status of the CCF according to state law and in training *damar* collectors, and when the security situation in South Aceh began to deteriorate in early 1999, the CCF concept still awaited effective development.

Nevertheless, the most significant obstacle occurred at the district level: WWF-LP's activities in Menggamat and across the district more generally challenged the networks of power and interest supporting logging, not only endangering the material base of the web of exchanges surrounding *cukong, tauke,* and local *oknum* involved in logging in South Aceh, but potentially threatening the foundations of the district government budget. As a WWF-LP report later noted, the district government agency had benefited from the proceeds of logging for many years. The agency believed that WWF-LP's conservation initiatives would "gradually reduce revenue either from legal or illegal logging retributions" (Perbatakusuma et al. 1997: 5). Even though the governor of Aceh signed the agreement establishing the Manggamat Conservation Forest, the major local politician, the *bupati*, made no steps towards achieving the goals set out in this statement. A WWF-LP paper reported that he told WWF project workers that "WWF is only an international NGO working to conserve the Park. It has no right to intervene [in] our resources management" (Perbatakusuma et al. 1997: 5). When WWF-LP management attempted to extend the project beyond the original two years, they were required to produce a letter of support from the *bupati* and the governor, and this support was not forthcoming.[114]

In Menggamat, WWF-LP successfully developed the CCF proposal and succeeded in gaining interim recognition for this initiative at the provincial level. However, effective authority in South Aceh was not vested in the official agreements approved by the governor; the balance of authority lay with local officials and powerful networks of exchange involved in logging. Neither WWF-LP nor YPPAMAM had effective authority to control access to forest resources. As former WWF-LP staff recognized, only sustained attention from higher authorities—the governor and the Regional Forestry Office (*Kanwil*)—in support of the CCF initiative would be able to curtail local clientelist arrangements supporting the logging.[115] Even though attempting to nest the CCF institutional arrangements in wider state regulations was important, changing the status of resources in South Aceh under state law would not guarantee the desired outcome. The key factors were distribution of power at the local level, as well as the clientelist arrangements and the socio-economic conditions in which they were embedded.

"Informal constraints"—the informal ways by which human beings structure human interaction, including norms, conventions, and values— "will not change immediately in reaction to changes in the formal rules" (North 1990: 45). Despite changes in the formal status of the Menggamat forests, the informal arrangements underpinning the logging network shaped forest outcomes. In 1997–98, fluctuations in the dominant socio-economic realities shaping local life, followed by political changes within South Aceh and beyond, altered the situation.

Krismon and *Reformasi*

During 1997–98 a range of changes altered the situation in Menggamat, causing the tide of uncontrolled logging to begin to recede. Logging continued across South Aceh, but for several reasons logging in Menggamat was now on a much smaller scale.

The first and most significant factor was the economic crisis (*krismon*), which led to wide fluctuations in the central economic elements determining village life. As I described in the last chapter, as the value of the rupiah sank, coincidentally the U.S. dollar value of patchouli oil skyrocketed. These twin influences led to a drastic increase in the local price of patchouli oil. Those villagers who had abandoned logging and, at the suggestion of WWF-LP and YPPAMAM, had switched back to *nilam* cultivation some months earlier, began to reap windfall profits. As the contagion of "*nilam* fever" (*demam nilam*) spread in Menggamat, farmers were now prepared to open new plots in distant forest areas. At this time the rate of opening land increased dramatically—mostly on steep land close to the village and in secondary forest near remote hamlets located far up the Kluet River. As long as the boom lasted, villagers involved in logging switched over to *nilam* farming. Many loggers sold their chainsaws to obtain capital for opening *nilam* plots. As wood supplies dried up, *cukong* now used their capital to buy *nilam* oil and trade it with wholesalers from Medan.[116]

However, the *nilam* boom was based on a commodity price fluctuation. The *nilam* crops would not be harvestable until seven months after planting. By mid-1998, *nilam* prices began to fall, just as many of the crops planted during the previous wet season began to be harvested.[117] Unfortunately, *nilam* production increased just as demand slackened. According to farmers, in 1998 values, *nilam* was worth growing only if prices were higher than 500,000 Rp/kg; by January 1999, *nilam* had fallen to 160,000 Rp/kg. As early as October 1998, villagers had begun to turn back to logging. In the meantime other factors had impinged, and logging was now on a much smaller scale.

The *krismon* also affected sawmill operations on a broad scale. While there had been ongoing raids against illegal loggers over the 1990s, they had never curtailed the illegal logging and the sawmills had operated with apparent impunity. However, the economic crisis initially led to a collapse in demand for wood from Japan, South Korea, and Taiwan, and the international price for plywood fell by 40 percent during 1997–98 (CIFOR 1998).[118] By October 1998, provincial officials estimated that half the sawmills in South and West Aceh had closed or gone bankrupt, due to the high cost of components for imported sawmill machinery as well as difficulty in obtaining wood supplies.[119] In a similar fashion, in North Kluet the number of sawmills dropped from the seven that had operated there at the height of the wood boom to only two, and even these worked sporadically—when there was wood. Clearly, the economic crisis had also hurt the *cukong*, who now lacked the capital to support large-scale logging operations. Before the economic crisis, a *cukong* might provide capital to several *tauke*, and each *tauke* would have several chainsaws and logging teams. However, following the economic crisis a *cukong* could afford the capital for only one *tauke*. At most a *tauke* would have three chainsaws, and sometimes only one.[120]

A second factor was also very significant: the difficulty of obtaining high-quality timber. As one informant explained, sometimes there was not enough wood even for one sawmill to function:

Logging is much reduced now. Now there is no wood. Operational costs are too high compared with the profit to be gained. Before, it just took one day; now it takes a few days to reach the forest. Previously, there was a lot of timber close to the river, and it didn't entail operational costs in getting it. Before, loggers only needed to put the wood into the river . . . Now they must pull it long distances as well. One log needs two to three people. The sawmills were right next to the river, and before it was much easier to float wood down. So there were no operational costs.[121]

Consequently, at the time of my last visit to the area in January 1999, the logging of Manggamat forests continued, but on a much smaller scale.

Although perhaps not as important as the first two in determining what had occurred in Menggamat, a third dynamic also affected logging operations in South Aceh. In the wake of *krismon*, the political changes known as *reformasi* led to significant shifts in the power of the local network of power and interest that previously drove the logging. In early 1998, South Aceh obtained a new *bupati*. According to local sources, the new *bupati* was a relative of a senior figure involved in Yayasan Leuser International (YLI).[122] YLI had long enjoyed influence at the center: in 1998 President Suharto issued a presidential decision clarifying the official legal standing of YLI/LMU's management activities in the Leuser Ecosystem, and the Minister of Forestry in the transitional Habibie government was a former member of YLI. Now, a district forestry official asserted, YLI had used their considerable influence in Jakarta to ensure the selection of the *bupati*. Accordingly, the new *bupati* supported YLI initiatives. For instance, according to the Medan daily *Waspada*, in 1998 the *bupati* oversaw the creation of a fact-finding team to investigate the "sins" (*dosa*) of logging concessions (HPH) operating in South Aceh. The team consisted of representatives of local agencies as well as critical NGOs such as *Rimueng Lamkalut* (Acehnese for "tiger meditation"), an NGO specifically concerned with the logging problem.[123] The *bupati* not only discontinued the previous policy of tolerating illegal logging in the name of raising local revenue but also attempted to have the leases of HPH terminated. Partly as a consequence, in 1999 the district government faced a large decrease in revenue.[124]

The fall of Suharto in May 1998 led to a more open political atmosphere and a movement against what Indonesians referred to as corruption, collusion, and nepotism (KKN). In the latter half of 1998, this change provided an opportunity for various student and other groups to campaign against KKN and illegal logging in South Aceh. In June, students met with police and military leaders in the presence of the *bupati* to protest against the behavior of officers involved in logging.[125] Responding to this mood, the *bupati* discussed replacing all district government officials who were considered to be involved in KKN.[126] Local officials could no longer so unashamedly accept payments for supporting logging operations. The police and army posts that formerly demanded payments from logging trucks in every sub-district were no longer tolerated.

During 1998, concerns about environmental conditions across South Aceh were more widely discussed. The new *bupati*, Sayed Mudhahar (a former *bupati* and at this time a senior YLI figure) and local groups discussed how the degraded watershed forests could no longer carry out their hydrological functions. After heavy rainfalls streams quickly flooded, and streams tended to dry up quickly after the rains. Over the previous months severe floods had hit South Aceh. Moreover, the *bupati* noted, twenty-nine irrigation systems in the district faced water shortages, affecting agricultural production and thus having "a serious impact on local farmers." In South and North Kluet areas, irrigation systems served by the Kluet River could not meet planting schedules due to water shortages.[127]

In Menggamat itself, as Effendi (1998) has written, local forests had significant ecological and economic value, providing essential watershed protection for local agricultural activities. However, illegal logging and agricultural encroachment into the watershed had diminished these ecological benefits. Since 1994, reports had described the outcome of the logging of forests behind the communities alongside the Kluet River. In 1996, severe flooding caused significant damage to housing and agriculture in Menggamat. In a study carried out in Menggamat, Effendi estimated that due to uncontrolled logging, flooding had caused damage in just one of Menggamat's thirteen villages amounting to Rp 85 million (Effendi 1998). A 1998 LMU survey also showed that older residents estimated that the average reduction in water levels in streams around Menggamat was 25 percent, while there had been a 50 percent decrease in water levels in the major river flowing through the sub-district, the Kluet.[128]

In essence, damage to the hydrology of watershed forests meant that forests could not retain water. With more rapid water flows during the wet season, there were frequent landslides on steep slopes, and the river became increasingly unstable. The erosion of riverbanks caused the river channel to widen and often unpredictably to shift course, washing away the gardens of local residents. On average, Kemukiman Menggamat now suffered flood damage three times each year. These floods affected farming lands and houses, causing serious economic loss. Decreased water flows in the dry season also caused water shortages for the irrigation systems supporting wet rice cultivation, in turn leading to diminished agricultural productivity.

Now vocal groups ascribed these problems to fifteen years of unsustainable logging by logging concessionaires and illegal logging networks sustained by local KKN.[129] Protesting against logging operations, the new groups went on the offensive, calling for the cancellation of logging operations (IPK) and timber concessionaires (HPH) in the district.[130] In August 1998 the most vocal group, Rimueng Lamkaluet, wrote to the

Minister of Forestry calling for the immediate cancellation of all HPH and IPK working in South Aceh. If they were not cancelled by October 5, Rimueng Lamkaluet threatened to "run amuck" and "burn to the ground" the base camps of IPK and HPH that refused to leave the area.[131] Local religious leaders, officials, and village heads supported the threats.[132] LMU was understandably comfortable with these developments and coordinated with these groups.[133]

As a consequence, in South Kluet, an area adjoining Menggamat, in late 1998 the forestry department imposed a temporary moratorium on the operation of one concessionaire, PT Medan Remaja Timber (discussed earlier), pending a field investigation. Since 1996 another NGO, Yayasan Leuser Lestari (YLL), had been providing detailed reports to the authorities on the illegal activities of this concession. Local people held the concession responsible for flooding and irrigation problems and complained about the building of a road "through a village's crops without talking to the villagers" (Newman et al. 1999: 39). Finally, in April 1999, the threats were carried out: "frustrated by the slow pace of negotiations after the team's inspection, 400 people and students burned the company's base camp to the ground." The government also withdrew the company's license.[134]

Nonetheless, the effects of the economic and political changes that had worked against the logging networks during 1997–98 proved to be temporary. A number of actors attempted to make the most of the opportunity for curtailing the logging networks, but in 1999 the balance of power moved back towards the logging networks. The demand for wood had again increased and, with few other economic opportunities available to impoverished villagers, widespread logging activities in the Leuser Ecosystem resumed. By the middle of 1999, "wild logging" of remaining areas of lowland rainforest was occurring in adjacent areas of South Aceh. In South Kluet in mid-1999, an investigative report by two NGOs noted that "while trekking the team saw areas of the National Park devastated by logging, heard chainsaws in most directions, witnessed the felling of trees within the park and learned the location of the sawmills receiving the stolen logs" (Newman et al. 1999: 35). The logging networks involved *cukong* and *tauke* following the *modus operandi* discussed earlier. By March 1999, logging teams had even moved into the Suaq Balimbing Research Area, an internationally renowned orangutan research site located within the boundaries of the Gunung Leuser National Park and a focus of LMU's forest protection efforts. "By July it was estimated that about 100 loggers were operating within the study area, based at 24 logging camps" (Newman et al. 1999: 37). The NGO's report observed that in a time of political instability, and with a separatist movement now operating in Aceh, law enforcement had deteriorated to the point that "authorities drop any pretence at even trying to deal with the problem" (Newman et al. 1999: 38). The report added: "The new reformation era of Indonesian politics has shifted some of the power to those who wish to oppose the law for political or financial reward . . . Coupled with this are local timber barons with military and police support, who are exploiting this current power vacuum" (Newman et al. 1999: 31). Once again, district networks of power and interest had proved capable of adjusting to change.

Conclusion

In many respects Menggamat conforms to the classical image of a frontier as a physical place in rapid transition. Here, as technology and infrastructure develop, and as new market opportunities emerge, there is a continual shift in the resources that actors can extract, in what is valuable, and in what is worth extracting. Consequently, as these cumulative changes are worked out on the ground, both the landscape and local patterns of resource uses are continuously transformed.[135]

On this forest frontier, villagers focus their livelihood strategies on farming and extractive activities to produce commodities that fetch high market prices. Historically, the inhabitants of Menggamat combined agriculture with the collection of forest products. The importance of forest products to local livelihoods always varied with fluctuations in their prices. Only during specific periods could villagers depend on the harvesting of non-timber forest products for livelihood. When prices were high, there were boom periods; the same products were ignored when prices were low. In the past agriculture was primarily a subsistence activity, but ever since the introduction of *nilam* by the Dutch and of other cash crops more recently, farming has been a primary source of cash income. With these changes, as in Sama Dua, rather than being devoted to staple food production, in the core area of Menggamat swidden agriculture became a part of the establishment of an income-generating tree crop agriculture.

In Menggamat farmers need to adjust to the rapidly fluctuating prices of forest and agricultural products during periods of boom and bust. Consequently, rather than pursuing a single agricultural strategy, farmers have to continuously adjust to the reality of a shifting frontier. Despite these rapid changes, in general, apart from timber, the trend has been for forest products to become a less salient part of local livelihoods and for agriculture to increase in importance. Consequently, even though *adat* assumptions and rules relating to access to and use of resources are long established, to adjust to this fluctuating environment, local institutional arrangements need to be dynamic, responsive to change, and subject to revision. Farming here occurs beneath rainforest-covered hills subject to erosion, flooding, and landslides. Over time *adat* institutions have developed an awareness of the limiting conditions under which agriculture takes place—that the Kluet watershed maintains the water supply, that it guards against erosion and flooding, and that degradation here affects the productivity of agricultural land. As a consequence, villages had developed some rules to prevent irresponsible farming and environmental damage, such as forbidding the cutting of trees in certain places or at certain times of the year. In remote, closely knit villages, these rules mainly relied on the principle of shame: villagers were loath to go against village values if it meant loss of reputation and standing in the village. For on-going violations of *adat* rules, offenders ultimately faced sanctions enforced by a meeting of the *adat* council. As demand for land has increased in the core areas of Menggamat, and as the area has increasingly been integrated into the wider system of administration, the *adat* regime has tightened its rules governing the opening of land near the core village complex. Consequently, in recent decades *adat* necessarily has shifted to accommodate the pattern of agrarian transformation occurring in upland areas, where rather than retaining forest areas, villagers have sought to secure a better future by expanding the cultivation of trees crops (Li 2000; Sunderlin et al. 2000). In contrast, in outer frontier areas the *adat* regime remains loose, and farmers have continued to shift plots more freely.

Over the last decades, entrepreneurs and politico-bureaucrats have also found new ways to use the frontier area to pursue particular political or development agendas. For example, as technology, infrastructure, and markets for timber developed, politico-bureaucrats have used their power over permits and sanctions to gain benefit from timber found on the forest frontier. They have done this for self-enrichment, to find extra-budgetary support for their agencies, to reward political supporters, and to co-opt opponents.

In South Aceh the logging phenomenon can be understood as a system of exchange and accommodation that begins with entrepreneurs (*cukong*), brokers (*tauke*), and "rogue officials" (*oknum*). Initially it included forestry staff, army personnel, and other local functionaries at the district level, but then the system expanded to embrace *adat* leaders and village heads. To stay in the logging game, *cukong* and *tauke* needed to conform to a system of exchange of extra-legal gifts and favors to local politicians and state functionaries. *Oknum* occupying positions in the district and sub-district government engaged in what Scott (1977) has called "office-based patronship," using their discretionary power over licensing, permits, and law enforcement as a means of controlling access to forest exploitation.

At first the *adat* authorities failed to maintain control over access to and use of local forest territory. However, they did not want just to observe the depletion of local forests, and in alliance with villagers, they acted to affirm *adat* property rights over local forests to defend the "right of avail" over surrounding forest territory. Outside actors could extract timber from their territory only if they paid *adat* fees. As local villagers joining logging teams and *adat* heads extracted rent for their own benefit and for that of the village, village actors enjoyed a portion of the flow of benefits from *adat* property. Accordingly, a wider local pattern of exchange around illegal logging accommodated *adat* assumptions and authority structures.

The resulting networks produced significant rents, both for self enrichment and for support of district government budgets and political agendas. This revenue increased the popularity of the key local politician, the *bupati*, who could then use expanded provincial budgets to support projects and programs that offered opportunities to clients and followers. Over time these exchanges generated the institutional arrangements that governed access to the forests. Here, direct personal ties based on reciprocity substituted for state institutional arrangements that operated in accordance with the law or the *adat* institutional arrangements of the villages. In this way, the locus of control over access to and use of forest resources shifted from the village and *adat* heads up to the network of functionaries (and their business partners) officially responsible for implementing regulations.[136]

Ostrom has identified several principles characteristic of self-governing "successful" regimes that have developed institutional arrangements to ensure the sustainable management of common pool resources. She has found that, where successful governance regimes operate, appropriators with higher economic and political assets need to be interested in generating an effective governance regime: that is, they need to be "similarly affected by a lack of coordinated patterns of appropriation and use" (Ostrom 1997: 7). In the successful regimes she discusses, resource users usually have the autonomy to devise their own institutions, a right not challenged by government authorities. Institutional arrangements in governance regimes (successful in these terms) also tend to be organized in multiple layers of "nested enterprises." If this principle operates, less formal levels of organization will be nested within "externally organised political jurisdictions" (Ostrom 1992: 76).

Translating the Menggamat experience into these terms, previously the *adat* regime in remote Menggamat enjoyed autonomy in making decisions and crafting rules with respect to tenurial rights, the collection of forest products, and protecting the limiting conditions facing agriculture. However, a range of contextual factors changed. A new road and the advent of chainsaws, trucks, and sawmills made tree felling a viable and profitable activity as Menggamat became integrated into national and international markets for timber. As the local government structure changed, village leaders became more integrated into a wider administrative framework, and the institutional arrangements relating to village government altered. Local leaders became more dependent on higher officials. At this time, some powerful local *oknum* were backing *cukong* or acting directly as *tauke* or *cukong* themselves. Accordingly, the village became more of a subsystem within wider institutional arrangements that profited from logging. However, as outside loggers and *oknum* bore none of the negative externalities involved in unsustainable logging, they had little inclination to help craft institutional arrangements that would guard the ecology of the Menggamat watershed. At the same time, since they had a limited period of office, officials had a different discount rate to local resource users. In other words, they undervalued the future benefits that Menggamat's villagers might receive from sustainable use of the forest, and overvalued the rents to be extracted from logging during their period of office (Ostrom 1992: 34). Through allowing illegal logging, these office-based patrons displaced the role that *adat* heads had enjoyed as gatekeepers of resource use. In this situation, the local *adat* regime was unable to prohibit logging, and the forests became subject to widespread and unsustainable logging.

Viewed from whichever perspective, an NGO trying to implement a community-based forestry initiative in this context faced significant challenges. WWF-LP activities combined community development and income-generating activities as well as building upon aspects of the *adat* regulations that had related to the forests. The later involved using *adat* to craft CCF regulations in a form that could nest within the wider state legal framework. The aim was that a new community conservation regime would guard the limiting conditions under which local agriculture operated while generating income for local villagers. As it turned out, given the deep roots of the timber interests in South Aceh, the WWF Leuser Programme represented a courageous—if somewhat quixotic—tilt against deep-rooted interests. If this intervention faced several obstacles that were ultimately insurmountable, it also raised particular problems.

To pursue a community-based natural resource management initiative, WWF-LP attempted to find an indigenous logic for community management, to build community management upon *adat* intuitional founda-tions, and to justify its conservation agenda in terms in Menggamat *adat*. "Revitalizing" *adat* in these terms involved a particularly simplified, ahistorical reading of *adat* that ignored aspects of the local working of *adat* (cf. Eghenter 2000).

While the CBNRM narrative finds a foundation for conservation in community-based property rights, in two particular respects Menggamat *adat* suggested that assuming a direct correspondence between *adat* and sustainable management would involve misreading *adat*. First, in South Aceh *adat* regimes have long encompassed levying fees on commercial uses of forest products without particular regard to sustainability. Pro-vided that local property rights were respected, the logic of these regula-tions could easily be extended to facilitate extractive logging. Second, WWF-LP and YPPAMAM sought to show a direct link between manage-ment of forest areas under *adat* and conservation outcomes by contrast-ing the relatively intact forested hills surrounding the core Menggamat area (*hutan adat*) and the state forest area (*hutan negara*) degraded by loggers (see figures 12 and 13). This contrast served to support the argu-ment that *adat* management should be extended to the *hutan negara*. The forested area surrounding the core villages in Menggamat indeed remained comparatively intact, albeit for rather complex reasons. In the core area the establishment of income-generating tree crops over recent decades had subsumed the older swidden practices, leading to the gen-eration of an agro-ecological system that involved tree cropping. This helped make the area look forest-like. Moreover, when logging teams moved into Menggamat's forests, villagers would not tolerate logging in forested areas adjacent to their gardens (*kebun*), where tree felling on steep land might damage crops or create landslides. Moreover, log-gers preferred to work in the state forest, not only because of the lack of clear property rights there, but because the most valuable species of timber still grew there. The contrast between the area where an *adat* regime persisted and the degraded state forest area remained interesting and pertinent to developing a conservation strategy. However, attributing these different outcomes to a simple difference between an "ecologically sustainable *adat* regime" and "a flawed state management regime" was misleading: the reasons underlying this contrast needed to be explored more thoroughly in order to find out how this could help project strate-gies. Otherwise, while *adat* could be used instrumentally to serve a con-servation purpose, doing so inevitably left aside aspects of village *adat* practices: the creation of a "conservationist *adat*" only partially repre-sented the values and practices of Menggamat villagers. It represented a particular definition of the local socio-legal order, a definition that was inevitably contested by a range of village actors.[137]

In Africa the colonial state guided the process of according particular ideas and interests the status of "customary law" (Chanock 1985). In contrast, in this case an outside conservation project and its local partners oversaw the transmutation of a "revitalized" Menggamat *adat* into the officially approved rules for the CCF. The problem was that WWF-LP had to work in the space between the state legal discourse and the local *adat* order. To do so involved a double act: using the authoritative language of existing government statutes while embracing the complex, flexible, and somewhat ambiguous guidelines for social relations found in *adat*. But rather than providing WWF-LP and its local partners the authority to craft rules that suited the local situation, at this time existing policies failed to allow sufficient space for local-level rule making. This points to a problem with the attempt to "nest" *adat* within state law. In attempting to gain state approval for the CCF initiative, the project had to accord with the state legal framework. Because this process had to be mediated by restrictive national legal forms, it forced the CCF concept to move away from existing *adat* practices. It also extended the reach of bureaucratic assumptions and state control over local resources at the expense of an alternative, autonomous, and (in local eyes) more just *adat* tenurial regime.[138]

The advocates of "*adat* law" in Indonesia often use the term "customary law" as if it designates "a straightforward set of traditional rules." However, here, as in Africa, it actually refers to "a cultural construct with political implications" (Moore 1986: xv). Without explicitly working with such an understanding of the term, attempts to "revive" *adat* have proceeded from the flawed epistemological assumption that "customary law" is always there, waiting to be found and turned into an inventory of rules (Chanock 1985: vii). To an extent this project intervention was built on a one-sided and simplified interpretation of identified and "recreated" *adat* rules. Therefore, it failed to deal with the diversity of opinion within "the community"; it also failed to build a consensus within a diverse "community." These failures led to significant resistance to the community forest initiative.

We could argue that the CCF initiative involved locally situated actors using state laws to extend state control over themselves, and to this extent it created state-sanctioned authority in another "community space" (Agrawal 2001). Seen from this angle, community-based conservation might become a vehicle for realigning the relationship between the state and its rural subjects (Doolittle 2001; Li 2002). Like colonial processes of creating "customary law," CBNRM conducted on these terms risked incorporating localities "into a different world on very disadvantageous terms" (Chanock 1985: vii). Yet, as the last chapter suggested, for many years *adat* and state orders have constantly made mutual adjustments and accommodations; by exploring the affinities of *adat* with state legal forms to protect their own *adat* order, as in Sama Dua, Menggamat actors may be able to use the symbolic capital associated with the state law to support *adat* property concepts that contradict state legal norms. Contested socio-legal contexts are more unpredictable than is often suggested: the emasculation of the *adat* order was not a forgone conclusion.

Although the project involved community development, income-generating activities, and institutional changes, these initiatives could not offer better alternatives to unsustainable logging, especially given the lucrative opportunities offered by logging and the instability of local livelihoods. As the director of WWF's national program later acknowledged, "the commitments from local community that were secured through community development activities often washed away with the increased incidence of legal and illegal logging activities by non-local people."[139] Part of the problem was that, as outside forces locked on to new opportunities on the forest frontier, the livelihood strategies left over for marginal villagers and propped up by conservationist interventions tended to be "opportunities that have no other claimants." These were "unviable development alternatives" that are of "nominal value" to the local people themselves (Dove 1996).

In Menggamat, the most significant problem was that key officials and even some village heads supported the logging while the *bupati* valued its contribution to district budgets. These *oknum* and their village allies collectively had a tacit but effective veto over law enforcement and the working of the community forest initiative. Without their support, the leaders who backed the WWF-LP plan could not effectively implement the new regime. At best, *adat* heads could defend *adat* property rights over surrounding forests by charging the loggers a fee. Furthermore, many villagers had become economically dependent on logging, and successive, rapid economic shifts continued to drive local villagers to mine forest resources. In the short term at least, the proposed conservation regime could not guarantee village livelihoods and therefore it struggled to get off the ground.

Across South Aceh the economic and political shocks of 1998–99 led to the temporary demise of the networks of exchange and accommodation. This demonstrated how these networks were embedded within the Suharto regime; they depended on patrimonial relationships sustained within the regime. These relationships were as stable as the regime and the economic and political structures on which they depended.

Within Menggamat itself, the fluctuations over this period also confirmed that patterns of unsustainable resource use have complex, multidimensional causes. To remedy the conditions leading to environmental decline would involve slackening the demand for timber; providing an economy that offered stable economic livelihoods and viable alternatives to logging; supplying steady revenue streams for local district government budgets in the absence of timber taxes; building greater local awareness of the long-term impact of unsustainable logging; developing an accountable political system with an active civil society able to control the behavior of local officials and politicians and ensuring effective law enforcement; renegotiating customary rules governing resource access and use to take ecological values into account; and reconciling those customary rules with legitimate and enforceable state rules. The shifting dynamics over the 1996–99 period suggest that a change in one factor without simultaneous changes in several others would not sufficiently alter the dynamics driving unsustainable resource extraction. Further, for any change to amount to more than a temporary fluctuation and truly make a difference, it would need to be sustained over the long term. The scale of change that would be required lay beyond the reach of this community-based intervention.

4

Power and Interest in Badar

The Alas land (*Tanah Alas*) lies some two hundred and fifty kilometers northwest of Medan, the capital of North Sumatra (see maps 1 and 8). Before reaching the area, the road winds through the fertile market gardens of the Karo uplands north of Lake Toba. Then, after traveling over a range of steep hills, some covered in candlenut (*kemiri*) forest-gardens but many denuded and badly eroded, suddenly the bus turns a corner and a river valley appears—a fertile plain of intensely farmed rice fields (*sawah*). Here the Alas River meanders through a valley ten kilometers wide and sixty kilometers long surrounded by steep banks of mountains covered in the dense forest of the Leuser Ecosystem.

Traveling alongside the Alas River, the bus passes the regency capital of Kutacane and continues north towards Blangkejeren, the main town in the Gayo Lues area. As the road climbs and the valley gradually narrows, the country becomes increasingly mountainous. Close to the entrance gates of the Gunung Leuser National Park, the verdant rice fields of the valley floor give way to candlenut forest-gardens. Then the road enters the national park, which it bisects for the next forty kilometers. As the bus climbs, for some kilometers thick tropical forest hangs over the road before once again the view opens to reveal simple village houses and forest-gardens cut out of the forest. This is Badar sub-district, and here Alas people, who have moved north onto what is technically national park land, live alongside Gayo people, who have migrated down from the highlands of Gayo Lues (see maps 3 and 4). Badar, the sub-district at the

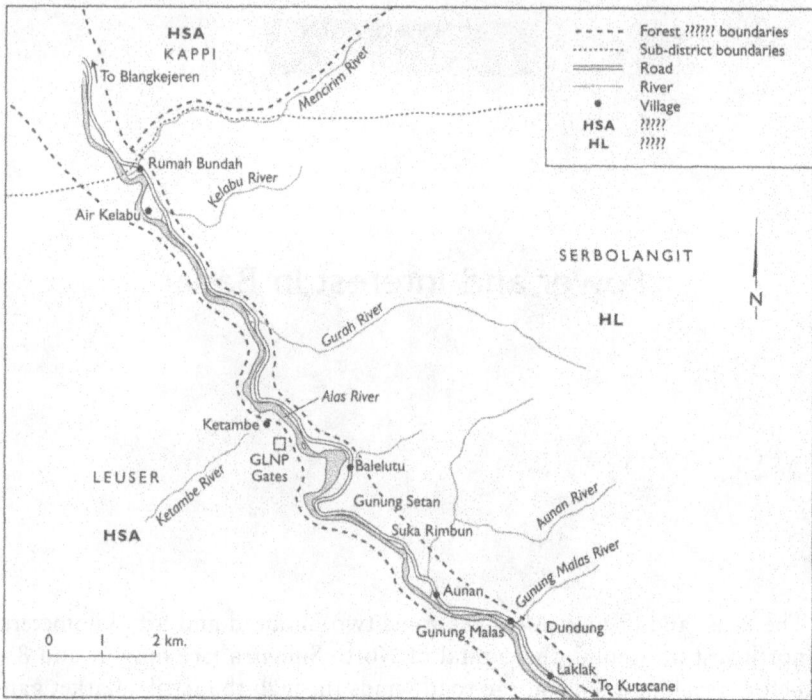

MAP 8. Jambur Lak-Lak village, Badar Sub-district, Southeast Aceh: forest boundaries and hamlets.

northern end of the Alas valley (*lembar Alas*), and its hilly margins are the homeland of two isolated and long-ignored ethnic groups, the Alas and Gayo Lues people. This area is the subject of this chapter.

In contrast to the two areas I discussed earlier, Badar is distinguished by a long history of conservation activity, and this has particular consequences. Since the colonial period, and more particularly since the 1970s, researchers, ecologists, conservationist NGOs, and the state forestry agency have vied with villagers, logging networks, and local government to determine patterns of resource use. Consequently, institutional arrangements here are more highly contested. Since the "rules of the game" are subject to dispute, the institutional arrangements operating here are more difficult to pin down. Moreover, as this is a site with a history of both latent and overt conflict, the actors involved are engaged in on-going struggles to control access to resources. Actors seek to strengthen their position by either highlighting certain details or hiding others. The records that researchers, ecologists, and conservationist NGOs and the state forestry agency have created articulate this history of disputes.

This chapter is divided into four sections. In the introductory section I discuss the Badar area and its agricultural and forest history. In the second section, I consider the evolution of village institutional arrangements in Badar. In the third, I discuss forest outcomes in terms of the actions and interactions of village institutional arrangements with logging networks, field foresters, and other officials, including district politicians. Finally, in the fourth section, I discuss the struggle over forest access generated by conservation interventions (particularly the high-profile ICDP) and its resultant effect, on local institutional arrangements.

Badar

Tanah Alas is the homeland of the Alas people, an ethnic group indigenous to the Alas valley. According to oral history, the Alas are predominantly the descendants of migrants from the Batak lands, from the Gayo highlands, and from the Singkil and Kluet areas of South Aceh. However, Minangkabau and Acehnese also migrated into the area at various times (Iwabuchi 1994a: 9). In 1985 Iwabuchi estimated that there were approximately seventy thousand people of Alas ethnicity (Iwabuchi 1994a: 8). To the north of the Alas valley lies Gayo Lues, the native land of the Gayo people. The Gayo Lues are a sub-group of the larger Gayo ethnic group found in the highlands of Central and Southeast Aceh (Hidayah 1997). Bowen (1991) reported that there were approximately forty thousand people belonging to the Gayo Lues sub-group. Although the Gayo and Alas people "had been sometimes regarded as identical by ethnologists," further investigation has revealed considerable differences between the two groups, especially in regard to hereditary succession, modes of marriage, language, *adat,* and agricultural methods (Iwabuchi 1994a: 248). In 1974 an administrative change fused Gayo Lues with Tanah Alas to form the district of Southeast Aceh.

Badar sub-district itself contains thirty-three villages and extends from the long-settled heartland of Tanah Alas to the more recently settled lands bordering with Blangkejeren sub-district. The settled area extends alongside the river Alas, forming a wedge shape inserted into the mountainous Gunung Leuser Ecosystem (see maps 2 and 8). Some twenty-five thousand people live in Badar, working rice fields alongside the river or cultivating candlenut (*kemiri*), coffee, and other products in gardens cut out of the hillsides.

Agriculture and the Collection of Forest Products

In the heartland of the Alas valley, as in the other two sites, the Alas people historically followed a partial system of shifting cultivation. Villagers cultivated wet rice (*sawah*) along the floor of the valley, and for many years the Alas valley produced a rice surplus. To supplement rice production, Alas people also practiced shifting cultivation, opening dry land plots (*ladang*) in the surrounding areas. In these gardens the Alas grew tobacco, chilis (*cabe*) and other spices, durian, pineapples, bananas, rambutan, and other fruits, and a variety of stimulants such as marijuana (*ganja*), coffee, and *sirih* (*Piper* spp.) for their own consumption (Paulus and Stibbe 1921: 27; Iwabuchi 1994a: 29–31; Rijksen and Griffiths 1995: 142). Generally, after a few years vegetable gardens were abandoned, and farmers then opened a new plot in another place. The forest regenerated on the old field, and after a number of years the plot could be used again. Since the early 1980s farmers have also developed candlenut and coffee smallholdings, colonizing surrounding hillsides, and even opening up enclaves within TNGL.

During the 1980s economic forces and cultural practices alike drove the conversion of hillside forest land to candlenut gardens. In older villages on the floor of the valley, land was growing increasingly scarce. Population growth contributed to the fragmentation of landholdings over time and helped stimulate agricultural pioneering.[1] In addition, certain cultural patterns also encouraged conversion. Alas families customarily paid a high bride price to marry off their sons. Alas families had to sell land to migrants from North Sumatra to meet these costs, and thus Alas families continued to lose control over land. Until recently, they could acquire land through forest pioneering, and young villagers would open areas of forest on the margins of the valley for cultivating cash crops such as coffee and candlenut.

Iwabuchi has shown that the poorer villages on the floor of the Alas valley could be distinguished from the more affluent villages on the margins of the valley.[2] Agricultural figures indicate the underlying causes of these differences. On the floor of the valley, the predominant crop, rice, gave a gross return of Rp 2.5 million per hectare at 1996 prices. In comparison, candlenut—the main cash crops in hillside gardens—yielded Rp 3.7 million per hectare, by far the highest price among the main crops grown in the valley at the time (Jordan and Anhar 1997: 38–39). Consequently, on the margins of the valley where land was available for candlenut gardens, average cash incomes per hectare tended to be higher. Until recently, young Alas couples lacking land in the core area could secure a better future by forest pioneering along the margins of the valley.

Around Badar itself, at least up until the economic crisis, many plots of land remained abandoned and covered in weeds. This "sleeping land" (*lahan tidur*) represented a vestige of an *adat* regime that accommodated shifting cultivation; it recognized the continuity of property rights over unused land as long as at least one perennial tree planted by the owner continued to stand.

A migrant from South Aceh related that, compared with the Gayo, the Acehnese, and the Aneuk Jamee people, for the Alas people hunting and gathering in the forest were not particularly important. For instance, the Jamee opened forest gardens some distance from the village and slept out in the forest. In comparison, in the immediate vicinity of the Alas River, a satisfactory life could be sustained by agriculture and fishing.[3] There were people with esoteric forest skills (*pawang*) amongst the other ethnic groups of Southern Aceh, but the Alas people were generally fearful of penetrating far into the forest (Iwabuchi 1994a: 34).

An older villager described how in the colonial period villagers collected forest products to supplement agriculture. When the market value of forest products was high and buyers visited the Alas area, villagers collected *damar* resin and large-diameter rattans from the surrounding forest. This villager remembered that even a child could collect enough *damar* resin in one day to buy one *kaleng* of rice—enough to supply a family for a whole week. Collectors would go into the forest only so far that they could return at night.[4] Villagers also collected pandanus, vegetables, fruits, and other products in the forest (Iwabuchi 1994a: 34).

In the 1970s, as the Alas area was increasingly integrated into the market economy, and as the villagers experienced greater needs, they increased the harvesting of forest products. In the late 1970s, according to van Strien, villagers collected materials mostly in small quantities for personal needs including herbs and spices, fruit, and other foodstuffs. However, collectors would also penetrate deep into the forest to gather those products with a commercial value in large enough quantities to be traded. These included rattan and fragrant oils and resins for perfume and incense. For instance, collectors tapped resins (collectively known as *damar*) from trees of the Dipterocarp species. Careless treatment killed the tree, and van Strien reported that people penetrated far into the forest to find these trees. The result was that "over large areas all specimens have already been killed" (Strien 1978: 51).

When I visited the Alas valley during 1996–97, the collection of forest products continued on a small scale. However, as collecting within GLNP was prohibited, villagers were reluctant to discuss this issue. However, sometimes a man, exhausted and soiled from a day in the forest, could be seen wandering back to the village with coils of rattan slung over his shoulder, and small piles of rattan were left sitting outside village houses. In Kutacane, a trader bought rattan in the market, laying out his stock near the fruit stalls. Through 1996 and 1997, people collected rattan mostly for their own use—for example, using it to make fish traps. At that time, villagers rarely collected *damar*. At Rp 200 per kilogram, the price was considered too low. Villagers would not consider it worth gathering until prices reached Rp 300–400 per kilogram. All the *damar* in the immediate vicinity of the villages had been collected, however. "A collector would have to go a long way," an informant said, "and at these prices it is hardly worth collecting."[5] Another villager mentioned that people collect it only during times of scarcity, for example when they were waiting for the candlenut trees to bear fruit. "Every few weeks a *damar* trader (*tauke*) passes through the village," he said. "If the police find out, then he will have to pay off the police."

By December 1998, during the economic crisis, villagers had begun to collect rattan more intensively. In Badar sub-district, along the Blangke-jeren-Kutacane road, rattan wholesalers had opened depots for buying rattan from collectors. The "subsidy from nature" provided by these extractive products was still a very important source of additional income for the rural poor, especially at times of crisis (Hecht et al. 1988). On 11 December 1998, I witnessed a pickup truck loaded with several tons of rattan pulling out of a rattan warehouse. A few days later, I saw five villagers returning from the forest loaded with rattan. Groups of villagers were taking to the forest, returning to the village in the evening, or even sleeping overnight in the forest. Collectors then sold the rattan to wholesalers who had obtained licenses, despite regulations forbidding collection in the park. With the rampant inflation, the failure of the candlenut harvest following the drought and pest infestations of 1998, and the collapse of *nilam* (patchouli) prices, poor villagers became increasingly dependent on small-scale extraction of forest products. Rattan prices had increased to between Rp 1,200 and 1,500 per kilogram, while *damar* fetched Rp 700 per kilogram. By collecting twenty-five or thirty kilograms of rattan, villagers could obtain "just enough to eat and send their children to school."[6]

Village and *Adat* Institutional Arrangements

An abundance of forest resources has been the historical experience of the Alas and Gayo people.[7] Dutch colonial sources reveal that the Alas and Gayo existed in small, scattered villages located on ancient footpaths that connected their highland communities (Kreemer 1922; Kreemer

TABLE 4.1
Administrative Structures in the Alas Valley

Pre-colonial (before 1904)	Colonial (1922)[i]	New Order (1998)
Kejuren—local lord[ii]	*Zelfbestuurders van de Landschappen* (self-governing territorial head): each ruling over four *mergo*	*Bupati*—district head
Raje-mude—vice-headman	*Pengulu* or *raje*—territorial headman (head of *mergo* area)	*Camat*—sub-district head
Pengulu si-empat (four heads in *Si-empat* council of state)	*Pengulu si-empat*	*Kepala Mukim*—head of village league
Pengulu suku—clan head	*Pengulu*—village heads	*Kepala Desa*—village head

[i] As described by Kreemer (1922). This is a schematic rending of administrative structures: there were several administrative restructures during the colonial and post-independence periods.

[ii] When the Dutch came in 1904, they found two small kingdoms, each ruled over by a local lord or *kejuren*. Under each *kejuren*, there were five lower chiefs: a vice-lord (or *raje mude*) and four territorial chiefs (*pengulu si-empat*). Each of these local lords exercised direct control over important villages and tribal headmen (*pengulu suku*) (Iwabuchi 1994: 15). However, at this time the basic unit of organization was not the village, but the patrilineal descent group or clan (*marge* or *suku*). This meant that the local lords did not rule over continuous territory but over descent groups whose villages were sometimes geographically scattered.

1923; Rijksen and Griffiths 1995: 125). Even today, according to official statistics, the Badar sub-district is still heavily forested, with 86 percent (362,887 hectares) of the area classified as state forest land. Clearly, as in the other two sites, the abundance of forest has had implications for the structure of local resource-related institutions. Other case studies have also shown that local institutions tend to evolve to protect the most valuable and scarcest resources. Accordingly, where the resource that was most salient differed for neighboring communities (even where the basic customary concepts tended to be similar), local institutions tend to evolve along different lines (Bowen 1988; Benda-Beckmann and Benda-Beckmann 1994; Otsuka 1998a, 1998b).[8] Communities organize to create institutional arrangements that control access to and use of resources when those who benefit from utilizing a resource observe substantial scarcity. Unless resources are relatively scarce, "there are few reasons for appropriators to invest costly time and effort in organising" to control common pool resources (Ostrom 1997: 9).

Historically, villagers in Tanah Alas have not immediately depended on the maintenance of the forest as a resource system. Once again villagers were primarily farmers: they produced rice, vegetables, fruit, and cash crops for consumption and for sale. As Ostrom noted, dependence on a resource system increases the likelihood that self-governing institutions will form to manage access to and use of the resource (Ostrom 1997: 6). Consequently, village regimes were not principally concerned with managing the sustainable harvesting of forest resources. On the contrary, village institutions were more interested in governing access to and use of productive farming lands—the fruit of human labor.[9]

In the past, according to Alas *adat*, lands adjacent to a village were considered to be under the corporate ownership of the village and subject to the *adat* authority of that village. In the core villages of Tanah Alas, this land was known as village land (*tanah kuta*), and consisted of an area covering a radius of about two kilometers. Although the borders of the village land were unclear to outsiders, for villagers and people from the surrounding areas, the borders were clear and were usually marked by a natural feature such as a stream (Koesnoe and Soendari 1977: 23–24).

Originally, each village (*kuta*) "accorded" with a clan (*marga* or *mergo*) controlling an area "for habitational purposes and the business of daily living" (Koesnoe and Soendari 1977: 21). A village tended to be administered collectively by the village elders, who divided up responsibilities according to their respective capacities. An elder who was a senior head of one of the family branches acted as head of the *adat* administration and was known as the *simetua* or *petua*. After the colonial conquest of the area in 1904, as in the other areas, in the process of setting up a territorial system of administration the colonial government rearranged the indigenous authority structures, determining that each village should have a person responsible to the government (see Table 4.1). The village councils then had to choose a head to act as chairman of the village council. This head was henceforth known as the *pengulu kuta* (or *pengulu kampung*). The other *adat* posts in the village were retained (Koesnoe and Soendari 1977: 21; Iwabuchi 1994a: 61).

Within the village lands, tenure pertained to the property rights gained over individual plots obtained with the permission of the village head (*pengulu*).[10] If the land was suitable for irrigated wet rice agriculture, villagers cultivated the lands continuously, thereby securing permanent tenurial rights. However, in addition to wet-rice fields (*sawah*), as I noted earlier, farmers opened dry-land plots to plant vegetables for their own consumption. Before clearing forest for agricultural use, a villager would inform the village head, who would then inspect the area to ensure that it was suitable and that it did not belong to anyone else. After some years, farmers tended to leave their vegetable plots fallow; under a shifting agricultural system, they would open a new plot close by. Koesnoe and Soendari (1977), writing before candlenuts were widely grown, found that once an area had been worked, this land then became the personal property of the person involved. Clearing the land was a mark of property rights, and farmers reinforced these incipient tenurial rights over fallow areas by planting fruit trees before abandoning an area.[11] During the 1920s, the area alongside the Alas River that was subjected to intensive wet-rice agriculture extended to Jongar, a village lying a day's walk from Kutacane, the local administrative center. In 1922, Kreemer reported that the land under tenure did not extend beyond there (Kreemer 1922).

Territorializing the Alas Valley

Following the Dutch conquest of the area in 1904, as in the other two cases, the colonial authorities gradually began their process of "territorialization." This amounted to drawing boundaries, and controlling what people did and their access to natural resources within these boundaries. In a process that in many ways parallels that which Vandergeest (1996) described as having occurred in Thailand, the colonial state further specified Alas territory, at the same time creating new categories of property rights and land uses. First, as elsewhere, the Dutch began setting up a system of indirect rule through "self-governing heads" (*Zelfbestuurders*). To this end, the colonial authorities territorialized local administration by mapping out the territories responsible to each head.[12] The mapping process also set the border of Tanah Alas with neighboring Gayo Lues at Lawe Mencirim.

As in Sama Dua, this process consolidated Alas territoriality. An elderly informant from the Alas village of Jongar (in southern Badar) reported that villagers from Jongar collected forest products in the surrounding rainforest and grazed their cattle upstream as far as Balelutu (thirty-one kilometers from Kutacane), which is now the site of the GLNP entrance gates. According to other informants, the Alas territory extended northwards along the Alas River beyond here, to the stream known as Lawe Mencirim (now located within the national park). This claim was based on a story that historically Alas people had opened land in this area. Furthermore, the lords (*kejuren*) of the neighboring Gayo and Alas peoples had come to an agreement that set the border between their two territories here. Subsequently, the Alas lord (*kejuren*) had asserted territorial control up to this point, and this still remain as the northern boundary of Badar on state maps.[13]

Second, in accord with the "domain declaration," which allowed that all land not under settled agriculture now belonged to the state, the colonial authorities began to make an area available for commercial use. In 1906, shortly after gaining control of the area, the colonial authorities allocated a seventy-five–year lease for developing a rubber plantation to the company CV Penang Kultur. The lease extended over a large area of 1,853 hectares of unopened forest in Badar.[14] The plantation lease extended 1.5 kilometers along the east bank of the Alas River and included foothills to the east of the present-day road.[15]

Third, at the time of the creation of the Leuser reserve complex in 1934–35, the colonial authorities set about classifying areas for specific resource uses. In this process, the colonial authorities left the existing village lands in the Alas heartland under the authority of the village heads. With the nominal agreement of the Alas "self-governing heads," the colonial forest authorities divided the Alas valley into *adat* land (*tanah adat*) and state forest reserve. The *tanah adat* was an area left for the expansion of local agriculture under the administration of the territorial head, while the remaining area comprised the Leuser reserve complex. According to official forest maps, the boundary lines of the Leuser reserve ran along the foothills to the east and west of the Alas valley, meeting where a hamlet now stands (Rijksen and Griffiths 1995: 40). In the late 1930s, the authorities marked the boundary between the *tanah adat* and the state forest reserve with a pathway. This boundary lies to the east of the present-day road, while to the west the river Alas formed a natural boundary, directly opposite where the village settlements now stand. Local people could open land within the *tanah adat* with the permission of the *kejuren* (local lord or "territorial head") concerned. Though opening land beyond the pathway or across the river was forbidden, villagers continued collecting forest products in the reserved forest areas.

At the northern end of the rubber plantation, the colonial government built a troop depot and a jail. Here during the colonial period lay the *pintu rimba* ("forest door"), the point where the footpath connecting Tanah Alas and Gayo Lues territories left the inhabited plain and entered the forest. The *pintu rimba* marked the end of the area under cultivation. The Alas people considered the area beyond to be tiger-infested wild jungle unconducive to human habitation. In the past, before proceeding past the *pintu rimba*, villagers would carry out a small ritual, placing a branch next to the path and asking protection from the guardian spirit of the forest (*aulia*) before proceeding into jungle inhabited by potentially malevolent spirits. Nonetheless, in 1936 the *kejuren* allocated an area of land for local smallholder agriculture just beyond the *pintu rimba*.[16] In the 1930s, a small settlement existed here—until villagers abandoned the area following an epidemic of measles that killed all the hamlet's small children (Rijksen and Griffiths 1995: 275).[17]

This territorialization process consolidated the northern boundary of Alas territory with Gayo Lues (at Lawe Mencirim), but it left much of the surrounding area to the east and west within the Gunung Leuser National Park. The steep, jungle-covered hillsides of the Bukit Barisan range soon cut off possibilities for agricultural expansion or easy passage in either of these directions. Prior to the colonial government's forest-mapping exercise, Alas people had entered the surrounding forest to collect forest products. However, beyond the valleys and hillsides immediately surrounding the Alas River, the forest had not been subject to tenure. As the area was unsettled and not under contention with other groups, the Alas people had never needed to specify the extent of Alas territory to the east and west in precise terms. Moreover, as Li (1999: 12) has pointed out, "the direct attempt to regulate the relationship between population and resources was not a feature of rule" in pre-colonial political systems. There was a gap here for the state mapping exercise to clearly articulate its territorial claim over the area.

The territorialization of the Alas valley begun in the colonial period resumed when international conservation agencies such as IUCN and WWF rediscovered the conservation value of the Leuser area during the 1960s. At this time Dutch scientists and conservation organizations, such as the Netherlands Commission for Nature Protection and the Netherlands Appeal of the World Wildlife Fund (WWF), began a special campaign to raise funds to conserve the Leuser area and supported a series of ecological surveys. Colonial conservation plans for the area were dusted off, and new activities planned. In 1971, the Dutch ecologist H. D. Rijksen arrived to carry out research into the status and ecology of the orang-utan and other primates. Rijksen helped establish a research and orang-utan rehabilitation center at Ketambe in the Badar area of the Alas valley (WWF 1971/72: 126). Under the patronage of Dutch universities, IUCN, and WWF, the Ketambe research area eventually enclosed 491 hectares of pristine rainforest on the west bank of the Alas River. In the 1970s the Dutch zoologist N. J. van Strien assisted the Indonesian forestry department in developing a management plans for the Leuser forest area. Their proposals came to realization when, in 1980, the Minister of Agriculture joined together several nature reserves gazetted in the 1930s, creating Indonesia's first national park, the Gunung Leuser National Park (GLNP) (Wind 1996).

TABLE 4.2
State Forest in Southeast Aceh

Land Use	Area (ha)	Percentage
Protection forest (*Hutan Lindung*)	168,000	17
Limited production forest (*Hutan Produksi Terbatas*)	135,000	14
Nature reserve/National Park (*Hutan Suaka Alam/ Taman Nasional Gunung Leuser*)	500,000	50
Forest reserve (*Hutan Produksi yang Dapat Dikonversi*)	77,000	8
Total forest area (*kawasan hutan*)	880,000	89
Total district area	995,990	100

Source: Aceh Tenggara Dalam Angka Tahun 1997

During the 1980s, the provincial governor oversaw the creation of a consensus forest land use plan (*Tata Guna Hutan Kesepakatan,* or TGHK) for Aceh.[18] In the process of classifying areas as "protection forests" and "nature reserves" and "conservation areas" the state forestry mapping planning (TGHK) classified some 89 percent of this district and Badar itself as state "forest zone." This process locked up 50 percent of the district within the national park (see Table 4.2) and left only 135,000 hectares of land (14 percent) for non-forestry uses.[19]

Resource planning during the colonial era attempted to balance local land uses and to balance plantation development or forest exploitation on the one hand with forest conservation requirements on the other. The same balancing act affected state planning during the New Order period, with adjacent areas set aside for conservation, logging, and plantation development. In contrast, in his management plan for the Leuser reserves, van Strien suggested that the boundaries of the Leuser reserve complex be revised to fit natural boundaries such as rivers and watersheds (Strien 1978). In a similar fashion, the management plan he developed for the Leuser ICDP pointed to the "ecological inadequacies of the conservation area." Taking a bioregional approach to conservation, the master plan argued that the national park covered too small an area to preserve the predominant ecosystem types and the mega-fauna typical of northern Sumatra. To overcome these problems, a team of ecologists selected a more extensive conservation area, known as the "Leuser Ecosystem," in the process creating a new territorial concept that aimed to include important areas, such as lowland rainforest and other significant wildlife

TABLE 4.3
Land Use in Badar Sub-district

Land use	Area (ha)	Percentage
Rice land (*Tanah Sawah*)	2,530	1.4
Dry land agriculture (*Tanah Kering*)	10,128	5.7
Built up area (*Bangunan*)	485	0.3
State forest (*Hutan Negara*)	159,222	89.5
Other	5,585	3.1
Total	177,950	100

Source: Kecamatan Badar Dalam Angka Tahun 1996

habitat that lay outside the national park (Rijksen and Griffiths 1995: 30). According to the ICDP *Masterplan* the Leuser Ecosystem would be subject to the authority of the newly established conservation agency, the Leuser Management Unit (LMU), which would obtain a "conservation concession" from the state to manage it. In the following years LMU engaged in a complex process of territorialization, dividing the Leuser Ecosystem into bufferzones and core conservation areas. This zoning was a strategy that aimed to attenuate the control of access to natural resources within specific areas.

Protection will be facilitated because access to the "concession" area will, as a matter of principle, be prohibited for everybody, except for persons holding a special license. Sustainable harvesting of particular quantities of forest-produce will be strictly regulated and controlled and can take place exclusively in the outer "bufferzone" fringes of the Leuser Ecosystem. (Rijksen and Griffiths 1995: 180)[20]

The colonial history is particularly important here. To gain the consent of local leaders to the first gazettement of the Gunung Leuser Reserve (*Wildreservaat Goenoeng Leuser*), in 1934, the then-governor for Aceh, Van Aken, convened the primary local actors in the colonial system of indirect rule in the area, the "self-governing territorial heads" (*Zelfbestuurders van de Landschappen*). There is evidence that at this point local leaders in the Gayo highlands made a strategic decision to support colonial conservationists to obtain protection forest status for forests areas to prevent the opening of more colonial plantations or other exploitation (Rijksen and Griffiths 1995: 37). Yet, it is debatable whether the agreement between the colonial state and the heads meant that the Alas people gave their consent to locking up such an extensive area of forest for conservation for perpetuity.[21] In any case, in the Alas valley the original reserve most likely had very limited impact on the livelihood activities of the Alas people. The reserve boundaries left sufficient land outside the reserve for opening new agricultural lands. As the population was so low compared to the extent of the forest, "self-use" amongst a small population had a limited impact on the Leuser forests. This reserve status only became an issue when the population grew and the demand for land could no longer be met outside the reserve boundaries.

Nonetheless, the status of the area as state forest reserved for conservation left later generations with fewer rhetorical means of asserting a territorial claim over these areas. As described by Doolittle (2001), when the British colonial state systematically organized native property rights in Malaysian Borneo, the official delineation of customary rights seemed to strengthen local rights at that time. However, this dispensation has subsequently constrained present-day understandings of native customary rights. In a similar fashion, the *tanah adat* created by Dutch colonial authorities in the Alas valley has restricted local autonomy and decision making in the contemporary period. Even so, during the 1990s, the Leuser Management Unit frequently used the agreement with the self-governing territorial heads of the colonial period as a rhetorical means of arguing that the protection of this area was legitimate because it was based on an agreement signed by the indigenous "territorial heads" of the area on behalf of the "community."

Gayo Migration

Over the last thirty years, significant numbers of Gayo have moved down into Badar. Consequently, the dynamics driving this migration as well as the land-use patterns and customary arrangements predominant in Gayo Lues are relevant to the study of forest management in Badar. While several sources recorded the northward movements of the Batak peoples into the southern end of the Alas valley (Koesnoe and Soendari 1977; Iwabuchi 1994a and 1994b), unfortunately little information has been available about the corresponding southward migration of the Gayo. Indeed, there has been very little information available concerning Gayo Lues.[22]

However, a few reports concerning Leuser do contain brief discussions of the Gayo Lues area. In 1978 Van Stein wrote that

the typical landscape in most of the mountain areas [consists of] peaks still covered with forest, below a belt where the forest is cut and burned and the soil planted, and there under a belt with completely exhausted soils covered with hard tall grasses. These zones are steadily creeping upward till the forest has disappeared and the tall grasses (*alang-alang*) cover the whole mountain. In . . . the area around Blangkejeren . . . the peaks have not yet been reached, but there is already a wide belt of infertile grassland between the rice-fields in the valley and the dry fields on the slopes. (Strien 1978: 48)

A forestry department report also described the sub-district of Rikit Gaib in the uplands of Gayo Lues in similar terms (Department Kehutanan and Universitas Syiah Kuala Darussalam–Banda Aceh 1993: 40–58). Here the slope of the lands, the prevalent soil types and the shortage of water favored shifting agriculture over wet rice agriculture, the agricultural model preferred by state planners. Farmers practiced shifting agriculture erected temporary shelters and opened plots on the less sloping lands close to rivers. Roads in the area were poor, and forest farmers could either walk the sixteen kilometers to the *kecamatan* centre of Rikit Gaib or catch a four-wheel vehicle serving as public transport. In the nearby sub-district of Terangon, shifting agricultural practices have converted large areas of forests and left large areas covered in pine trees. Here, as elsewhere, Indonesian state policies have long attempted to bring what they believed to be order, control and "development" to upland regions where upland populations have practiced extensive shifting agriculture (Dove 1985; Li 1999).[23] Accordingly, the Social Affairs Department classified the shifting agriculturalists of Rikit Gaib as an "isolated community" (*masyarakat terasing*) and targeted them for a development program that attempted to change the nature of local agriculture and compel them to lead a more settled way of life (Department Kehutanan and Universitas Syiah Kuala Darussalam–Banda Aceh 1993: 40–58).

During my brief visit to the Gayo Lues highlands in September 1997, informants described local land use patterns. In the mountainous valleys around Blangkejeren (the capital of Gayo Lues), some farmers cultivated *wet rice* in the valleys while also growing cash crops such as coffee and tobacco on the hillsides. However, land suited to wet rice cultivation was scarce in many upland areas of Gayo Lues, and the cultivation of dryland rice (*padi ladang*) was integrated with various cash crops in shifting plots. Because the roads were poor and the distances from market formidable, farmers faced significant expenses marketing bulky agricultural products such as candlenut. As tobacco was of high value and low bulk, upland farmers chose to plant it as the predominant cash crop. After the first annual crop of tobacco farmers would plant chilis (*cabe*) or dryland rice (*padi ladang*), and then tobacco again.[24] After the second crop, the land had to be left fallow for at least three to four years. According to farmers, villagers could continue to do this up to three times, but then they would leave the area uncultivated. In the uplands of Gayo

Lues a species of pine (*Pinus merkusii*) native to the area spontaneously colonized worked-over areas. Farmers maintained that if this pine took over, it was not possible to replant the land. And so gradually the land became secondary forest, scrub or covered in pine, as villagers progressively moved further and further out from the village settlement in search of new land. In the past the Gayo used to manage the extensive pine area under a sophisticated system of traditional *adat* rights, deriving turpentine, firewood, timber, and other products from these semi-wild pine stands. However, in the late 1980s "the stands to the North of Blangkejeren were given out for exploitation to feed a paper-production plant" (Rijksen and Griffiths 1995: 137).

To find new areas for wet rice production, early in the twentieth century the Gayo began to pioneer the lands surrounding the Alas River north of Tanah Alas. At the time the Dutch created the Leuser reserves, Gayo farmers had already started moving down the footpath connecting Kutacane and Blangkejeren. Some years earlier, with the permission of the territorial head of Blangkejeren, Gayo people founded villages along this footpath. When the Dutch established the Leuser reserve, these areas were mapped out as enclaves within the boundaries of the reserve. After independence, some Gayo people moved further south into Alas territory, settling in the village lands and *tanah adat* in Badar, just south of the rubber plantation. In the 1960s, with the permission of the descendant of the Gayo who held the original tenurial rights issued by the *kejuren* in 1936, newcomers opened land north of the former rubber estate. Following the unification of Gayo Lues and Tanah Alas into Southeast Aceh district (*Kabupaten*) in 1974, larger numbers of Gayo people began to migrate down to Badar.[25] As Iwabuchi (1994a and 1994b) has noted, the Alas people were accommodative of newcomers, and to bolster Kutacane's status as a district capital over an area including Gayo Lues, the Alas people had to allow for Gayo in-migration. In-migration was facilitated by the fact that the Alas people were the descendants of migrants from many parts of northern Sumatra. The Alas people generally held that there was plenty of land, and unlike the more closed nature of Acehnese or Gayo communities, the Alas had a tradition of readily accepting newcomers (Iwabuchi 1994a: 32). If a settler wished to begin farming in a village, he would be given a piece of land that had not been opened already (Koesnoe and Soendari 1977).

An elderly Gayo villager described moving down to a hamlet of Badar in 1968. At that time there were only five families living in the hamlet, and the settled area ended further south. When Gayo moved into this area of Badar sub-district, there were few residents and only incipient village institutions. "Back then," he said, "we had to open the land. The trees were enormous and it might take two men two days just to cut a tree. There were no chainsaws then. We planted dryland rice (*padi ladang*), *cabe* and coffee." Before the first crop matured, he made money whatever way he could—cutting trees and selling the wood, collecting rattan and damar.[26]

Seeing the in-migration of Gayo into this area, in 1975 influential Alas figures from the core Alas area of Natham and Jongar staked a claim over a flat area of abandoned plantation land within Badar suitable for wet rice cultivation just north of the last settled village. As this area was still subject to a seventy-five–year lease issued in the colonial period, these Alas figures used their influence within district government circles, ultimately obtaining permanent tenurial rights (*hak milik*) over this area. Consequently, Alas people gained tenurial rights over one of the last significant areas in the northern Alas valley suitable for *sawah*, establishing an enclave of Alas people in two hamlets within the other predominantly Gayo hamlets.[27]

After 1976, the building of the road to Blangkejeren further facilitated the in-migration of Gayo people, who came to occupy unused lands in Badar. In accordance with the Village Government Law (UU No. 5/1979), several hamlets that stretch along the east bank of the Alas River next to the Kutacane-Blangkejeren road were joined to form a composite village (*desa gabungan*) here in 1987.[28] To the west of the composite village, across the Alas River, rises the mountainous rainforest of the Gunung Leuser National Park (TNGL), and to the east lies the Serbolangit Protection Forest, an area that has now largely been turned into agricultural land.[29] Gayo people also opened gardens in areas earlier classified as forest reserve and now forming a part of the Gunung Leuser National Park. Nowadays, outside of two hamlets within the composite village, the majority of inhabitants of the composite village were in-migrants (or their children) from Gayo Lues.[30]

Adat and Village Regimes in Badar

To understand how customary institutional arrangements are constantly being re-negotiated as conditions change, it is necessary to appreciate the changing political and economic contexts where rules are invoked, challenged, and restated (Moore 1989). Changes tend to be particularly pronounced in frontier situations. As Bowen (1988) found in a comparison of two communities in Central Aceh, Gayo customary practices in a pioneering situation differed considerably from those in a long-established area of origin. In a similar fashion, the incipient customary institutions developed by the new Gayo communities of Badar differed from those operating in the Gayo Lues heartland to the north.[31]

First, Gayo people tended to move south as individuals or in small groups. As they were still young when they left, they tended to be poorly acquainted with the *adat* regulations regarding resource use in their native villages. The new settlements lacked elders skilled in the dispute resolution and consensus building fundamental to the operation of an *adat* regime. Thus, in the frontier villages of Badar and Gayo, *adat* institutions failed to develop strongly. This situation was complicated by the fact that *adat* regimes tended to vary within the Gayo Lues homeland. As the newcomers were from different areas, they didn't share a single vision of what constituted *adat* Gayo.[32] Time would be needed before stable institutional arrangements could evolve.

Second, in the pioneer area agro-ecological conditions differed from those prevailing in the core area of Gayo Lues. As the agricultural system in Badar evolved in a different direction from that in the Gayo highlands, this helped create different priorities and contributed to variation from the original institutional forms. In Gayo Lues an extensive form of swidden agriculture predominates, but in Badar sub-district, in the 1970s, the Gayo migrants began combining temporary plots for food production (such as those developed by Alas people) with more permanent coffee gardens. Later, from the early 1980s, in accordance with government policy that sought to end shifting cultivation, the *bupati* and village heads encouraged farmers to plant candlenut trees, and the residents of Badar began opening permanent candlenut gardens. In the past forest had been cleared to open temporary swiddens, but now pioneer farmers enjoyed permanent rights over land covered in coffee plants and candlenut trees. The Gayo left behind the extensive swidden system and secured more permanent tenurial rights over permanent plots under tree crops. Subsequently, to discourage farmers from practicing shifting cultivation, the village head threatened to rescind tenurial rights over plots abandoned by farmers.[33] Thus, compared with Gayo Lues, the village regime was less tolerant of shifting agriculture.[34]

Administratively, the Gayo hamlets were under the sub-district head (*camat*) of Badar and the *kepala mukim* of Lawe Mencirim, administrative units headed by Alas leaders.[35] As the Gayo migrants were new to the area, they could not readily establish a territorial claim over an area already marked out as state forest within Alas territory; undoubtedly they needed to accommodate Alas influences. Yet, Gayo and Alas informants in Badar consider the differences between Alas and Gayo *adat* to be significant and prefer to live in hamlets dominated by their own ethnic group. This meant that, as I have noted, the Alas people have tended to settle in two hamlets within the composite village. Here, where people of Alas origin predominate, Alas *adat* rules have prevailed. However, in other hamlets of the composite village, hamlets *(dusun)* were run according to Gayo norms. Therefore, when members of the two groups interacted, for instance, when a dispute arose, mediators used the *adat* predominating in the hamlet concerned. Disputes were handled at the hamlet level first; if they couldn't be resolved at this level, they were then taken to the village level. In this way, the Gayo have nested their *adat* regime within a wider Alas area.

Under these conditions, in the composite village, according to several villagers, in recent years the village government and customary institutions had often faced significant problems. Village institutions had been a site of on-going conflict. Elected leaders of hamlets and the composite village—as well as the village resiliency council (LKMD) appointed by the village head—had frequently changed. For example, although the period of office was eight years, in one particular hamlet, hamlet heads *(kepala dusun)* had rarely lasted more than two years. According to one informant, the main problem was that the *kepala dusun* misused funds, showed favoritism *("anak kiri, anak kanan")*, or "paid insufficient attention to the people." As a result, eventually villagers lost patience and agitated for his removal from office.[36] In a similar fashion, the village head *(kepala desa)* of the composite village often was felt to lack credibility. One informant said that the position could be bought, and in five years the incumbent would enrich himself through commissions on village business.[37] In November 1998, a village meeting dismissed the village head *(kepala desa)* for corruptly misusing Rp 800,000 of government agricultural assistance funds (KUT) that were meant to help develop corn production at a time of economic crisis.

As the village's hamlet heads and the village council (LKMD) had responsibility for mediating village disputes, their lack of credibility has hampered the functioning of the *adat* regime. This problem has been compounded by the weak basis of *adat* institutions in the new villages in Badar, including the lack of agreement concerning *adat* practices and the lack of *adat* leaders with the skills for resolving disputes to the satisfaction of both parties. As a consequence, at times disputants took village problems to the police. However, where police demanded payments from the disputants for solving the conflict, people with financial resources could engineer a resolution in their favor. If most of the payment went to the police, the person wronged would fail to obtain adequate compensation. Alternatively, the accused person might claim to be offended *(sakit hati)* because the offended party reported the dispute to the police rather than subjecting it to *adat* mediation in the village. Subsequently, the accused party might take revenge in repeated offenses against the accuser. Accordingly, taking a dispute to the police usually led to one party's feeling wronged; police involvement did not lead to effective resolution of disputes but rather created a village riven with festering quarrels.[38] If the stability of village leadership and the ability of *adat* institutions to solve conflicts were indicators of the capacity of the local *adat* regime, in many respects the *adat* regimes here were functioning inadequately.[39]

In the past, village heads had greater authority over access to the village area. According to a former hamlet head, previously even soldiers or tourists entering the area would first report to the *kepala desa*. According to the *adat* regime then operating, if anything happened to a visitor in the village, the *kepala desa* would then take responsibility and help deal with the problem. However, if a visitor failed to report, even if the visitor was wronged, he or she would be fined. This regime ensured that visitors asked permission from the village head before gaining access, opening land, or even taking timber. Yet, over the years, with the emergence of other authority structures, the village head lost this discretionary power, and outsiders visited the village without asking permission.[40]

The management of land and forest resources was linked to the status of forest lands surrounding the village. As I noted earlier, the Gayo and Alas in-migrants entered an area where, building on colonial claims over the forest, most of the land was already formally part of the state forest zone and thus subject to the Ministry of Forestry. Yet, in the absence of a state forestry agency capable of monitoring compliance to forestry laws and sanctioning those who broke them, up until the time when the Gunung Leuser National Park was created in 1980, these laws and sanctions had little meaning. The forest boundary set by the colonial foresters had disappeared and there were no clear boundaries. Moreover, there were virtually no forestry officials on site to control access to and use of forest resources.

As we will see, after the advent of the national park, these hamlets had competing authority structures: one was associated with the national park and in theory dedicated to implementing state forestry regulations; and the other associated with village authorities and bound to the village's normative order. Thus was created a situation of competing and overlapping authority systems, each one associated with its own (state or *adat*/village) property claims without any overarching arena for resolving disputes. In this context, covert uses of resources have operated largely within an institutional matrix working contrary to what might be identified as a state forestry or an *adat* normative order.

Clientelist Accommodations and Local Patterns
of Resource Extraction

By 1978, in North Sumatra, large concessions held by clients of powerful figures at the center—often several tens of thousands of hectares—were already operating, while in Aceh itself logging had only recently started (Strien 1978: 49). In the Leuser area significant logging concessions had also been issued, with virtually all areas on the borders of the national park included, and several concessions including areas within park boundaries. In addition to these large concessions, at the district level the local forest authorities also allocated local concessions of not more than one hundred hectares in forest reserves. These included areas that needed to be set aside as "protection forest" if forestry regulations were to be strictly applied (Strien 1978: 49). Allocation had begun in 1974, when Southeast Aceh obtained district status. By 1978, the forestry agency had granted several small local concessions on reserved land in Badar. As van Strien put it, "Several sawmills operate with a licence, or operate on a large scale outside their concessions, often in the [Leuser] reserve. There is even one medium-size sawmill which has been operating for several years on a 100 ha concession (PT Aceh Tenggara) but actually has logged more than 3,000 ha and processed many logs stolen from the reserve by roving bands of independent log poachers" (Strien 1978: 50). Two years later, a WWF consultant reported that there were seven sawmills operating in the Badar area, supplying logs from protection forest and the TNGL. Most of the logs were extracted from a concession operating in the Serbolangit protection forest, an area just to the east of the former plantation land (Blower 1980).

In a pattern mirroring that occurring at the national level, logging operations were clearly connected to local agents of the state working beyond their legal responsibilities (*oknum*) and their families or favored clients. Natsir, the son of Syahadat, the *bupati* of Southeast Aceh, for instance, had obtained a concession from the head of forestry for Aceh province for the Serbolangit protection forest, adjacent to the composite village and the Leuser reserve. Natsir's sawmill, PT Aceh Tenggara, obtained its license directly against the recommendation of the Nature Conservation Service (PPA).[41] However, in addition to the *bupati*'s family, forestry department staff were also directly engaged in logging. A PPA official, Abu Mukim, operated a sawmill while still receiving a PPA salary. Another forestry department official working in the Kutacane Office named Simanjuntak owned another sawmill under his wife's name (Blower 1980: 14). Clearly officials working for a state forestry agency responsible for both forest protection and resource exploitation, and with ample opportunities for self-enrichment, faced a conflict of interest that worked against the implementation of forest policy. As Blower put it, "Though some forest officers pay lip-service to conservation, they appear in general uninterested and uncooperative, providing no assistance to PPA in law enforcement (e.g. in connection with illegal logging in the National Park), and actively promoting timber concessions in conservation areas even when specifically requested not to do so" (Blower 1980: 5). From the mid-1970s, logging in the heartland of the wildlife sanctuary and, as of 1980, the national park, proceeded most visibly along the Kutacane-Blangkejeren road and in the Kappi area just beyond the Ketambe research station. Village informants described how forest police (*jagawana*) were heavily involved in the logging, with several military figures also owning sawmills.[42] Bands of loggers cut swaths of forest alongside the road, leaving the banks of the Alas River unstable and subject to erosion. During the late 1980s, loggers even began to move into the forest surrounding the Ketambe research area.

Villagers described how the situation in the Alas valley began to change. Increasing numbers of forest guards were posted to protect the reserve. It was now officially illegal to collect forest products, cut wood, or open forest gardens in national park forest. But local people did not accept these rules, and the divided loyalties of the forest guards living amongst them created a new situation. In a pattern reminiscent of that described by Peluso (1992) in Java's teak forests, some guards would accept payment in return for turning a blind eye. To supplement their small government incomes, it was not unknown for guards to become directly involved in poaching and illegal logging. In this situation local villagers wishing to gain access to forest might be negotiating with those officially responsible for forbidding its use. Instead of the village access rules, new informal understandings had emerged concerning the state forest reserve. In the struggle over access to land and forest resources, new institutional arrangements emerged that accommodated local demands for land. Under this regime, use of the forest was no longer mediated through local *adat* institutions but rather through field officers using their discretionary powers in an accommodating (albeit corrupt) fashion.

In 1996 a villager explained how this operated. "If someone wanted to cut wood, they would go and ask the PHPA (forest guard), and he asks for a payment or a share of the wood. He manages the wood. If he is caught, his superior says 'OK, that's enough for now, but don't do it again.'[43] Then he turns around and does it again behind the superior's back." "Nowadays the regulations forbid cutting wood in the park or bombing the river [to catch fish]," he continued, "yet it is forest guards who cut wood and the police bomb the river. If I cut wood or bomb the river, I'm arrested, but if the forest police do it then nothing happens."[44] Consequently, as Rijksen and Griffiths have noted, the image of the Ministry of Forestry as "the representative of the state guarding the common good has been tainted in the local mind." As Blower reported in 1980, "there appears to be considerable hostility among local people towards [the conservation agency]" (Rijksen and Griffiths 1995: 175).

As the logging frontier moved into the area during the late 1970s, land speculation—often promoted by wealthy patrons living elsewhere—also began to emerge. People began opening forest-gardens on the pathway between Kutacane and Blangkejeren. By planting a few coffee plants in a newly opened plot "for speculative purposes," speculators or their agents could claim tenurial rights. Upon the completion of the road, which was then under construction, they could sell the land to in-migrants (Strien 1978: 48–49). Shortly after clearing the land, those with the financial means were able to secure rights over the land under state law by registering their claims with the agrarian office (now known as BPN). Once these official rights were obtained, they could be used as collateral for borrowing money from the bank.

After the inception of the national park, the presence of the forestry apparatus affected the operation of local customary institutions. Once national park status was granted, forest guards were placed in Badar and gained field-level responsibilities. Whereas in Menggamat, *oknum* officially acting as gatekeepers of the forest used their privileged position to extract rents through illegal logging from afar, in Badar forestry officials were stationed in the hamlets surrounding the national park. These personnel were poorly paid and resourced. Moreover, given the history of collusion between forestry staff and timber networks, it was hardly surprising that some forest guards used their position to extract benefits from the park—either turning a blind eye to exploitation in return for payments or even participating in timber extraction. Ironically, as one villager who had previously worked for the WWF project in the area observed, the creation of the national park had the perverse effect of hastening timber extraction in the area.

From the 1970s, as the Ketambe area drew researchers and conservationists from abroad, it became the focus of conservation efforts in the Alas valley. When the Gunung Leuser National Park was created in 1980, PPA (Perlindungan dan Pengawetan Alam, later PHPA), the Nature Conservation and Wildlife Management Service, built the park headquarters in the district capital, Kutacane. Subsequently, visitors to the area judged the success of the park by inspecting efforts to "protect" the highly visible forest alongside the Kutacane-Blangkejeren road. As the appearance of the forest here served as an indicator to traveling conservationists of

Figure 16. The southern end of the Alas valley, in Southeast Aceh. John Mc-
Carthy.

Figure 17. Badar Sub-district, Southeast Aceh. Candlenut gardens line the
banks of the Alas River in the Gunung Leuser National Park bufferzone. John
McCarthy.

Figure 18.   Badar Sub-district, Southeast Aceh. Opening a new *kebun* within the national park's "enclave." John McCarthy.

Figure 19.   Deforested hillside, southern end of the Alas valley, Southeast Aceh. John McCarthy

Figure 20.   A Leuser Development Programme field office, Southeast Aceh. John
McCarthy.

Figure 21.   Ketambe Research Station, Gunung Leuser National Park, Badar
sub-district. The signs prohibit entering the area, taking forest products, and using bombs or poison when fishing in the river. John McCarthy.

Figure 22. An enclave village within Gunung Leuser National Park, with rice, candlenut, and coffee being dried on the Kutacane-Blangkejeren road. Badar sub-district, Southeast Aceh. John McCarthy.

Figure 23. Cooking *nilam* leaves to produce patchouli oil. Badar sub-district, Southeast Aceh. John McCarthy.

Figure 24. A tobacco *ladang*. Gayo Lues, Southeast Aceh. John McCarthy.

Figure 25. Buying rattan in the market. Blangkejeren, Southeast Aceh. John McCarthy.

the park authorities' ability to protect the TNGL, logging and forest pioneering here became an embarrassment to the national park management
and donor-sponsored conservation projects.

In 1989, deforestation within the park gained international attention,
and IUCN declared the park to be one of the ten worst-managed parks
in Asia (Wind 1996: 4). Because the local state agents were heavily involved, the authorities called in an army unit from outside the area. In
1990 an army mobile brigade (*Brimob*) from Banda Aceh clamped down
on the logging. *Brimob* arrested many people, several sawmills were shut,
and the scale and visibility of illegal logging in the "heartland" of the
park alongside the road were curtailed.[45] Afterwards, conservation efforts increasingly involved trying to keep forest resources safe from timber networks and local farmers. Since 1990, the forest authorities have
more strictly applied regulations in the forest area alongside the Alas
valley road in the heartland of the national park than anywhere else in
the park. There have also been successive attempts to relocate so-called
"forest encroachers" who settled in the area in the 1980s. However, the
pattern of resource extraction continued, albeit in a more discreet form.

Few people in Southeast Aceh were prepared to discuss the practices
surrounding illegal logging. Yet, in late 1998, a resident of Kutacane described how, even though there were no official logging licenses in existence within Southeast Aceh, "there are several sawmills working, but
none have permits. Officials shut their eyes and take payments."[46] When
there was a logging raid, he said, the officials involved warned the loggers in advance. Alternatively when a patrol caught loggers in the act,
there was a settlement (*perdamaian*) on the road, and the loggers were
allowed to go after paying a fine. Other sources alleged that to proceed
with illegal logging, loggers paid police and forestry *oknum* a monthly
fee, with payments passed on to some senior forestry officials working
beyond the law.[47] At the apex of the network were four key business
figures, predominantly from the Desky clan (*marga*), who were originally
based in Mbarung village, Kecamaten Babussalam. These figures dominated Southeast Aceh politics, and even the *bupati* was enmeshed in this
network. Those who upset this group would be excluded from the webs
of patron-client relations running Southeast Aceh.

A former WWF-LP worker familiar with Southeast Aceh confirmed
this account, describing the intimidating nature of the clientelist network
operating in Kutacane. He reported how a "timber mafia" controlled logging operations and other activities that extended beyond the bounds of
legality.[48] An atmosphere of secrecy and latent threat surrounded Kutacane life, he told me; people were too frightened to talk about certain
activities, and journalists were afraid to report them. "If someone asks
too many questions," he said, "they could disappear."[49]

Therefore, as in South Aceh, binding obligations structured the processes of negotiation and exchange surrounding many activities in the district, including timber extraction from protected forests. Those who violated these rules would be excluded from the webs of personal ties and obligations that structured exchange within the district. Key figures in the district, including the *bupati*, were enmeshed in a social order that extended to involve forestry staff working for the national park, police (*Polres*) and army personnel (*Kodim*), local government officials, the judiciary, and local religious leaders (*imam*).[50]

In the charged atmosphere following the fall of Suharto in July 1998, a group of students carried out a series of protests and demanded the resignation of officials in Southeast Aceh involved in corruption (KKN), including the *bupati*. The students alleged that the manipulation of projects in the district had led to a loss to the state of a billion rupiah due to activities involving the *bupati*.[51] However, the *bupati* secured his position by galvanizing support amongst his political clients, and thirty-nine village heads made a statement in his support.[52]

Then in late 1998, student groups and local NGOs began to focus on the logging mafia, particularly on logging operations in the Serbolangit area east of the road. Here, a Korean acting as the financial backer (*cukong*) had supplied a bulldozer to open a road going ten kilometers into the Serbolangit protection forest and into the national park beyond. The road facilitated logging operations in the area by people in his employ and other villagers, with trucks taking the logs out by night. According to a witness quoted in *Waspada*, the Korean was working with the collusion of local forestry officials: "if there was no collusion, it would be impossible to carry out this sort of activity."[53] Despite this attention in the press, logging activities continued in Badar and elsewhere.

Finally, in November of 1998, six members of a local NGO, Leuser Lestari, took matters into their own hands. They captured a truck carrying thirty cubic meters of wood taken from the Serbolangit protection forest, taking its occupants, including a security official backing the operation and the Korean *cukong*, and surrendered them to the police.[54] This action did not turn out to be conclusive. The next month, *Waspada* reported that forest police (*jagawana*) confiscated the Korean *cukong*'s bulldozer inside the TNGL in Badar, along with seven tons of wood. However, apart from confiscating the wood, while "awaiting instructions from the *bupati*," the *jagawana* failed to arrest a single person.[55] Informants noted that it was rare for this sort of case to end in court.[56] Typically cases of illegal logging that end up with the police elude legal sanctions. It is often reported that the police fail to take the case to court, for instance, because the evidence has been "lost" or because they "lack witnesses." Even when a case proceeds to the prosecutor's office, the prosecutor usually returns it to the police because, due to a lack of supporting evidence, the prosecutor cannot process the case further. Unofficially, in most cases there is a suspicion that failure to prosecute occurred because those facing sanctions had their "back-up" within the apparatus close down the case or otherwise paid money to ensure that the case failed.[57] In an earlier case, *Waspada* reported that forty people had been caught red-handed with eleven chainsaws inside the TGNL. However, "Up to this time not even one person had been arrested as a suspect and the TNGL side had given no indication of the time needed to ascertain who is behind the logging of the National Park . . . 'We will investigate this problem to the very bottom,' [an official] said."[58] The forestry department had set out operational rules as specified in government regulations, ministerial decrees, and laws regarding the state forest lands. Under this regime, *jagawana* were charged to patrol state forest land including the national park, monitor and collect information about what might be occurring there, and arrest and enforce the law against those infringing regulations. However, where the *jagawana* operated, the state forestry regulations had very little legitimacy among villagers and local government, and powerful local politicians and business figures supported illegal logging. Law enforcement requires concerted negotiation with affected parties (Wilshusen et al. 2002: 25). In this situation, in the course of such negotiations, local state agents made decisions regarding enforcement of regulations in accordance with village-based notions of justice, and the localized networks of exchange and accommodation. It was hardly surprising then that decisions according with the interests of powerful local patrons, and at times local state agents, allow for logging operations in contravention of the law.

Extractive Uses and Local Coalitions

While so far this chapter has considered the operation of clientelist networks, timber operations, and village regimes separately, in many ways they intersect. The extractive activities of loggers were compatible with the transformative activities of pioneer agriculturalists wishing to open new plots. For forest pioneers, clearing huge rainforest trees was a formidable task. Therefore, villagers have been happy to see illegal loggers extracting timber and creating secondary forest areas that were then easily opened for agriculture. In Badar, as in other areas of Indonesia, forest pioneering often followed in the tracks of logging operations (Sunderlin and Resosudarmo 1997). The former head of one hamlet described how in the 1970s an ethnic Chinese *cukong* approached the village: he wanted to open a sawmill and harvest the wood from the adjacent Serbolangit protection forest. By asking permission first and paying "compensation" to the village and the local government, the *cukong* respected local claims over the surrounding area. The hamlet head allowed the logging, and when the large trees were extracted farmers could readily open *kebun*. During the 1980s, in the enclave hamlets north of the national park gates, forest pioneering also followed in the footsteps of loggers.[59] Then, in 1998, villagers also supported the road-making activities of the Korean *cukong* in the Serbolangit protection forest just to the east of road. As a former hamlet head explained, in Badar there was no other work besides farming, and if people wanted to work as farmers, they needed to open land in the Serbolangit protection forest. "If the Korean takes the wood," he said, "this will enable farmers to open there."[60]

Local government-supported agricultural expansion and the logging that accompanied it helped to increase the revenue generated for local government coffers, thereby serving to consolidate the district's administrative position (Rijksen 1995: 132). Moreover, many villagers, officials, and businessmen within the Alas valley resented state attempts to control local forests. In the late 1970s, after wide consultations with local leaders, an Indonesian expert on *adat* law found that the Alas people hold that local land and forest resources "belong to the right of disposal [i.e., tenurial domain] of their community for the interest of their living and their survival" (Koesnoe and Soendari 1977: 73). In Southeast Aceh district, over 80 percent of the population were farmers (Rijksen and Griffiths 1995: 132). By supporting the wish of farmers within the Alas valley facing land shortages to continue the longstanding practice of opening agricultural plots within Alas territory, local leaders could gain local popularity. Leaders define and articulate the myths and symbols of a corporate group; they interpret a group's identity in relation to a wider society and this ability generates a leader's influence and power in a group (Brown 1994: 113). Through populist politics, district leaders could shore up their local standing by distancing themselves from or even criticizing an unpopular policy of the central government. Accordingly, the local government often sought to expand the area of the Alas valley subject to agriculture at the expense of state forest land. However, during the Suharto period, local officials were formally subject to a strong central government. There were limits to how far they could openly resist state policies. Even so, at least they could avoid the opprobrium that might be attached to fully implementing them. They could turn a blind eye to access to state forest in breach of state regulations, or patronize these activities for personal gain. Alternatively, at strategic moments such as this, local officials could allow local expressions of opposition by local actors.

In addition, besides this purely rhetorical resistance, local officials have also tacitly allowed activities that encroached on the state forest. While the central government consistently refused requests to excise agricultural land from the state forest, the regency had surreptitiously managed to almost double the areas subject both to *sawah* (91 percent increase) and to dry land agriculture (96 percent increase) since the 1970s (Rijksen and Griffiths 1995: 133). For instance, in 1989, the director-general of the State Conservation Agency (PHPA) wrote a letter to the local government agreeing in principle to carving a bufferzone from the national park consisting of an area of one kilometer in width alongside the Blangkejeren-Alas road north of Ketambe. The local government took this letter as a green light for opening the area. Within six months, loggers moved into an area classified as national park, taking the rich tropical hardwoods to local sawmills. Villagers followed, moving into the area and opening gardens in the foothills (Rijksen and Griffiths 1995: 134). At this time, migrants settled the most northerly hamlets of Badar as well as enclaves within Blangkejeren sub-district just to the north. In the run-up to the 1999 elections, local politicians had further opportunities to play a populist tune. In late 1998, a regional meeting (*Muspida*) addressed the land shortage faced by local farmers. A meeting of leaders of the local ruling Golkar party involving the *bupati* asked the central government to make GNLP lands in north Badar and Pasir Luk-Luk (Lawe Alas sub-district) available for agriculture.[61]

In a study from Latin America, Rudel and Horowitz (1993) described what they have called a "growth coalition." Such a coalition consists of actors from different classes who work together to open up regions for settlement and deforestation. According to Rudel and Horowitz, these forms of alliances are more often "tacit than explicit." For instance, affluent individuals, patrons or institutions, and a peasant population might develop a "symbiotic relationship," without acting together in pursuit of a common good (Rudel and Horowitz 1993: 29). In Southeast Aceh, the continuities and affinities between village interests in forest pioneering to support their livelihoods, the rent-seeking interests of logging networks, and the political and economic interests of their patrons and partners inside the local state apparatus led to the formation of an implicit coalition of interests that spanned several political levels. The implicit affinity of interests connecting village- and district-based interests effectively resisted the rhetorical claims over local forests advanced by conservationists and national state forest authorities while continuing to push back the boundaries of the forest reserve.

The Struggle over Forest Access:
State Forestry, Integrated Conservation and
Development, and Local Actors

Although the interests of pioneering farmers, loggers, and local networks of power and patronage coalesced, their activities were irreconcilable with wider international interests in non-extractive uses. As I noted earlier, from the 1960s on, international conservation agencies such as IUCN and WWF concentrated resources in the Leuser area. Indeed, for many years WWF's Leuser project comprised the organization's single largest project in Indonesia. Among other things, WWF provided for surveys, ecological research, orang-utan rehabilitation, the development of management plans, ecotourism and education center development, as well as the equipment and deployment of forest guards (WWF 1979/80: 252). The WWF's approach was consistent with the managerial and technocratic assumptions of international environmental discourse and practice at the time. Within this discourse, state bureaucracies had primary responsibility for managing the environment. Representatives of state agencies were invited to high-level conferences and summits, and they subsequently ensured that their nations became parties to international environmental conventions. As state environmental agencies developed policies, crafted laws, and gazetted natural areas in line with this international environmental discourse, multilateral and metropolitan donor agencies entered into agreements with recipient states to support the implementation of these laws and policies. As new policy initiatives and project interventions filtered down from distant policy-making arenas to the district and village level, they aimed to change local behavior in the hope that doing so would have desired environmental consequences. The implementation of the normative model found in state policies was supposed to lead to the desired situation in an instrumental fashion (F. von Benda-Beckmann 1989). But by the time the IUCN declared the park one of the ten worst-managed parks in Asia, the conservation programs in Leuser had failed to forestall deforestation. Despite the park's "protected status," the Leuser rainforests were deteriorating. The problems faced in the Leuser area exemplified the problems resulting from overreliance on authority tools—the "fines and fences" approach, which involved protecting areas legally and enforcing this protection via policing—criticized by advocates of the ICDP approach (Wells et al. 1992). As we will see, subsequent conservation initiatives in the Alas valley had little success overcoming these same problems.

In concert with the critique of state-centered approach to forest management, Indonesian state planners also recognized the failure of the previous policy approach. By the early 1990s, with numerous claimants competing for increasingly scarce resources, clearly state agencies lacked the human and financial capacity to manage the vast state forest zone. Consequently policy makers re-examined the role government could play in conserving biodiversity and recognized the limitations facing government agencies. State policy documents, such as the *Biodiversity Action Plan* and *Agenda 21–Indonesia*, advocated a range of "mixed instruments"—policy tools involving varying levels of state and private provisions with differing levels of involvement from other actors. At this time, responding to international concerns, donor agencies, including the World Bank, Asian Development Bank, European Community, and USAID, became involved in funding park protection and biodiversity conservation in the Gunung Leuser National Park (GLNP) area. In 1989 a World Bank-funded technical assistance project commenced in the Leuser area, which eventually led to a European Union–sponsored Integrated Conservation and Development Project.

Following the inception of this project in December 1995, the Leuser Management Unit became the key advocate of the international conservation community in the Leuser area. In outlining the rationale for the ICDP intervention, the *Leuser Development Program Masterplan* discussed the failure of regulatory policy instruments, arguing that it followed from the failure of the state forestry agency to establish an effective property regime in the Leuser state forest area. Proposing a rather radical approach to this capacity problem, the *Masterplan* proposed to strengthen management of the conservation area by creating a new agency to take the place of the existing conservation agency (PHPA). This new actor would effectively enforce state policy in the state forest estate by harnessing "control and enforcement" capacities already existing in the state apparatus, but "rarely, if ever, concerned [with the] protection of wildlands" (Rijksen and Griffiths 1995: 180). This plan would ultimately involve using authority tools: the *Masterplan* envisaged improving monitoring and surveillance, improving communication, creating anti-poaching squads and a well-organized guard force, and effectively mobilizing existing law enforcement agencies.[62]

According to one study, in Indonesia road construction, mining and logging concessions, and sponsored immigration projects threatened protected forest areas such as Leuser more than the activities of local communities (Wells et al. 1999). These types of threats are best addressed through influencing planning and investment decisions. For this purpose, LMU could draw on the highest level of the New Order. President Suharto himself officially launched the Leuser Development Programme in Banda Aceh on 21 December 1995.[63] Later, during 1997 and 1998, the Minister of the Environment and Minister of Forestry also visited the program area to support LDP activities (Leuser Management Unit 1996/97). With such influential patrons, LMU and YLI were able to en-sure that some development projects that would have facilitated "forest encroachment" were blocked. For instance, the president halted planned transmigration projects in the Lawe Bengkung and Singkil areas. Then, after the Minister of Forestry visited the ecosystem in 1997, this ministry refused to extend three timber concessions within the Leuser Ecosystem area, also knocking back several requests for conversion of ecosystem forests to plantations (Leuser Management Unit 1996/97). Through this top-down approach, therefore, the project enjoyed some significant victo-ries, putting an end to several planned developments. On such occasions LMU "demonstrated political clout that very [few], if any, conservation initiatives in Indonesia have had" (Barber 1997: 12).[64]

While LMU's power lay in its ability to deal with external actors not in the immediate vicinity, for LMU to attempt to implement state policies protecting the Leuser area entailed trying to change the *de facto* local rules of the game. In the next few years this attempt generated a series of struggles. LMU soon confronted antipathy in local government. As one observer noted, "LDP's decisive actions to stop a number of local projects in their tracks have fanned a good deal of resentment among lo-cal officials, one of whom likened LDP to 'intelligence police' who 'come here, snoop around, and then go back to Jakarta and report—then the project is suddenly cancelled.'" This dynamic led to widespread displea-sure and a backlash against LMU (Barber 1997: 13). In a similar fashion to the activities of WWF-LP, LMU's activities threatened the networks of power and interest engaged in unsustainable mining of Leuser forests. Nevertheless, while this "outside" interference incensed some members of the local elite, local figures had to pay lip service to conservation: LMU/YLI had such high-level support that they could not openly resist state policy.

During 1996–99, this dilemma was often illustrated by newspaper re-ports from Southeast Aceh. For instance, in 1994, *Serambi Indonesia* carried a report under the headline "TNGL: Gift or Curse." The report described a meeting in Kutacane to co-ordinate the funds for underdevel-oped villages (IDT), which local officials used to express resentment of the park. Presumably with the support of local public sector directors, a district youth committee (DPD II KNPI Southeast Aceh) asked the gov-ernor to support the excision of a section of the TNGL for mining and industrial development. According to the report, local opinion saw the national park as an obstacle to regional development. "Until now," a youth leader said, "for the residents of Southeast Aceh, over time, it is becoming less clear if the presence of the TNGL is a gift or is a curse."

A youth group associated with the Golkar party, the Pancasila Youth Group, had written a letter to the president expressing this wish. Abdullah Moeda, the head of the district parliament, criticized "a concept of development that prioritises the environment rather than considering humanity." The head of the regional planning board (*Bappeda*) also joined in, complaining of the problems facing those wanting to "free up" forest land for development (Serambi Indonesia 1994a). Despite EU promises of extensive development assistance, some months earlier *Serambi Indonesia* reported that "the existence of TNGL, according to the *bupati*, does not promise anything. It even seems to slow the process of this region advancing from 'miserable Aceh' to 'happy Aceh.' 'A large part of our region has been dedicated as the lungs of the world, but why hasn't the world dedicated anything to Southeast Aceh,' Syahbuddin said" (*Serambi Indonesia* 1994b). For several years, local newspapers have also carried stories of a planned road between Titi Pasir, in the Alas valley, and Bohorok in North Sumatra. By cutting a short route across the mountains to Bohorok, the proposed road would enable travelers to avoid the long and winding journey to Medan. Local people argued that it would cut some hours off the arduous trip to Medan, help local producers gain access to the large Medan market, and facilitate tourism in the Alas valley. Consequently, the road became a cause célèbre in Southeast Aceh. However, the new road would bisect the national park. As the paving of the Kutacane-Blangkejeren road had advanced logging and forest pioneering in Badar, the road would also have facilitated access to untouched forest for the Southeast Aceh logging mafia and farmers looking for new land. Therefore, LDP/YLI and the forestry department had repeatedly asked the central government to block this project. According to newspaper reports from March 1994, the local head of Golkar openly criticized this decision, while the *bupati* showed his local loyalties by stating that he still supported the road. Both lamented the role of unnamed local officials in the failure of the plan (Waspada 1994a and 1994b).[65]

The problems raised by disagreements with local government intensified as LMU came into conflict with the forestry department. As envisaged in the *Financing Memorandum* between the Republic of Indonesia and the Commission of the European Community, LMU would gain full management rights over the Leuser Ecosystem, displacing the Department of Forestry as the primary landlord of the Leuser state forest estate.[66] However, according to a Minister of Forestry decree, LMU/YLI management rights over the Leuser Ecosystem would be "based on a plan approved by the Minister of Forestry" and implementation of this plan would occur under the "overall guidance and control" of the provincial forestry office (*Kanwil*).[67] The president resolved the lack of clarity over roles and responsibilities in 1998 with a presidential decree. While this decree extended YLI's management agreement to thirty years, by reducing UML/YLI's role to "assisting (*membantu*) the Government in carrying out area management," it represented a victory for the Ministry of Forestry (*Keppres No 33/1998*, paragraph 3).[68] As we will see, this competition over roles and responsibilities clearly created coordination problems between Leuser and the forestry department, which affected project activities in the field.

LMU lacked the authority to enforce the law and physically depended upon forestry department, police, and military efforts. LMU instigated and funded integrated operations involving the police, army, and forestry personnel. As LMU law enforcement efforts faced problems at the district level, LMU reported on-going infringements of forestry regulations to higher-level provincial and state authorities. At times these authorities sent teams to monitor what was occurring at the local level.[69] Higher-level officials then instructed their local forestry representatives to enforce regulations preventing habitat destruction within GLNP forest. Yet these activities hampered local clientelist networks, loggers, and forest pioneers only temporarily, as was demonstrated in May 1998, when a provincial inspection team (*tim pansus*) found eleven sawmills operating in Southeast Aceh without valid permits (*Ijin Pemanfaatan Kayu*, or IPK). In response, the vice-governor of Aceh, in his capacity as head of provincial anti-logging operations (*Tim Pengamanan Hutan Terpadu*, or TPHT), issued a letter instructing the *bupati* to shut down the local sawmills by the end of May 1998. However, in July, when the provincial team returned to the area, they found that the sawmills were still operating, apparently due to the on-going need for wood in Southeast Aceh.[70] This situation stimulated a journalist to write: "The irony is that there are enough people with the power, responsibility and equipment for securing protection forest. But forest destruction and wood theft in the protection area continues to happen more and more."[71]

Finally, in August, now with instructions from the governor, district authorities shut down ten sawmills operating around TNGL in Southeast Aceh. These sawmills included those working in Badar and those operated by prominent businessmen, members of the local Desky and Selian clans.[72]

With actors within the law enforcement agencies involved in logging, powerful local interests shaped the context in which state agencies used their discretionary powers, and the implementation of law reflected local power configurations. Due to corruption and collusion, LMU law enforcement work floundered, despite the large amounts of project funding provided for this end. Still, LMU continued to support law enforcement efforts against villagers who opened plots or logged within the Leuser forests. As we will see, in village eyes this sharpened the LMU's legitimacy problem.

The *Leuser Development Program Masterplan* had outlined a strategy for confronting what it identified as "rural pressures"—the extensive socio-economic factors that stimulated the rapid conversion of the forest to other land uses. In keeping with the ICDP concept, the Leuser Development Programme would employ incentive tools that set about creating inducements, charges, sanctions, or even using force to encourage compliance to LMU and state policy objectives. Amongst the armory of incentive tools, the *Masterplan* emphasized the development activities designed to reduce pressure on the reserve. LMU sought to link development programs and conservation and find incentives to induce local communities to change their activities within Leuser Ecosystem boundaries. LMU entered into "quid-pro-quo agreements" with village leaders whereby LMU offered villages micro-project developments in return for villagers' agreeing to cease logging and opening new land within the Leuser Ecosystem.

LMU implemented its project through or in cooperation with government agencies, and local NGOs criticized the way this worked out. They alleged that because community development projects tended to be implemented by clients of government agencies, little money reached the target village communities.[73] One informant in Southeast Aceh even alleged that around 70 percent of funds were allocated this way.[74] Critics argued that thus, despite the best intentions of LMU staff, target villages rarely obtained the benefit of the European Union funds. During one interview, a forestry official acknowledged this problem, observing that one could hardly blame LMU/YLI for this—it was a problem within the state agencies themselves.[75]

Irrespective of LMU's good intentions, the dedication of its staff, or the amount of funding at its disposal, the population of the Leuser area proved to be incommensurate with the amount of funding available for micro-projects. As WWF-LP found in Menggamat, LMU funding for individual micro-projects was not enough to counterbalance the lucrative and direct economic benefits local actors derived from logging. Unless these micro-projects could eventually become self-sustaining, or LMU could secure eternal funding, the economic assistance would dwindle away. In a similar fashion to other ICDP interventions, when faced with increasing market incentives that limited the capacity of its economic development strategies to contribute significantly to sustainable livelihoods, LMU proved unable to have a significant impact (K. Brown 2002). The micro-projects were hardly significant when compared to opportunity costs forgone by local actors locked out of the ecosystem. In any case, it is not safe to assume that the provision of public goods directly changes private behavior (Newmark and Hough 2000).

Unresolved issues of legitimacy and social justice underlay these problems. Although ICDP activities involved (nominally) increasing development options of rural communities for conservation aims, villagers interviewed in Badar said that there were no economic benefits from LMU activities. One LMU employee also admitted that most villagers did not gain from the micro-projects. Meanwhile, villagers watched the LMU staff driving around in imported cars and the villagers employed by LMU using motorcycles and expensive equipment. They were aware of a gap between the facilities of local staff and the simple nature of the micro-projects. These micro-projects were unable to deliver significant benefits to villagers. At the same time, the law enforcement activities that LMU supported, by criminalizing the livelihood activities of poor forest pioneers, attempted to prevent the rural poor from gaining access to forest resources. Consequently, LMU's activities created resentment and jealousy, generating conflict and resistance to LMU objectives. Ironically, while LMU success depended upon building alliances with villages against loggers, LMU field workers and law enforcement officials came to be perceived as the "common enemy." As a result, forest encroachment and illegal logging continued despite the large budget spent by the LMU. "In the end the community felt cheated and hated LMU," a former LMU field worker noted, and "we were even ostracized (*diusir*) by one Southeast Aceh community."[76]

The Leuser ICDP approach derived its legitimacy from the state law and relied heavily upon legal assumptions and regulatory policy tools. The Leuser ICDP might have tried to renegotiate customary rules governing resource access to ensure they effectively took ecological values into account and to reconcile those customary rules with legitimate and enforceable state rules. Instead, however, in concept it merely envisaged local communities signing contracts that guaranteed the integrity of the state forestland in exchange for economic benefits. The implicit assumption was that local uses of the forest amounted to trespassing on state land. Contrary to the complex process of territorialization that divided the Leuser Ecosystem into bufferzones and core conservation areas, however, "state forest" remained a state planning concept, found on state maps that either contradicted local notions of territoriality or worked against village livelihood interests. Law enforcement actions based on a territorialization process that villages could not accept would only lead to protracted conflicts with local communities. Project activities in the field proceeding from this basis were hamstrung by a conceptual framework that failed to provide appropriate tools for facing the fundamental conflict between state and village institutional orders.

The Institutional Terrain in Badar

In Badar the uncertain institutional terrain that emerged at the village level can more readily be understood in terms of the actions and interactions of village institutional arrangements with field foresters than from LMU activities. Here, following Peluso (1992), we can distinguish three conflicting roles or identities of a lower-level forest guard, each associated with its own set of priorities: He could act in accordance with the role of professional forester, a small-time patron, or a householder trying to support his family. The *jagawana*'s role implementing forest conservation laws reflected the interests of this wider conservation community; at specific moments and to a limited degree these obligations led to actions that affected the activities of loggers and forest pioneers. To varying degrees, the role that *jagawana* played (as protector of the forest) also affected the ability of village heads to allocate tenurial rights over surrounding lands. At the same time, the *jagawana* and other forestry personnel were embedded in the clientelist context of Southeast Aceh and could profit from this context. Simultaneously, they might also act in the interest of their clients in the village, who had cultivated their support over time. In thrall to these contradictory orientations, forest guards could readily shift perspective as the situation required. Consequently, the rules were not fixed.

As I have already noted, according to customary tenurial practices, those wishing to open land must ask permission from the village or hamlet head (*kepala desa*, or *kepala dusun*). However, in the hamlets of Badar, the ability of the *kepala dusun* to grant land has varied over time. According to an Alas villager who settled in the enclave hamlet just prior to 1990, the *kepala dusun* gave him permission to open land, which he thought guaranteed his right to settle there. Even though the land was included in the national park, it lay in the bufferzone area that the *bupati* was happy to see opened. At that time, forestry officials were openly involved in logging the area. Besides warning him that it was state land, they made little effort to stop him from opening a plot there. From this farmer's perspective, his rights were confirmed when taxation officials began to levy land taxes on his farm lands.[77]

According to an older resident, at this time "those wishing to open new land might pay *jagawana* to look the other way. Before anyone noticed, the candlenut trees are big enough and no one can do anything about it." This meant that to obtain land, he added, "in other villages it may be enough just to let the *kepala desa* know, but here it is not like that."[78] If the candlenut trees were already bearing fruit, it would be difficult to move the farmer: he would have established permanent tenurial rights, according to *adat* principles, and the government would have to pay compensation. As a result, *jagawana* were unable to move farmers: they would tell those with mature candlenut gardens, "Don't go any further."[79]

An older villager had found several years earlier that lower-level officials had discretion over state rules, that these rules may not be applied consistently over time, and that this situation has consequences for the security of tenurial rights. During the 1970s, he had opened GLNP land across the Alas River from one particular hamlet. Realizing that the oral agreement of the *kepala desa* was insufficient to support his tenurial rights, he had successfully obtained an official letter from a district government official. However, the validity of the letter was challenged after the official who had issued it moved, and he was forced to abandon the land. "Other officials would not recognize it as valid," he said, "so I could not hold on to it, and three years later I had to move."[80] Due to these overlapping and unsettled authority structures, village institutional arrangements remained ambiguous, and villagers had to eke out a living in an uncertain context.

Moreover, during struggles between forest pioneers and forest officials, the problems undermining law enforcement and the lack of coordination between the forestry agency and LMU further compounded this uncertainty, for, as I noted earlier, in the early 1990s there was a move towards tightening the implementation of state regulations regarding GLNP. The extent of logging and forest pioneering in the Kappi area of GLNP beyond Ketambe had become too visible a reminder of the failure of the park to enforce state regulations. The loggers and smallholders opening gardens here represented a threat to the state territorialization that set aside the area for biodiversity conservation. In keeping with government actions elsewhere, senior forestry officials sought to impose stricter control over TGNL land, including over those who had settled in Kappi (Li 1999: 16). The official discourse held that villagers who opened candlenut gardens here were *perambah hutan* ("forest encroachers" or "squatters"). Following a coordination meeting between forest authorities and local government agencies in 1991, the authorities had resolved to relocate the "forest encroachers" who had settled on TNGL land (Department Kehutanan and Universitas Syiah Kuala Darussalam–Banda Aceh 1993).

The forestry department commissioned a report into the socio-economic, anthropological, and physical condition of these "forest squatters," aiming to choose a site and to plan their re-location to it. The report identified 1,077 families subject to relocation, including 672 families in the Badar section of the park. Approximately 70 percent of these families were from Gayo Lues. In surveying average land holdings here, the researchers found that most of the people had moved into the park to improve their livelihoods, attracted by the availability of fertile land suitable for growing food crops in addition to tree crops such as candlenut. In the enclaves, the surveys found that land parcels varied between 0.5 and 3.5 hectares in size, with average land holdings of around 2 to 2.5 hectares. According to the report, "forest encroachment taking the form of uncontrolled cutting of wood, shifting agriculture and opening of settlements is frequently observed along the Blangkejeren-Kutacane road" (Department Kehutanan and Universitas Syiah Kuala Darussalam–Banda Aceh 1993: 3). The report declared that the "forest encroachers" were having a negative impact on the conservation of the national park, causing erosion along the steep hillsides next to the road, causing landslides and affecting the functioning of a transport link vital to the local economy.

Having adopted the North American concept of national parks, which establish protected areas that exclude human uses, when forest authorities in developing countries have designated national parks, they have forced indigenous and local people to move, with little attention to the impact on local communities (McLean and Straede 2003). In Indonesia, forest authorities have also identified and inventoried "encroachers" into forest areas such as national parks zoned by state planners for particular uses, in many cases subjecting these people to relocation in an official transmigration settlement.[81] Following this established pattern, the transmigration department, in collaboration with other agencies, formulated a plan to move the "forest encroachers" to Lawe Bengkung, in a remote corner of Southeast Aceh.[82] Villagers recounted that the authorities visited their enclaves to announce that only those who had recently opened *kebun* would have to move. As it turned out, forest farmers who had been resident in the enclaves for over twenty years—before the Kappi area had officially become a part of TNGL in 1980—were also required to move. One farmer described to me how, following the investment of years of work creating a productive candlenut garden, he like other villagers in the hamlet found himself classified as an illegal "forest squatter" and subject to transmigration. The villagers were moved to Lawe Bengkung, to the south of Kutacane, the area identified by the report (mentioned earlier) that the Department of Forestry had commissioned (Department Kehutanan and Universitas Syiah Kuala Darussalam–Banda Aceh 1993).

This site was located in the midst of a logging concession, and the only transport available was the logging trucks that passed through continuously.

In September 1997 a villager critically assessed the consequences of state forest policies implemented amidst the contradictions between the suffering of threatened nature, a state territorialization that invoked the international conservation discourse, and a village-based sense of natural justice. In an interview laced with irony, she described how the transmigration site was better suited for tourists than the enclave north of Ketambe from which the "forest encroachers" had been removed: "There are a lot of elephants, tigers, and orang-utan [in the remote transmigration site]. It is complete—that is the real place for PHPA—they should take the tourists there . . . Where do you see elephants, tigers, and orang-utan here [in the national park enclave along the Kutacane-Blangkejeren road]?" Some mornings in the wet season, she explained, a herd of elephants would turn up, usually in a group of six or nine, but occasionally as many as sixteen. "They would trumpet outside the house, and put their feet forwards and their trunks up," she imitated an elephant with its legs and trunk in a supplicating position. Then they would make a special roar. "So, we would bring out some rice, and say: '*Datuk* [lord], please take this and go a long way from here. I have children, and grandchildren, and we are scared.'" She said this in a placatory tone. "And then they would roar again. So we'd give more rice, and then they pick it up by their trunk and swing it on to their back and move off." She imitated an elephant stretching its back, and continued:

In the zoo they have elephants, but they are enclosed by a fence. Here they were just wild. It was an elephant village, and we were living in the midst of it. They are just like human beings. How would we feel if someone came and wrecked our village . . . When the forest is finished, where are the elephants to go? They saw us, so they were mad. They thought we were destroying their forest. Then they would wreck the crops or move right into the house—wrecking it completely . . . and so in the end we had to come back here.[83]

Poor planning and implementation have plagued this "local transmigration" effort. The Leuser Management Unit also resisted the plan, not because it angered villagers or led to hardship, but chiefly because it involved shifting local people into an area within the Leuser Ecosystem that LMU ecologists had identified as being important for elephant populations. Subsequently, the families who had moved there were forced to return to the enclave. Back in Badar, they found their gardens overgrown and their houses fallen into disrepair.

During 1996–98, disputes between the forestry agency and local villagers continued without any sign of resolution. According to local villagers, when *jagawana* caught villagers cutting timber or opening areas within the forest, they compelled them to sign agreements stating that they would not do it again. One informant noted that such agreements enabled the *jagawana* to wash their hands of responsibility, even though local people would most likely continue to log or open lands.[84] However, if *jagawana* arrested illegal loggers or squatters and the case ended up in court, the villagers found that if they pled that poverty had forced them to "encroach" into the park, the court would release them. Forestry staff complained that this would happen even if the illegal activities were sponsored by outside logging networks. Nonetheless, villagers resented what they considered to be continuous harassment by *jagawana*. One *jagawana* I interviewed in this enclave area during late 1998 noted that the dispute had reached a level such that, due to the risk of violence, *jagawana* would only go into the area with four colleagues.[85]

In 1997, interviews revealed that villagers who had been forced to move to Lawe Bengkung and been impoverished by the experience were particularly bitter. One told me:

I have now opened six hectares of candlenut trees, and now they come and tell me I have to move. They said that they would pay compensation, but I haven't seen this. Perhaps it would be better if they shot me. Every day I have gone out and worked in the *kebun*—for many years only eating once or twice a day while I established this *kebun*. People come here as just-married couples wanting to start a new life. No one wants to be rich, but just to have a livelihood and send their children to school. We don't receive a salary like a government official.

Another explained: "No compensation was paid when we came back. But it doesn't matter what Yayasan Leuser, PHPA, or the government say, I won't trust them again. It is enough to make you wild (*bandel*). I am sure we could get arrested for speaking like this but we have children. PHPA care more about the elephants. Poor things (*kasihan*), they say [about the elephants]."[86] Following this dispute, hamlet heads (*kepala dusun*) recognized that, within the land classified as state forest, especially in the national park, they had no authority to grant land and were unwilling to take responsibility for villagers opening land there. According to a village informant, in 1998 a villager who wanted to open a *kebun* was still obliged to report to the *kepala dusun*, even if the area lay beyond the boundary of the GLNP. However, the *kepala dusun* would grant land only if the area fell under his authority. "If he can't, he will say: 'it is up to you.' Some villagers are brave [enough to open TNGL land], while others aren't."[87] Nonetheless, in the less protected Serbolangit protection forest to the east of the road, villagers have continued to open forest for a considerable distance.[88]

Yet, when the economic and political crisis struck Indonesia in 1997–
98, the ability of the state agencies and their conservationist backers to
implement state law declined even in more visible areas alongside the
road. At the beginning of 1998, villagers responded to the crisis by
opening new land. With a boom in patchouli oil (*nilam*) prices, villagers sought out forest areas to plant this crop. In some areas of the Alas
valley, during a period of drought and forest fires, areas of bufferzone
forest were deliberately burnt to facilitate forest pioneering; in hamlets
within Badar, farmers opened plots directly across the Alas River. As the
river acted as the park boundary and these plots were in clear sight of a
main provincial road, doing so represented a challenge to the authority
of TNGL officials. The forestry authorities and police arrested several
people. However, the villagers argued that this was unjust: the activities
of companies in other areas of the park were much more destructive,
and local villagers were only trying to make a living. For this reason, village leaders protested to the district assembly (DPRD), requesting that,
because of land shortages, villagers be given permission to cultivate this
land, even though it was within the park boundaries. A session of the
regional assembly (DPRD) discussed the case in the presence of forestry
officials. However, the assembly failed to come to a formal resolution
to stop these activities. Subsequently, the forestry authorities and villagers had a tacit understanding (*tau sama tau*) about these activities, and
jagawana no longer challenged the new plots.[89] The problem presented
to the authorities was that, in the face of the crisis, villagers could claim
that they needed to take desperate measures to feed themselves—that
they needed to open new land irrespective of its status. "Before the crisis,
forestry could stop people opening here, but now they have to accept
this."[90]

Conclusion

In this chapter I have analyzed the gradual emergence of shifting, unstable institutional arrangements governing access to and use of natural
resources in Badar. The roots of this process emerged during the early
decades of the twentieth century when a variety of non-state actors including scientists, scientific organizations, and environmental pressure
groups both in the Netherlands and in the colony itself provided the impetus for environmental regime formation in the Netherlands East Indies
(Paulus and Stibbe 1921; Dammerman 1929: 22). As in other colonial
contexts at that time, pressure groups and the wider intellectual environment prompted the colonial administration to see the need to create

new environmental regimes for controlling access to and use of forest and other resources (Groves 1996). The process of environmental regime formation in the Alas valley followed the earlier process of developing a territorial system of administration involving indigenous authority structures arranged by the colonial administrators. In setting up an environmental regime in the Alas valley, the colonial state fixed boundaries between nature reserves and territorialized indigenous administrative areas, at the same time providing space for the expansion of local agriculture. Although this environmental regime fell into abeyance in the post-independence era, the legal concepts that provided for it were incorporated into the post-independence state, and continued to provide the dominant notion of how to protect the environment.

In discussing the parallel evolution of a village institutional order in Badar, I have explored how customary (*adat*) rules have changed, been renegotiated, and even been enveloped by wider social and economic forces. In this frontier region a range of influences ensured that the institutional arrangements governing access to and use of land and forest resources would differ from the customary patterns that had operated historically in the Alas and Gayo heartlands. The relatively recent settlement of pioneer agriculturalists from different areas of origin and of two distinct ethnicities acted to complicate the evolution of strong *adat* arrangements. As these pioneer agriculturalists had opened land in an area where most of the forest had a pre-existing status as state forestland, the role of state forestry agency is particularly important.

By the 1960s and 1970s, the creation and management of a national park and protected area system had become an essential element in a nation's development planning within developmental discourse. Consequently, Indonesia, alongside other nations in the developing world, tried to lock up rich areas of biological diversity in protected areas. As in the past, the idea was that, in the midst of a sea of underdevelopment and rapid economic growth, biological diversity could be preserved by setting aside large islands of wildlands for nature conservation (Quigg 1978). With international NGOs, conservation agencies, and donor agencies supporting this process, in Indonesia this idea was developed initially by renewing efforts to conserve forest areas that had first been gazetted during the colonial period. When the Indonesian government created Indonesia's first national park, they chose to revive the colonial state's conservation regime in the Leuser area. The government created the Gunung Leuser National Park (GLNP) by joining together nature reserves in the area that had first been gazetted in the 1930s.

After the revival of the protectionist paradigm in Leuser, the state forestry agency picked up the task of regulating the relationship between population and resources in the area. Despite the classification of most of the area as protected forest and later national park, for some years the national park primarily existed on paper, as there were virtually no forestry agency personnel present to monitor compliance to state rules or sanction those who transgressed them. Eventually, the state forestry authority placed forest guards in the area to extend state control over access to the forest. After this time, forestry officials emerged as a source of authority that worked in the same area as village and *adat* leaders who, according to village practices, had previously controlled access to surrounding lands. Particularly during law enforcement campaigns, field foresters conformed to their role as professional foresters. However, the poorly paid, under-committed forestry officials stationed in the hamlets surrounding the national park often used their discretionary powers over the enforcement of state sanctions to extract rents from loggers and forest pioneers. As one villager argued, the creation of the national park may have had the perverse effect of hastening timber extraction in the area. Such inconsistencies in the intent of forest field officers compounded the existence of overlapping authority structures and conflicting property claims associated with village conservation agencies, forestry officials, and logging networks, ensuring that institutional arrangements pertaining to the forest remained shifting and uncertain.

During the 1970s, as forest pioneers from neighboring areas continued to move into the area, logging networks had emerged and began extracting timber. Local elites involved in these networks rejected the national park and protection forest status of the majority of the Alas lands, at least partly because this status obstructed the timber operations of district entrepreneurs. To contest the state's territorial strategy that set most of the district aside for nature conservation, they asserted that the Alas people had a "right of avail" over these lands. At the same time farmers under economic pressure and facing land shortages challenged the state territorialization that prohibited local uses of surrounding forests. Despite the weak legal basis for *adat* territorial claims over surrounding forests, local people maintained that, in the absence of alternatives, they were entitled to meet their subsistence needs by continuing to open surrounding forest for agriculture and to collect forest products. Thus timber extraction by district logging networks proved to be compatible with forest pioneering by villagers, and the affinities between these groups led to the formation of an implicit coalition of interests that spanned several political levels. As the different actors—including timber entrepreneurs and brokers, village leaders and pioneer agriculturalists, and the local agents of the state—entered into exchanges and reached accommodations, they effectively established the key institutional arrangements determining access to and use of forest resources.

In attempting to intervene in this situation, the Leuser Development Programme tried to apply the ICDP approach. ICDPs work on the premise that some appropriate set of incentives exists to induce local communities to change their practices. Brandon and Wells (1992) have distinguished three main strategies employed by ICDPs: (1) strengthening park management, (2) providing compensation or substitution to local people for lost access to resources, and (3) encouraging local social and economic development. Considering the second and third of these strategies first, LMU's attempt to compensate local people for lost access to forests and otherwise encourage development faced the same problems as WWF-LP intervention in Menggamat: despite European Union funding, LMU micro-projects spread out across such a large area proved incapable of providing stable economic livelihoods and viable alternatives to logging, whereas the demand for timber offered village and district actors just such lucrative returns.

Many of LMU's problems in furthering village development in part derived from the structural relationship between government agencies and a donor project working within state structures during the New Order. The Leuser ICDP grew out of an inter-government agreement between the European Union and the Republic of Indonesia, and it primarily worked through government structures. For the first few years of the project, at least, this relationship was symbolized by the fact that LMU located its district offices within Bappeda offices.[91] LMU/YLI also employed seconded or former high-level New Order officials in key decision-making roles. In the field, LMU tended to implement community development projects through government development agencies and their clients. Due to inappropriate project design and "leakages," little of this development funding benefited target village communities. Moreover, despite use of the rhetoric of participation in project planning, this "participation" tended to amount to consultations regarding the achievement of preset conservation goals.

LMU's efforts to strengthen park management involved attempting to ensure that state regulations protecting the state forest were consistently applied. However, LMU lacked independent means of enforcing conservation laws and depended on state agencies. But, with powerful but unspoken collusive relationships encompassing these agencies, logging networks and forest pioneering continued to contravene the official forestry laws. In the face of the combination of interests working in the district, even with the advocacy of a high-profile EU-sponsored ICDP with high-level support within the state, the state authorities responsible for implementing the rules have lacked the ability or the will to implement the law in a consistent and dependable fashion. Consequently, park management has been unable to do much more than periodically tighten up enforcement and briefly constrain the behavior of resource appropriators; they have not been able to fundamentally alter the dynamics of this situation. Only at certain moments, when LMU activities combined with more rigorous media reporting and the emergence of a more active civil society during the reform era, did the balance shift somewhat towards state law enforcement.

In this case the outcome has been an on-going and largely covert struggle over access to and use of forest resources. With overlapping institutional arrangements relating to forest resources, there have been no clear decision-making or conflict resolution arrangements. In the absence of clear authority structures willing or able to consistently govern access to and use of the forest, the intertwining of local government, *adat*, state forestry institutions, and international conservationist agencies has created a complex, dynamic state of affairs. As wider economic and political forces affected the situation, the transformation of the forest has been the inexorable outcome of a conflict in which there were no clear winners.

Conclusion: Institutional Arrangements in Southern Aceh

To conclude, I return to the questions that animated this study: How do local institutional arrangements governing resource use emerge and evolve under various conditions? How do particular institutional patterns lead to certain environmental outcomes? Let me begin to reflect on these institutional conundrums by discussing the understanding of customary *adat* institutions that has emerged from this study: How can we best conceive of customary institutions? How do they change? How do they manage natural resources? Do *adat* arrangements provide a basis for sustainable resource use? Then, in the next section, I consider the long history of state action in the area: What has been the impact of intermittent state attempts to remake local institutional arrangements and impose a state forestry regime? And, in theoretical terms, how can we best understand the nature of the relationship between the state and *adat* institutional orders? The final section discusses how the most salient institutional patterns found at the district level have led to environmental decline.

Adat Institutional Arrangements: Adapting to Local Agro-ecological and Socio-economic Constraints

Since the time of first settlement, Southeast and South Aceh have been frontier areas. Dutch colonial records from the late nineteenth and early twentieth century describe a sparsely populated mountainous region covered in rich tropical rainforest. Here, narrow valleys and small coastal plains suitable for human settlement lay between the steep mountains of the Bukit Barisan range and the sea. By the time of colonial occupation, different waves of migration had poured over the shores of South Aceh and down the Alas valley to the east. The isolation of each valley and narrow coastal plain fostered the evolution of discrete sets of villages each with its unique heritage, blend of ethnicities, customs, language, and identity. Though the structure of village government varied somewhat in the different settlements, there was a general pattern in which a village head and other *adat* leaders worked under the supervision of a council of elders (in Acehnese known as the *petuhapet*). This village governance structure itself functioned beneath a primary local headman (such as the *kepala mukim* or *datuk*). Under these structures, communities developed dynamic and evolving institutional arrangements regarding access to and use of resources in their area.

While agricultural practices have varied considerably in this area, historically the general pattern was for villagers to engage in what Conklin (1975) termed a "partial supplementary" swidden system. Farmers opened temporary swiddens (*ladang*) for food or (at times) cash crop production, combined either with settled rice production (*sawah*) in the valleys or with the development of more permanent gardens (*kebun*) in the hills. Agriculture also depended on the ecological functions of adjacent forests—for example, the hydrological functions provided by adjoining watershed forests. However, this dependence was indirect: as forests were so extensive, the impacts of the forest clearance that occurred were often not immediately apparent.

Agro-ecological and social constraints lent themselves to the development of particular forms of property relations. As I noted earlier, a property regime gains its character largely from what the decision-making group believes to be scarce and valuable, and hence what needs to be protected with rights (Bromley 1989). Unless resource units are relatively scarce, there are few reasons for appropriators to invest costly time and effort in organizing to control access and use (Ostrom 1997: 9). In the past South and Southeast Aceh contained extensive, mountainous lands covered in forest. Forest was abundant, and the population was comparatively small. Thus, property regimes tended to be concerned more with what was scarce: labor and the productive agricultural land created by its deployment.

In all three cases that we have looked at, farmers engaged in swidden cultivation. The evidence available suggests that the *adat* principles relating to the residual property rights that a farmer might enjoy over formerly swidden areas varied somewhat both between communities, and within the same community over time. With the importance of permanent gardens producing cash crops increasing over time, customary rules became increasingly focused on protecting permanent property rights over productive land and trees in all three cases. *Adat* arrangements, such as the *seuneubok*, provided a framework for the transformative use of the forest: the clearing and conversion of native forest into gardens. In other words, village institutions became particularly attentive to tenure—the "rules and practices specifying who is to get access to land, at which time and in which place" (Dijk 1996: 18).

Nonetheless, surrounding forests remained important to village livelihoods. They constituted a reserve of agricultural land available to future generations of farmers. Moreover, villagers ranged across the landscape collecting valuable forest products. For centuries, forest products—such as incense wood (*gaharu*), resin (*damar*), forest rubber (*getah*), camphor, benzoin, rattan, wax, honey, rhinoceros horn, deer horn, and elephant tusks—were amongst the most valuable exports from the west coast of Sumatra, including southern Aceh. *Adat* arrangements provided for the collection of forest products and some timber extraction by local residents.

Yet, though forests were important sources of non-timber products, the characteristics of forest resources as common pool resources did not lend themselves well to management by instituting tenurial rights. In contrast to the agricultural products generated from cultivated lands, forest products were generally attained without the sustained investment of labor. Many of these resources were widely dispersed in the forest, they tended to be non-renewable or difficult to regenerate within a short time, or their regrowth and distribution were hard to predict. Even resources that could be subjected to a sustained management regime, such as *damar* trees, which were tapped for their resins, were widely scattered through the forest. Furthermore, as the market prices of forest products tended to be subject to wide fluctuations, villagers never altogether depended upon the gathering of forest products as a source of livelihood. Dependence on a resource system increases the likelihood that self-governing institutions will form to avoid the losses associated with open-access, common-pool resources (Ostrom 1997: 6). Thus, there was less incentive for communities to invest time and effort in developing a sophisticated property regime requiring monitoring and the imposition of sanctions.

However, communities did attempt to control access to and use of surrounding forests through the diffuse form of territorial control known in the Indonesian legal terminology inherited from the *adatrecht* scholars as *hak ulayat* ("right of avail"). With the permission of the customary leaders, villagers would ordinarily open land within *adat* territory. They also had free access to the forest surrounding their villages for their everyday needs. Dutch records reveal that commercial users, especially those from outside the villages concerned, had to ask permission for access and had to pay a tax (known as *pantjang alas* or *bunga kayu*) on products either harvested or purchased from the area. The leaders of small local principalities imposed such taxes on the export of forest (and agricultural) products from the ports and made use of them for their own personal income and for the upkeep of their administration. Local heads also collected these taxes on forest products on behalf of their corporate group, taking what they considered to be their due as administrators while passing some tribute on to higher authorities.

Advocates of indigenous rights may be tempted to romanticize the "environmental knowledge" of traditional cultures, finding in customary institutions a basis for sustainable resource management. To be sure, in specific circumstances protecting village livelihoods and the property rights upon which they depended might involve attending to local environmental conditions. For instance, Menggamat villagers cultivated land in a steep, forested area subject to flood, landslide, and pest infestation. Over time they developed a consciousness of the limiting conditions under which agriculture takes place—that the degradation of forests maintaining hydrological conditions would affect the productivity of agricultural land. Consequently, villages here had (to some degree) developed rules to prevent irresponsible farming and environmental damage, such as cutting trees in an inappropriate place or at the wrong time of the year. However, as I discussed in the Sama Dua case, collective action tended to be motivated by the defense of property rights rather than by the need to stabilize environmental conditions. Despite the attraction of looking at customary resource management "through a green lens," this study found that in these areas village regimes were neither primarily concerned with nor organized to ensure environmental outcomes—such as the sustainable use of forest resources over the long term.[1]

In studying customary institutions, as other scholars have noted, it is necessary to avoid a second set of temptations. If the term "customary law" (or "*adat* law") is taken literally, observers may understand *adat* as merely meaning rules or rule-like formulations. As colonial scholarship suggests, researchers can find socio-legal ideas analogous to an analytical concept of law in *adat*. Furthermore, such concepts and principles must necessarily to be taken into account by the state legal system. Yet, as has long been noted, the concept of "*adat* law" emerged from colonial intervention during a specific historical and political context (Benda-Beckmann and Benda-Beckmann 1985). *Adat* varies considerably across

Indonesia, and in some contexts *adat* has many of the characteristics of law.[2] Yet, despite its rule-like character, analysts need to resist the appeal of reducing *adat* to "law." *Adat* encompasses both implicit, deeply held social norms and more explicit rules. As I am not the first to argue, customary (*adat*) institutional order(s) can be conceptualized as shifting ideas and patterns of social ordering. *Adat* arrangements are constantly subject to mediation, compromise, and change; because they are constantly renegotiated, they are both resilient and dynamic.

Bowen's (1988) study of property relations across two sites in Central Aceh demonstrated that, as agro-ecological conditions begin to vary, *adat* institutional arrangements tend to diverge. For instance, when farmers settled in new areas, although they carried the same semantic *adat* categories from their homeland, the property relationships underlying these *adat* concepts tended to change to reflect the new conditions (Bowen 1988).

In the three localities studied in this volume, a major economic crisis exacerbated fluctuations in the prices of commodities upon which villagers depend; a changing political situation affected the capacity of the state to enforce resource management laws; as was most apparent in Sama Dua, changing agro-ecological situation also affected conditions on the forest frontier. The advent of timber entrepreneurs backed by powerful local interests, and the presence or absence of forestry department and conservationist actors also affected local conditions.

Villagers and other actors found themselves caught between short-terms needs (feeding their families) and long-term considerations of ecological integrity and economic security. As villagers reevaluated their livelihood strategies, they reassessed the opportunities offered by local conditions. This situation also affected how actors conceived of their identities and interests, the meaning of *adat*, the use of state law, and the resultant patterns of resource use.[3] In other words, local institutional arrangements in a particular location typically evolve to deal with the specific problems facing the corporate group concerned. Consequently, the evolution of institutional arrangements is intimately tied up with the problems created by specific agro-ecological and socio-social conditions, and *adat* institutions need to be understood in this context.

The *adat* order at the same time incorporates legal, social, political, and supra-mundane religious functions to generate a village socio-ritual order that encompasses local notions of identity and associated notions of appropriateness. Although clearly individuals calculate what is in their best interests and at times act accordingly, *adat* institutional arrangements are not built solely on the logic of rational choice.[4]

This point raises a corresponding problem for institutional analyses

that conceptualize institutions as "rules in use." Recently, more sophisticated approaches have aimed to identify "the multiple and overlapping
rules, the groups and individuals affected by such rules and the processes
by which particular set of rules change in a given situation" (Agrawal
and Gibson 1999: 638). As in this study, this form of institutional analysis provides tools for understanding local processes. Nonetheless, to a degree the caveat just mentioned also holds: institutional analysis needs to
avoid the flawed "epistemological assumption" that "customary law" or
local institutional arrangements are "always there, waiting to be found
and turned into an inventory of rules" (Chanock 1998).

State Interventions, Local Institutions, and Resource Control

State attempts to remake local institutions have had a critical effect on
adat institutional arrangements, once again demonstrating how state
formation involves the creation of community identities recognizable by
the state (cf. Agrawal 2001). This process began when, shortly after the
Dutch colonial forces subdued the area during the Aceh War, the colonial
authorities began constructing a legal and administrative regime according to a familiar colonial pattern. As in other parts of the colonized world,
the valorization of custom was a political technique of colonialism; by
means of it, the colonial state sought an alliance with an indigenous elite,
who gained a vested interest as the "customary" authorities administering
an acceptable form of "customary law" under a system of indirect rule
(Rouland 2001; Holleman 1981; Warren 1993; Li 1997: 6).

Colonial scholars studied local practices and customs, finding in them
the concepts to construct *adatrecht* (*adat* law); administrators used social
and geographical features to fix particular communities within particular boundaries, organizing sets of villages into "self-governing territories" or "*adat* law communities" under specific "self-governing heads"
(*Zelfbestuurders van de Landschappen*). This process created apparently
homogeneous territorial and social units of organization under the one
appointed head, developing a "ethnic spatial fix" that helped consolidate each set of communities' sense of social and geographical identity
(Li 1997). "Self-governing heads" were authorized to use "*adat* law" to
regulate and to administer the internal affairs of their communities, under the paternalistic guidance of colonial administrators.

It is certain that this process encroached on remote areas of southern Aceh, but there is a lack of evidence regarding its precise impact on
village affairs. Given the isolation of the villages in those times, apart

from the colonially constructed "*adat* law" found in colonial texts, it is likely that village headmen were able to administer local customary rules "more or less in the manner of their forefathers" (Sonius 1981: lxiii).

After independence, in a clear break with colonial policies, as in the newly independent African nations, the Indonesian state embarked on a nation-building project that entailed establishing the primacy of the state's institutional arrangements. In accord with the evolutionist assumptions of high modernism and in the name of economic development and national unity, the diversity of "customary law," which was associated with the colonial period, had to be replaced by a modern state system. This "negation of pluralism" entailed setting up a unitary administrative system across the archipelago, abolishing the territorial administrative regimes that operated under the colonial system of indirect rule, and re-shaping institutions down to the village level (cf. Rouland 2001). The New Order's Village Government Law (UU No. 5/1979) represented the apogee of this process, leading to the establishment of a model of village government set out on a national scale according to a uniform format. In Aceh, the new law took away the power of important local *adat* headmen (such as the *kepala mukim*) and marginalized village *adat* councils. Henceforth, rather than villagers' being subject to an *adat* council controlled by village elders, the village head oversaw the village assembly (LMD) and village councils (LKMD). Because the village head became a government official accountable to an appointed official (the sub-district head, or *camat*), village government became increasingly responsive to instructions from above.

State territorialization processes in the area evolved in parallel with the remaking of village institutions. By the time of the Dutch conquest of Aceh, the colonial state had well-developed policies for controlling the use of natural resources. In the 1870 agrarian law, the "domain declaration" (*Domein verklaring*) allowed that all land not under constant cultivation was to be classified as state domain, enabling the colonial authorities to set aside areas for either exploitation or conservation. The colonial authorities gradually operationalized this principle through a territorialization process that set the boundaries for areas that had been set aside for either exploitation or preservation. Given the history of violent anti-Dutch resistance in Aceh, colonial territorialization here proceeded carefully. As in Tanah Alas, when setting aside areas for plantation agriculture and forest conservation, the colonial authorities made allowances for the agricultural needs of villagers (under *hak ulayat*) by carefully setting aside areas for agricultural expansion.

During the Suharto period, the state built on and even extended the colonial territorialization process through the TGHK forest-mapping

process. This process classified most of South and Southeast Aceh as
state forest estate and placed it at the disposal of the forestry department.
Whereas the earlier colonial territorialization process had attempted to
establish a careful separation of state and *adat* domains, the TGHK process created contradictions between state forest classifications and pre-
existing notions of territoriality and tenure associated with indigenous
systems of authority and local notions of justice.

This process had particular effects on specific sites. Indeed, of the three
sites studied in this volume, until recently, two of the areas were so inaccessible that they remained marginal to both forest exploitation and
conservation. In the third area, Badar, after driving a road through the
valley, the colonial authorities first allocated an area to colonial planters.
After creating a category of *adat* land (*tanah adat*), the colonial authorities then allocated land along the Alas valley to customary uses, at the
same time reserving a large area of forest for nature conservation. This
territorialization process proved critical to later developments in Badar,
sowing the seeds for conflict when, after 1970, conservationist actors
returned to the area just as growing local populations in Badar were expanding into surrounding forest.

In essence, the colonial state forestry regime and its post-colonial successor shared the same ambition; both wished to control the revenues
derived from resources exploited in the forest estate over which the state
claimed domain control. In the colonial period this process of extending control over forest resources began as colonial officials attempted
to understand and then take over revenue generation that had been in
the hands of local *adat* leaders. As colonial officials developed forestry
ordinances, they reinterpreted *adat* rights such as the concept of "right
of avail" to the point that *adat* almost lost any effective meaning. Consequently, colonial policy here cut both ways: whereas the colonial administration set up *adat* authorities and the mapped territories over which
they were granted authority within the colonial dispensation, colonial
forest ordinances attempted to emasculate *adat* control over the only
valuable assets over which *adat* heads exercised control.

Despite these attempts by colonial administrators to co-opt *adat* authorities, emasculate *adat* controls over the forest, and reinterpret *adat*
concepts to serve the interests of the parallel state order, here, as elsewhere, the dominant colonial legal order failed to penetrate fully and
encountered pockets of resistance (Merry 1991). Notwithstanding the
extension of its legal and administrative apparatus, village actors continued to gain access to and use of resources according to long-standing
adat assumptions. Although the state rules empowered only the appropriate state agencies to regulate use of or levy taxes on state forest, colo-

nial times *adat* heads had continued (and still continue) to collect taxes on specific forest uses within the area subject to a local *adat* territoriality. Moreover, on many occasions, local actors have asserted the primacy of local *adat* property rights, attempting to ensure that, if forest exploitation is to take place, it should do so only on the terms of the corporate group concerned. In the many instances described in this book, this pattern has continued into recent times.

Reflecting on this history, it is clear that the relationship between the state and *adat* orders is more complex than that suggested by this history of dominance and resistance. For instance, in Sama Dua, when *seuneubok* institutions began to be undermined by the extension of state institutions, *seuneubok* heads were able to accommodate or even capture the state order. At particular strategic moments, *seuneubok* leaders could mobilize the state regime to support an *adat* order whose intentions were at odds with the objectives of the state forestry regime. Similarly, at other moments, such as during the resolution of village disputes, state agencies have also needed to make use of some aspects of *adat*. Despite the failure to acknowledge *adat* rights over forest areas in state law, to some degree state mapping processes have recognized the concept of a "right of avail" by attempting to leave room for the agricultural expansion of local farmers into surrounding forests. Yet, when state decisions threatened *adat* territorial claims over the surrounding area, the Sama Dua constituted a "community" capable of collective action to defend its territory. Consequently, over time, *adat* regimes and the state order have opposed one another, co-existed, overlapped, and worked together. In many ways, *adat* and state regimes have a complex relationship that involves "shifting patterns of dominance, resistance and acquiescence" that occur simultaneously (Wilson 2000).

District Networks of Power and Interest

In the course of research I discovered that a logic of exchange and accommodation between actors at the district level established the institutional patterns driving unsustainable resource extraction. Taking an actor's perspective, I have sought to understand how this logic of exchange operated.

Over a long period of time the official state budgetary system has failed to meet the economic needs of local political actors, either to maintain the district governmental apparatus, to sustain political loyalties at the local level, or to meet the personal ambitions of politico-bureaucratic actors.[5] To work around this, local agents of the state could use their discretionary powers over permits and the application of sanctions to enforce state regulations. At the same time, they could utilize these powers and their control over access to district forests to generate significant rents, both for their own personal use and for political purposes. For instance, local politicians could increase their popular support by expanding district budgets to support projects and programs that offered opportunities to clients and followers.

Of course, state agencies working with nationwide organizational rules are often unable to accommodate the interests of diverse groups and the variety of variables involved in local settings. Consequently, it is unlikely that externally imposed rules will reflect the circumstances of resource users or that they will enhance the development of effective governance regimes (Ostrom 1990). Indeed, as this study has revealed, the official administrative system created difficulties for local actors in other respects.

For example, the state forestry regime set up difficulties for local actors wishing to gain access to the forest through official legal channels. In the wake of the state terrorialization process carried out by New Order forest mappers, with the permission of the central government outside elites operated large concessions (HPH), gaining privileged access to local territory. These local territories included areas which were considered to be subject to local customary rights and included some areas where local entrepreneurs operated small-scale timber concessions. At the same time, largely at the urging of foreign conservationists, the state forestry regime locked up large areas of forest mapped for conservation. In effect, the New Order oversaw a system permitting industrial logging by corporate interests that damaged the environmental foundations that supported local livelihoods and curtailed farming livelihoods, but it also prohibited poor farmers from opening other forest areas set aside in protected areas.

In this way the formal system helped legitimize the emergence of an informal system of resource extraction that circumvented the "contradictions inadvertently created by the formal political and economic systems" (Oi 1989: 229); it also allowed for "the flexibility needed to survive the myriad rules of a bureaucratic state that has failed to satisfy the needs of its citizens" (Oi 1989: 228). On the one hand, the extra-legal system generated opportunities for local entrepreneurs who were jealous of the privileges of corporate actors with connections to the bureaucratic elite in Jakarta, wished to gain access to resource opportunities and might otherwise feel blocked by the regime. By creating labor opportunities in local forests, it offered livelihood strategies to villagers, and ways around a forestry regime that local people otherwise regarded as unjust and inequitable. On the other hand, the informal system of resource extraction served the interests of the local state apparatus, and fitted into a wider pattern. Scholars have analyzed the New Order system of governance in terms of horizontal and vertical networks of power and interest largely financed with extra-legal revenues, including those derived from the timber sector.[6] Under this system civil servants pursued entrepreneurial activities and extracted rents where possible, either for self-enrichment, or to command the loyalty of others both within and outside the pecking order, or to sustain the political interests of the institutions in which they were embedded. Within this system, politico-bureaucratic actors were free to extract rents from lucrative areas as long as they remained politically loyal to the regime and provided payments up and down the chain of command.[7]

Within the wider patterns of exchange and accommodation, local entrepreneurs were allowed to operate quasi-legal timber operations in exchange for payments and political support. If entrepreneurs and their agents within and outside the local state apparatus wished to maintain a timber operation, they had to conform to a system of exchange of extra-legal gifts and favors with certain local politicians and state functionaries. Clearly, diligent officials who contested such accommodations would face significant risks to their careers (or worse). Therefore, to survive in a district, at the very least officials needed to avoid coming into conflict with evolving networks of exchange and accommodation.

In the Menggamat case, initially it appeared that the wider district interests supporting logging had displaced local *adat* authority. However, closer examination revealed that, as external networks supporting logging entered the village, they found ways to accommodate local leaders and *adat* territorial claims. At first the *adat* authorities were not able to maintain control over access to and use of the forest within their territories. However, rather than merely watching the depletion of local forests from the sidelines, village heads entered into exchanges and compromises; loggers could cut in *adat* forests if they agreed to pay (*uang pembangunan*) taxes to village leaders along the lines of the customary *pantjang alas* tax of colonial times. In this way, village leaders attempted to reassert *adat* property rights over local forests and defend the village's "right of avail" over surrounding forest territory. In imposing taxes for the extraction of timber from *adat* territory, they were also extracting rents for their own benefit. Local villagers also participated in the logging by joining logging teams. In this way village actors enjoyed a portion of the flow of benefits from what in their view was village property. Accordingly, a wider local pattern of exchange around illegal logging accommodated *adat* assumptions and authority structures.

In Badar, the system was even more encompassing. As different actors—including entrepreneurs, *adat* leaders, pioneer agriculturalists, and the local agents of the state—entered into exchanges and reached accommodations, they effectively established a complex institutional matrix, a social field with its own rules and means of inducing compliance. In this way, these exchanges and accommodations effectively established the key institutional arrangements determining access to and use of forest resources.

Indonesians are familiar with the encompassing nature of such corrupt webs of exchange and accommodation that exist beyond the law, referring to the phenomenon as a "vicious circle" or a "devil's circle" (*lingkaran setan*). When asked to explain this term, an official volunteered that "this circle [or wheel] does not have an end: it turns constantly so that it is difficult to tell the head from the tail. They are all like devils, but it is unclear who is the boss."[8]

Because the vicious circle involved forestry staff, army personnel, and other local functionaries, the state's legal sanctions could not readily be applied. Direct, personal ties based on reciprocity substituted for the legal arrangements found in state law. As these networks grew to embrace *adat* leaders and village heads, *adat* sanctions were neither brought to bear on loggers nor villagers violating *adat* principles. Rather, as *adat* shifted to allow such practices, local *adat* authorities allowed rapid exploitation, sacrificing the long-term value of the forest for short-term gain. Yet, if control over forest access apparently slipped beyond either the *adat* or the state regulatory order, these orders continued to affect the emergent institutional arrangements. As the *adat* order shifted to accommodate loggers, and as local agents of the state used discretionary powers attained under the state legal order, both of these orders played a part within co-incident and shifting patterns of resource use. As these refracted orders interpenetrated with each other, and with the institutional order that emerged from reciprocities between actors, they helped to shape the institutional arrangements that governed access to the forests. Consequently, the legal and "illegal" domains—together with *adat* and state orders—came to constitute parts of a single system.

In the northern hamlets of the Alas valley, there were additional dimensions. Here, the surrounding forests had protected or national park status. Because forest guards (*jagawana*) were stationed in this area, village leaders could not necessarily assert territorial control over the forest on behalf of their corporate group. During the 1970s and 1980s at least, some forest guards (acting as *oknum*) took over this role, extracting rent from loggers and pioneer agriculturalists in return for turning a blind eye to these activities. Most villagers were new to the area and wished to extend their gardens, and the activities of logging networks facilitated the clearing of land for pioneer agriculture. Here, affinities between the interests of villagers wishing to improve their livelihoods through opening new forest gardens, the interests of logging networks in seeking rent, and the political and economic interests of their patrons and partners inside the local state apparatus shaped an implicit coalition of interests that extended over several political levels. These parallel interests worked together, supporting the on-going transformation of the forest frontier, while checking the activities of foreign environmentalists and national policy elites interested in biodiversity conservation.

These reflections suggest that, despite the New Order state's near monopoly on violence and policy, the "tyranny of the center" so often discussed by commentators and regional politicians since 1997 has been somewhat overstated: there has always been a gap between formal authority and control (Niessen 1999). At the district level, informal decisions were often made outside the legal framework and largely beyond the scrutiny of higher-level government agencies.[9] In other words, even at the apogee of state power under the New Order, the capacity of the Indonesian state to implement laws, especially in remote districts, was limited. Even before *reformasi* and the recent regional autonomy initiatives, in some respects the management of local resources was already localized: as long as they remained politically loyal to the New Order, district elites were allowed to operate with a large degree of independence from central government supervision.

Although these arrangements corresponded with neither legal nor *adat* arrangements as they are generally understood, their vibrancy and resilience reflected their ability to accommodate the interests of a wide circle of actors. Even so, they existed outside of any clearly accountable or transparent governance system—either understood in terms of state law or *adat*.

The predominance of institutional arrangements that emerge from reciprocities and a shared but tacit understanding among actors has significant implications. These institutional arrangements help structure the processes of competition, negotiation, and exchange within these particular settings, determining resource access and use to a significant degree. Though they stand in complex relation to *adat* and state institutional structures, they are typically exterior to the forms of accountability and transparency that are available within the *adat* and the state institutional orders. Yet, due to their very flexibility, they are able to accommodate the interests of a diverse range of actors. Given their negotiated, fluid, and constantly evolving character, they tend to be emergent rather than fixed and stable. There is something to be said for understanding such institutional arrangements as "rules in use," but to do so may risk reifying rules that can be identified and overlook the role of evolving interests and strategies.[10] Analysts will tend to misread local institutional arrangements if they overlook these institutional forms; yet they are all too readily ignored in analysis. Their very nature makes them particularly resilient and exceptionally difficult to change. As I will discuss below, they create significant problems for outside projects that choose either the state system or *adat* institutional orders as the key site for intervention.

Parallelism

In conclusion, this study has shown how institutional arrangements of a specific nature have structured action and determined resource outcomes in these localities. Institutionalist approaches may assume that a researcher can clearly locate the "rules-in-use" governing a particular resource, but the issue of power can confound the task of identifying "rules-in-use." Rules can be seen as legitimizing discourses: the ability to define or apply a rule involves the application of power and is influenced by the capacity of an actor within a specific social field (cf. Mosse 1997). There are usually competing views of what should constitute the rules-in-use in a particular setting. Actors pursue ever-changing interests as prices of commodities fluctuate, political and ecological circumstances wax and wane, and powerful outside actors intrude or withdraw. Over time, diverse actors can make different interpretative uses of parallel normative systems, including those corresponding to the prescriptions of *adat*, the regulations used by different authorities (be they sectorial laws of central government agencies or district regulations of local government), and the binding conventions generated by reciprocities between actors within a social field. At particular moments, in order to stake out or to defend a claim over a stream of benefits, parties invest in and make use of an institutional order at the level that advantages them. At the same time, other parties can contest that order at the (same or another) level where they have significant capability. In such situations, environmental entitlements—the ability to command resource bundles—tend to materialize from a process involving negotiations, accommodations, and conflicts between social actors that involve "power relationships and debates over meaning" (Leach et al. 1999). Insofar as they can be identified as rules-in-use, in situations where overlapping and competing arrangements exist, the actual institutional order will tend to be emergent, reflecting the shifting interests, ecological and economic conditions, and power relations that affect a particular context at a particular moment.

Consequently, an emergent institutional order comes into view at the intersection of parallel sets of institutional arrangements. Within this institutional order, the separate strands of underlying and overlapping but parallel normative or "legal" universes can still be seen. Yet, as these parallel orders buttress and oppose—intersect and separate—they give rise to "the shifting patterns of dominance, resistance, and acquiescence" characteristic of such complex settings (Wilson 2000). As the interests of actors shift with the changing situation, actors can seek new ways to control streams of benefits within such fluid arrangements. In circumstances where the interests of actors working across parallel sets of arrangements concur around resource exploitation (as they did in Badar and Menggamat), this emergent order proves to be a powerful force for resource depletion.[11]

As the project interventions in Leuser demonstrated, these characteristics also make sustainable natural resource management particularly difficult to achieve. In Chapter 6, I consider these project interventions related to three recent policy approaches that aim to ameliorate the situation: first, the integrated conservation and development project (ICDP) approach; second, the community-based natural resource management (CBNRM) model; and third, current experiments in democratic decentralization. Two project interventions have already attempted to apply the ICDP and CBNRM approaches in the Leuser area, and in 1999 Indonesia embarked on a decentralization program.

Epilogue: Alternative Policy Models: Lessons from Southern Aceh

If state law itself operates as a constitutive system that creates conceptions of order that frame, structure, and organize state action, the sources
of these conceptions of order are multiple. As the heirs of colonial legal
narratives, post-independent states have used them for their own purposes, retaining and even extending the conceptual models and territorial
modes of resource control they inherited from the colonial past. More
recently, transnational processes of environmental regime formation have
become interwoven with state rule-making processes. Particularly since
the 1980s, the United Nations system and the nongovernmental and
academic communities have inscribed environmental management ide-
ologies as well as sustainable development and environmental planning
discourses into various transnational policy documents and instruments.[1]
As nation-states have come to concur with these various international
environmental treaties and agreements, these transnational legal orders
have come to impinge on the way developing states have conceptualized and framed environmental and development policies. These interpenetrating processes have helped secure the legal narrative underlying
the state forest regime pertaining to nature conservation, particularly the
preservationist model of nature conservation applied in the Leuser area.

The deployment of a complex of apparatuses and institutions legiti-
mized by state and transnational legal narratives in the Leuser area forms
one chapter in this large tale. It occurred after international donors and

Indonesian state planners alike recognized the failure of state attempts to apply regulatory policy approaches to protecting tropical forest areas in Indonesia and elsewhere (Bappenas 1993; Barber 1995; KLH and UNDP 1997; Wells et al. 1992). As the discourse underpinning conservation shifted, donor agencies chose to intervene directly by supporting integrated development and conservation project (ICDP) and community-based natural resource management (CBNRM) project interventions. These interventions involved transnationally supported agencies such as the European Union's ICDP or nongovernmental agencies such as WWF. Each of these interventions involved specific ideas, practices, and normative visions, in an attempt to apply "a specific politics of the environment" that supported particular regimes of rules (Zerner 2000). In the following section, I reflect on the lessons to be learned from these attempts to support state institutional structures vested with the power to implement rules, to "revitalize" customary resource management arrangements, to find legal support for these arrangements, and to make these emergent management regimes seem reasonable or even natural and otherwise develop new institutional structures that regulate access to and use of natural resources in this area.

First, in assessing the complex issues associated with this ICDP, this chapter considers some of the principal problems working against the success of an ICDP approach that attempted to combine the use of incentive tools together with regulatory instruments to induce local communities to change their resource management practices.[2] Second, at a time when the need to find a place for customary (*adat*) orders within the unitary Indonesian state has emerged as a burning issue, I will consider questions raised by the CBNRM project in Menggamat. What are the key issues raised by efforts to devolve resource management responsibilities to "revitalized" customary (*adat*) orders? Is there congruence between local livelihoods, the management of surrounding forest areas by community institutions, and conservation outcomes? What problems are entailed in efforts to "revitalize" *adat* values and norms to serve as the basis for community-based natural resource management? What political problems are associated with such efforts? What are the implications of trying to nest *adat* rights within the state legal regime? Finally, I will consider the implications for the understanding of institutional patterns that have emerged from this study for an alternative policy formulation that is now being implemented in Indonesia—democratic decentralization.

Before proceeding further, we need to first examine the specific historical and political context in which these CBNRM and ICDP project models were formed, promoted, and institutionalized.[3] To provide this background, the next section briefly examines the development of conservation policy in Indonesia and the history of attempts by conservationists and their state partners to impose conservation regimes in South and Southeast Aceh.

Conservation Regimes

Colonial Origins

During the late nineteenth century an essentially moral movement in the West linked the appreciation of wild nature with the preservation of natural pristine areas (Callicott 1991). With the creation of the first national park in the United States, in 1873, conservationist actors successfully argued for the necessity of preserving areas of natural beauty and wildlife, combining this objective with the provision of pleasure grounds for "the benefit and enjoyment of the people" (Forster 1973: 15). In the following years, other countries established nature reserves and park systems largely following this model. In colonial contexts, typically doing so involved developing game reserves, wildlife parks, and nature reserves in which the land rights of the inhabitants were often erased (Neumann 1992, 2000; McLean and Straede 2003).

Even earlier, the observation that European colonization had wrought ecological destruction in some remote territories had created environmental anxieties. These experiences served as allegories of wider environmental disaster and influenced scientific and wider intellectual thought. As Europeans encountered cultures that had other environmental knowledges and philosophies, at times colonial policies had to reach compromises with these differing attitudes. Colonial states were also susceptible to the influence of scientific lobbies in the metropolis and in the colony itself, and colonial states also became pioneer conservationists (Groves 1996).

In the Netherlands Indies, during the nineteenth century Dutch foresters had already set up forms of state control over the teak and non-teak forests of Java that involved mapping areas classified as state forests and establishing police forces "to restrict people's access to trees and other forest products" (Peluso 1992c). Colonial foresters combined the establishment of profitable tree plantations with the conservation of forest areas for hydrological reasons, creating forest reserves as areas that could be exploited for timber but had to be left under forest cover to prevent soil erosion and other ecological damage. Then in 1889 the Dutch established the first nature reserve in Indonesia, at Cibodas in West Java;

this site was initially dedicated to scientific research rather than nature conservation (Dammerman 1929).

Even in the colonial period, however, non-state actors, including scientists, scientific organizations, and environmental pressure groups, provided the major impetus for environmental regime formation (cf. Litfin 1993). Already in the early decades of the twentieth century, some Dutch residents of the colony worried that "the primeval forest" was disappearing at an "alarmingly rapid rate" under the pressure of "unrelenting cultivation" (Dammerman 1929: 22). In July 1912, Dutch preservationists founded the Netherlands Indian Society for the Protection of Nature, an association devoted to preserving wildlife and to promoting "the institution of nature reserves" (Paulus and Stibbe 1921; Dammerman 1929: 22).[4] Influenced by American conservationist models, the society applied to the colonial authorities for grants of large pristine natural areas whose uses would have to now accord with conservation. In 1916 the colonial government enacted an ordinance that allowed for the establishing of nature reserves in areas "which for the sake of their special scientific and aesthetic value, would as a matter of public interest, be kept as intact as possible. In such nature reserves it is forbidden to collect plants or animals, to hunt, to keep cattle and in general to perform any action by which the existing natural conditions are altered" (Dammerman 1929: 23).[5] This was the beginning of state policies that construed particular areas as pristine wilderness areas to be set aside for conservation and applied regulatory policy approaches (or authority tools) of exclusion for nature protection. This conservationist model involved closing off access to areas set aside by state planners, criminalizing encroachers, and largely neglecting the adjacent area's socioeconomic dynamics or demands (Barber et al. 1995). Following further submissions from the Society for the Protection of Nature, by 1929 the Netherlands Indies could boast of seventy-six nature reserves, fifty-five of them situated in Java (Dammerman 1929: 24). In addition, this society and other pressure groups advocated the gazetting of forest reserves in Sumatra to prevent the extinction of the unique wildlife found there. This effort culminated in the creation of nature reserves in the Leuser region, a mountainous and remote forested area, one of the last areas of the archipelago subjected to colonial rule.[6]

Conserving Leuser

The Leuser area of Aceh fell under Dutch control at the end of the Aceh war, and during the first decades of the twentieth century, colonial planters converted many surrounding areas to plantations and other uses.[7]

To the southeast, the Dairi area of Northeast Sumatra became famous for its tobacco and rubber plantations. Around Takengon, in the Gayo highlands to the north, the Dutch established estates, transforming forest areas into plantations (Bowen 1988). In comparison, the mountainous area of Southern and Southeast Aceh was somewhat peripheral to colonial commercial ambitions. Although planters opened some rubber estates in the Alas valley, these plantations were not particularly successful (Berge 1934).

In the late 1920s, the colonial government issued a permit to the geologist F. C. van Heurn to explore for mineral and oil resources in southern Aceh. During van Heurn's expedition, local leaders sought out van Heurn to discuss his findings. In subsequent meetings, the local leaders expressed their fear of "a permanent colonial invasion to exploit mineral resources" (Rijksen and Griffiths 1995: 37). Van Heurn found that the Gayo people living to the north of the Alas valley in Gayo Lues considered the mountain area around Gunung Leuser a sacred site: ancient Gayo myth held that the mountain itself was "interpreted as the link between heaven and earth" and the "primeval forest-abode of their ancestral spirits" (Rijksen and Griffiths 1995: 37).

Van Heurn failed to find deposits of any significance in the area, and he decided to support the efforts of local leaders to protect the area. As a naturalist, he thought of obtaining nature reserve status for the area. For their part, the *adat* leadership could make use of the colonial conservationist discourse: via a selective allegiance with Dutch conservationists, they might be able to protect local resources from outside exploitation. In the Netherlands, on 29 August 1928, van Heurn submitted a detailed proposal for creating a reserve to a meeting of the Netherlands Committee for the International Protection of Nature Conservation. The proposed "Gayo and Alas Wildlife Reserve" would include 928,000 hectares covering all types of terrain, "from the coast to the highest mountains" (Strien 1978: 19). The area "would be ideal for all of Sumatra's wildlife and provide a cross-section of the fauna and flora of northern Sumatra" (Strien 1978: 19).

In 1929, the committee submitted a proposal to the colonial government. Initially the colonial government saw no reason to protect an area of virgin jungle that was under no particular threat. Eventually conservationist interests prevailed: finally in July 1935, the governor of Aceh signed the document establishing a reserve that enclosed both sides of the Alas valley but excluded "virtually all known lowland rainforest areas."[8] This land was considered "inappropriate for protection in the light of possible future claims for development" (Rijksen and Griffiths 1995: 39).[9]

At this time a small population lived among the extensive forested mountains, residing for the most part in villages located near the coast or along small stretches of rivers. Forest resources were still abundant, and the original boundaries left sufficient land outside the reserve for opening new agricultural lands (Koesnoe and Soendari 1977; Strien 1978). In any case, the Dutch forestry service had few resources to invest in monitoring resource use in areas with little commercial value to them (Poffenberger 1990: 15). Traditional uses—such as the collection of forest products—continued within the reserve (see Chapter 4). Consequently, it is unlikely that the designation of the area as *Wildreservaat Goenoeng Leuser* at this time significantly affected local patterns of resource use. Nevertheless, the creation of state forests set aside for conservation—albeit with the formal assent of the local heads—set the scene for later conflicts.

The Re-emergence of Conservationist Ambitions and the Advent of ICDPs

The Second World War and the Indonesian Revolution disallowed any opportunity for conservation activities. However, during the post-war period, expanded scientific interest in ecological issues worldwide corresponded with the growth of international NGOs and conservation agencies. As the 1972 United Nations Conference on the Human Environment attested, international development agencies began to see sound environmental management as a key component of development. This agenda was expressed in interstate treaties and later by intergovernmental organizations working on environmental issues. Over time, developing nations participating in international meetings and fora—such as those involving the United Nations system—became parties to these treaties and at least formally acceded to this agenda. The international agencies supporting it generally had a technocratic orientation towards development that saw environmental management as a responsibility of state bureaucracies. In countries like Indonesia, these agencies provided financial and other support for initiatives in universities and state agencies for improving environmental management (Mayer 1996). Over time, states came to employ the language of sustainable development, environmental protection, and biodiversity conservation in their policies and laws.

Within developmental and conservationist discourse, the creation and management of a national park and protected area system became an essential element in a nation's development planning. With the growing perception of a global environmental crisis, the designation of a reserve system became a condition for international support for conservation programs. Consequently, the nations of the developing world tried to

lock up rich areas of biological diversity in protected areas. The idea
was that, in the midst of a sea of underdevelopment and rapid economic
growth, biological diversity could be preserved by setting aside large islands of wild land for nature conservation. And so the number of national parks in developing countries mushroomed. By 1978, there were
fifteen hundred national parks or equivalent protected areas in more than
one hundred countries (Quigg 1978).

Over the next decade, the issue of tropical rainforest conservation
shifted to the front of the international stage. Scientists projected that the
rapid loss of biodiversity would lead to the disappearance of up to half
of the world's species (Soule and Sanjayan 1998). As the tropical forests
contained the majority of the world's terrestrial and freshwater biodiversity, conservationists began to focus especially on the loss of biodiversity
associated with tropical deforestation. The World Bank commissioned
a major task force on this issue, and leading conservation organizations
published volumes of new research. At the same time, a range of international programs were also established to address this issue. Governments, environmentalists, and development agencies began to explore
more intensely a variety of approaches to curb the loss of tropical forests
(Bowles et al. 1998). In response to this crisis, international commissions and nature conservation organizations called for the "near-term
protection of at least 10 or 12% of the total area of each nation or in
each ecosystem" (Soule and Sanjayan 1998: 2060). If these efforts were
successful, this campaign would double or triple the area designated as
national parks or strict reserves.

Efforts to conserve the Leuser area re-emerged in parallel with these
developments. In the 1960s, under the aegis of international agencies
(including IUCN, WWF, and the Netherlands Commission for International Nature Protection), international scientists had begun a series of
ecological surveys in Leuser to "investigate the status of orang-utan and
Sumatran rhinoceros" (Strien 1978). These activities led to the building
of an orang-utan rehabilitation station in the Alas valley (see Chapter 4)
and the preparation of a management plan for the reserve complex.

While at this time Indonesia had yet to create a national park system,
moves were afoot to renew the forest conservation project of the colonial period. Finally, on 6 March 1980, the efforts culminated when,
by joining together several of the Leuser nature reserves gazetted in the
1930s, the Indonesian government created Indonesia's first national
park, the Gunung Leuser National Park (Wind 1996). Subsequently, Leuser became the World Wildlife Fund's single largest project in Indonesia
(Rijksen and Griffiths 1995: 43). Working with the predominantly technocratic, ecological approach to conservation prevalent at the time,

WWF funded surveys, helped develop management plans and ecotourism and education centers; helped to equip and deploy forest guards, and supported orang-utan rehabilitation and ecological research.[10]

By the 1980s, it was clear that the conservation programs in Leuser had failed to forestall deforestation. Despite their "protected status," the Leuser rainforests were deteriorating. In 1989, IUCN declared GLNP one of the ten worst managed parks in Asia (Wind 1996: 4).[11] The problems faced in the Leuser area exemplified the predicament resulting from over-reliance on authority tools—the so-called "fines and fences" approach to conservation.

The concept of integrated conservation and development projects (ICDP) emerged as a key policy model during the 1980s at a time when participatory approaches to development planning prevailed (Chambers 1983; Whyte 1984; Little 1994).[12] Despite variation in terminology and subsequent changes in conservationist thinking, the ICDPs generally reflected a transnational policy formulation that depended upon a number of key assumptions. It represented an attempt to draw together the goals of human development and poverty alleviation on the one hand and biodiversity conservation on the other hand. By these means conservationists were able to draw on the public and private financial support for both development and conservation (Mogelgaard 2003). It followed the recognition that previous regulatory approaches had raised social justice issues: by criminalizing established resource uses but failing to deal with social disadvantage, such regulatory approaches took away time-honored resource rights and left communities living near protected areas to bear disproportionate social costs. By alienating and impoverishing neighboring villages, these policies lacked legitimacy and provoked resentment and resistance (Wells et al. 1992).

The ICDP approach also accorded with the neo-liberal view that tended to see the failure of biodiversity conservation policy in terms of institutional, market, and policy failures (K. Brown 2002). To address these various failures, ICDPs have called on a suite of measures. Working on the premise that some appropriate set of incentives exists to induce local communities to change their practices, ICDPs deploy "incentive tools" and "socio-economic investment tools" to encourage compliance to pre-established conservation goals. Seeing a linkage between the loss of local livelihood and the conservation of biodiversity within adjacent areas, ICDP approaches typically attempt to link conservation and development goals by compensating local people for lost access to resources with limited access to certain resources and/or by investing in alternative income-generating activities in "buffer zones" (K. Brown 2002).

ICDPs have also assumed that policy failure in the area of conservation is at least partly due to incapacity of state agencies in developing countries to pursue protected areas policies. Accordingly, project interventions have tried to improve park management (Brandon and Wells: 1992). As in the case of the Leuser ICDP, they have attempted to help state agencies to more effectively use regulatory tools, advancing territorial conservation strategies and even extending these into adjoining areas, now reclassified as buffer zones.

In Indonesia, following this shift in international conservationist discourse, state planners allowed that state agencies lacked the human and financial capacity to manage the vast state "forest." Policy documents from this period re-examined the role government could play in conserving biodiversity, advocating a raft of policies involving varying levels of state and private provision with differing levels of involvement from other actors.[13] For instance, one key policy document, *The Biodiversity Action Plan,* invited international donor agencies to assist with priority conservation activities and encouraged donor and NGO activities to support policy goals, envisaging that state agencies would primarily encourage, stimulate, and coordinate donor activity (Bappenas 1993; NRMP and Bappenas 1994).

By the late 1990s, ICDPs had proliferated across the world. By 1996 there were already more than fifty ICDPs in twenty African countries (Newmark and Hough 2000). By 1999, foreign donors had pumped US$ 130 million into ICDPs covering some 8.5 million hectares of Indonesia's "conservation estate" (Wells et al. 1999: 2).[14]

Lessons from the ICDP Intervention

During the late 1980s the interest of international donors—including the World Bank, Asian Development Bank, European Community, and USAID—began to converge on Leuser. It was one of the most extensive tracts of forest left in Southeast Asia and an area of outstanding biodiversity, and donor agencies identified it as a possible site for an ICDP.[15] Government to government discussions between the European Union (EU) and the Indonesian government during and after the UNCED summit at Rio de Janeiro, in June 1992, developed the Leuser ICDP plan further.[16] At the end of a planning exercise that produced the *Leuser Development Programme Masterplan,* in May 1995 the European Union agreed to provide US$ 40,625,000 over a seven-year period to fund this project; the Indonesian government also committed US$ 22,500,000 in local currency equivalent from the Ministry of Forestry's Reforestation Fund (The Republic of Indonesia and Leuser Development Programme 1995: 6). The Leuser project became one of the largest ICDPs in Indonesia.

The Leuser project's design document presented a seemingly far-ranging critique of state forestry activity in the Leuser area. This critique saw the failure of state policy as stemming from the way the protectionist paradigm had been implemented; it thereby absolved the protected area concept itself and the assumptions guiding state resource policy from reproach. The thinking of the ecologists who designed the ICDP was firmly rooted in a Western ecological tradition that tended to exclude humans from the study of ecology. Thus, in the project's design, the "core area" of the park was viewed as a pristine natural area free of defiling human touch, endorsing a preservationist approach to conservation that regarded local uses of the national park as "encroachment."[17] This project design concurred with the assumptions of the New Order's conservation policy, and the logic of the Leuser ICDP replicated key aspects of the state policy regime.[18] The ICDP design envisaged developing an actor—the Leuser Management Unit (LMU)—capable of stepping in where the state forestry agency responsible for the area had so far been unsuccessful. In this way, the project hoped to remake the forestry regime that protected the biodiversity and other ecological values set aside by state policy. With LMU either taking over the role of the state or otherwise working in partnership with state agencies, the Leuser ICDP involved trying to intensify state resource control strategies. This approach subsequently led to LMU's attempts to articulate state territorialization more clearly and apply it more effectively. For example, LMU created a new zoning system for the Leuser Ecosystem, supported by intensified para-military law enforcement. While the Leuser National Park encompassed around 905,000 hectares of mountainous forests, the Leuser Ecosystem territory extends over 1.8 million hectares of the surrounding area. In this way, the ICDP involved extending state resource control strategies.[19]

Although in its specifics the ICDP differed from earlier interventions, the continuities which the project shared with projects of the colonial period were striking. The colonial conservation project involved an examination of indigenous practices and property claims, establishing boundaries between newly created *adat* areas and state forest lands. Now, based on a study of extant colonial documents regarding this territorialization, the ICDP process extended the colonial territorialization into a more extensive Leuser Ecosystem. Thus, in a more intensive fashion, a conservationist intervention again began to impose a territorial strategy on an extensive area of Leuser.

While the issue of legitimacy was one of the factors leading to the shift from earlier regulatory approaches to more comprehensive ICDP approaches, ironically a legitimacy problem haunted the Leuser ICDP as well. The project derived from an agreement between a multilateral donor and the Indonesian government at the highest level. It reflected international conservation norms inscribed in international agreements and national conservation laws. On this basis, LMU assumed a strong authority for its activities. Even so, for the project to avoid provoking conflicts with local villagers, it also needed to derive its legitimacy locally during project execution. To do so, the project needed to work with local socio-legal understandings, including the village regimes that presided over the long-established property rights underlying village livelihoods. While in many areas local *adat* regimes prevailed to various degrees, these regimes were practically "invisible" to the state legal order. The large gap between the state natural resource regime—in which the Leuser ICDP was embedded—and the reality of many forest practices meant that all too often state laws would remain "so many black ink markings on paper" (J. Griffiths 1995: 213).

The Leuser ICDP was not based on local consideration of local problems: village understandings did not support locking up forest resources in the Leuser Ecosystem or national park. The ICDP management compounded this initial shortcoming by not thoroughly engaging with *adat* arrangements, which are invisible in state forest plans. Rather than negotiating nature protection and human development goals in context, the Leuser ICDP involved imposing *a priori* conservation and development aims through eliciting "participation" in the field.[20] Here, the Leuser ICDP tended to repeat the inadequacies of other conservation projects that engaged in "participation," but only as a part of a top-down management process that tended to involve people in passive forms of consultation rather than as active agents (cf. K. Brown 2002). This so-called participation only softened what remained a top-down approach that involved the continued subjugation of local values and concerns. Villagers remained removed from decision-making and policy-making processes, and the totality of their concerns, values, and systems of knowing remained peripheral to the management process. Meanwhile, professionally trained ecologists and resource managers (Indonesian and foreign) tried to set the parameters for management of local areas, without addressing power imbalances or conflicts. As a result, the project failed to establish local legitimacy.[21]

Law enforcement against local "encroachment" into the forest focused on poor farmers and loggers, for the most part ignoring the powerful sponsors behind logging networks resident outside villages. Soon the Leuser ICDP faced resistance from a wide range of village and district actors. Working against village understandings, LMU found itself associated with repressive state territorial strategies. LMU's structural dependence on law enforcement agencies—whose enforcement activities all too often were refracted by the collusive relations between law enforcement officials and local clientelist networks, where the real district power lay—compounded the legitimacy problem. This problem was most apparent in the Alas valley, where a wide group of local actors consistently challenged the legitimacy of the state forestry policy that enclosed the national park, arguing that it did not sufficiently accommodate local interests. In the face of resistance from such a wide range of actors, for the most part the LMU's implementation of state policies and laws protecting the forest became fruitless. Soon LMU found itself involved in the same kind of failure of regulatory tools that the original project design document had so much lamented.

The use of incentive tools by an ICDP depends on the assumption that if the project provides some form of benefit to the "community," reciprocity between a community partner and the project will lead to desired policy outcomes. In this vein, the Leuser ICDP entered into "quid-pro-quo agreements" with villages. The process of creating these agreements with local partners occurred in authoritarian circumstances distinguished by a large power imbalance between an international project and its state partners on the one hand and local leaders and villagers on the other. LMU worked with state agencies that chose local leaders to implement micro-projects in the clientelist context of a political system that failed to provide representative and accountable local leadership. All these factors affected the effectiveness of these agreements.

In addition, given the size of the population in the Leuser Ecosystem and the scale of the development problems involved, the benefits LMU micro-projects offered could scarcely be compared to the opportunity costs forgone by villagers who might be shut out of the Leuser forests. With socio-economic fluctuations driving forest pioneering and logging, the market incentives for unsustainable resource extraction intensified. As other researchers have noted, the assumption that improving living standards of adjacent communities will enhance conservation remains contentious (Wells et al. 1999). The provision of public goods does not necessarily lead to changes in private behavior, especially if the aspirations of villagers point in the other direction (Newmark and Hough 2000). Clearly, villagers aspired to better livelihood and saw that they could achieve this through involvement in the market, particularly through the expansion of commercial tree crops.[22] Consequently, because communities failed to benefit significantly from ICDP micro-projects, LMU could not provide sufficient incentives to support sustainable livelihoods or to alter attitudes and behavior on a significant scale.

Often, ICDPs have overlooked external threats associated with vested interests and government policies (Wells et al. 1999). In contrast, LMU attended to state policy, regional planning, and state institutional development. Making use of its political ties at the apex of the state, LMU tackled the direct threats arising from the granting of concession licenses, development planning, and investment decisions taken elsewhere. On a number of occasions LMU persuaded the central government to rescind development decisions that would have hastened the extraction of unsustainable resources while not necessarily benefiting villagers. LMU could affect spatial planning, public investment decisions, and development coordination by mustering support among powerful state actors against projects supported by other actors. What success LMU achieved in this area supported the observation that, despite numerous disappointments, a particular component of an ICDP may achieve a degree of success (Hughes and Flintan 2001).

Nevertheless, despite its dedicated staff and its large resources, over the period covered by this research the Leuser ICDP was incapable of fundamentally changing the situation. LMU could neither apply incentive tools to induce local communities to change their practices nor determine access to and use of forest resources by applying regulatory instruments. The lack of change in the culture of corrupt public institutions involved in resource extraction compounded these problems. Finally, since LMU could not find compatibility between ICDP objectives and the aspirations of village actors, it was unable to alter the village-level institutional arrangements that worked against conservation aims.

The problems faced by the Leuser ICDP were not unique. A review of ICDPs in Indonesia concluded that "under current conditions . . . very few ICDPs in Indonesia can realistically claim biodiversity conservation has been or is likely to be significantly enhanced as a result of current or planned project activities" (Wells et al. 1999). This report laid the blame on the managerial and technical problems facing implementation, including "flaws in basic assumptions and planning, and a failure to address the real threats and capacity constraints that conservation projects face in the field," as well as "broader constraints" such as lack of support within Indonesian political circles and the wider society.[23]

The report concluded that "the major problems do not seem to lie with the ICDP concept itself" (Wells et al. 1999). Yet, this conclusion begs a critical question underlying this first generation of ICDP interventions of the Leuser variety.[24] These earlier conservation projects involved strategies of resource control that were built by ecologists on colonial modes of resource management. They were driven and funded by outsider actors who justified the need for biodiversity conservation interventions according to an international conservation discourse. Over a long period of time, these actors had etched conservationist norms into international conventions and national law, eventually overseeing conservationist strategies of territorial control. They were also constrained by the confined political and legal space allowed by the authoritarian New Order. Irrespective of the worthy intent of those advocating them, project interventions based on this foundation typically fail to obtain legitimacy at various social and political levels—apart from eliciting local "participation" in externally derived program goals. Consequently, this particular project ended up pitting the conservationists and their allies (such as they were) within the state against the needs and interests of villagers and local elites. Given the constraints state agencies have long faced in achieving policy outcomes in such contexts, the question remains: How can conservationist strategies of this type ever hope to succeed?[25]

Returning to *Adat*:
The Community-Based Solution

The ICDP label has been attached to a wide variety of project approaches. With many projects varying from the Leuser ICDP, some even concurring with the community-based natural resource management (CBNRM) policy model (Mogelgaard 2003). The CBNRM policy model holds that local populations have a greater interest in the sustainable use of resources than does the state or distant corporations because, among other things, local communities better understand the intricacies of local ecological process and practices and are better equipped to manage these resources through local or traditional forms of management (Brosius et al. 1998).[26] The assumption is that the more a community depends upon the surrounding resource base, the more incentive they will have to protect it (Lynch and Talbott 1995). In a similar vein, it is sometimes presumed that, although a connection between sustainable use of resources and local livelihoods once existed, it has broken down under the impact of market and state.[27] CBNRM approaches try to be participatory, aiming to re-establish a link between local sustainable livelihoods and biodiversity conservation.

Although specific interventions vary, typically CBNRM projects work to strengthen local organizational capacity, recognize the value of indigenous resource management systems, provide tenure security, encourage the development of environmentally sustainable livelihoods, and devolve power and responsibility for natural resource management decisions to the local level (Utting 2000). As these approaches try to deal with the

Epilogue

distributive effects of natural resource policy, they entail an explicit environmental justice agenda.

At the same time as the Leuser ICDP planning process, the World Wide Fund for Nature also decided to continue its earlier work in the Leuser area.[28] In keeping with the CBNRM philosophy, the World Wide Fund for Nature Leuser Project (WWF-LP) took a more "participatory, multi-stakeholder, NGO-focused approach" (Barber 1997: 30). As I discussed earlier (see Chapter 3), this approach involved utilizing new policy initiatives that provided limited opportunities for basing resource management on local property rights and using customary *adat* regimes as a basis for creating new co-management models. In this section I consider a series of lessons drawn from the Leuser CBNRM experience.

First, I wish to consider problems arising from efforts to "re-establish" the link between the development of local livelihoods, the management of surrounding forest areas by community institutions, and conservation outcomes. In choosing a strategy in Menggamat, the renowned *damar* garden system of Krui, South Sumatra, helped inspire WWF-LP to develop a community conservation forest (CCF). As described by Michon et al. (2000), in Krui a highly specific and parallel set of agro-ecological and institutional transformations led to the emergence of a successful Krui agroforestry model. The agro-ecological transformation occurred over a long period of time as Krui farmers developed agricultural techniques and cultivation patterns, learning to cultivate *damar* trees in an extensive plantation that resembled a natural forest. The parallel institutional transformation involved changing a common property system—a community-owned forest—via "controlled privatization" to a forest plantation. Under the Krui system, *adat* institutions both guaranteed individual rights and restricted these rights in a fashion that secured the continuity of the agroforestry domain. Similarly, in Sama Dua during the successive shifts from pepper to clove to nutmeg cultivation, the establishment of income-generating tree crops and the eventual emergence of forest-like agro-ecological systems involved corresponding changes in agricultural strategies and institutional arrangements.[29] In other words, the development of these systems occurred in particular historical, socio-economic, and agro-ecological contexts that required changes contingent on a diverse range of highly site specific factors.

In Menggamat, WWF-LP and YPPAMAM could point to the stark contrast between relatively intact secondary forest areas under *adat* management in the core Menggamat area (*hutan adat*) and the highly degraded state forest area (*hutan negara*).[30] The WWF-LP and YPPAMAM used this contrast to argue that *adat* management should be extended to the state forest. Yet, as in Krui, the reasons for the relatively intact *adat* area were rather more complex than a simple contrast between *adat* and state management regimes might suggest. In Menggamat a range of factors contributed to the survival of the *adat* forest in steep areas close to the villagers. For instance, villagers with gardens in this area would not allow logging teams to cut in steep forests in a fashion that would damage their crops, and loggers preferred to work in the state forest, where large specimens of valuable timber species could still be found.

While the role of *adat* institutional arrangements in Menggamat and elsewhere is pertinent to positive outcomes in such areas, these outcomes cannot simply be attributed to the ecologically sustainable nature of indigenous *adat* regimes. Putting CBNRM assumptions aside, the particular circumstances in which positive outcomes may emerge need to be carefully explored (Eghenter 2000). Given the highly contingent historical, economic, and social factors leading to particular outcomes, it may be difficult for project interventions to engineer the particular agro-ecological–institutional elements underlying successful regimes.

Second, CBNRM raises a series of problems regarding the "revitalization" of *adat* institutional arrangements. These are particularly significant at a time when, following the end of the authoritarian New Order regime (1966–98), local groups have invoked *adat* customary rights in a large number of struggles over natural resources. Following the implementation of decentralization laws, in many areas there have been efforts to reinstate *adat* as an alternative source of meaning and legitimacy for local institutions. Simultaneously, a national movement of NGOs and other reformist elements has been calling for new laws that more closely reflect the customary (*adat*) order of local communities (Benda-Beckmann and Benda-Beckmann 2002; Warren and McCarthy 2002; Acciaioli 2000; Li 2001). In 2001, a formal decree of the Indonesian supreme parliament (MPR) called for the reform of natural resource and land tenure laws and policies in accordance with principles that "recognize, respect and protect the rights of *adat* law communities." While the decree may have failed to satisfy many reformist groups, the highest legislature had finally responded to the re-emergence of the *adat* as an alternative institutional order.[31]

While "*adat*" can take on a coherent identity in these discussions, the term itself has been applied to a wide range of varied institutional arrangements found among the diverse indigenous populations of the archipelago. Moreover, the precise nature and status of *adat* are associated with long-standing problems within the Indonesian polity that resonate back to the unsettled arguments of Dutch colonial policy (Kahn 1993; Burns 1999). These unresolved problems concern how we can best understand *adat*, the precise role *adat* plays in land tenure and natural resource management, the imputed role of the *adat* order as a foundation or impediment for national development, its status as "*adat* law" (*adatrecht*), and the relation of *adat* institutional orders to the state order. WWF-LP's efforts to build conservation according to "revitalized" *adat* principles raises many of these same issues.

WWF-LP's community-based conservation intervention had to find a site for intervention that would sustain an interpretation of *adat* and could justify a conservation agenda. In finding this site, the project needed to develop a cultural and legal reading of local institutional practices that rationalized CBNRM practices: this rationalization needed to be based on *adat* elements found in the site. These *adat* elements could then be reworked in a fashion that suited their strategic deployment in the service of conservation (cf. Zerner 1994).

In colonial central Africa, the creation of a codified "customary law" involved interpreting custom. This required crafting rules out of a fluid system in which the kind of rules and the way they applied depended on circumstances. Rather than merely discovering rules, doing so required the actual creation of rules. The rules created under this process reflected the anxieties of local witnesses and the moral predilections and administrative purposes of the colonial officials involved (Chanock 1982).

The legalization of custom entailed taking some local claims of what might be considered "local law" and transforming these into legal rules, or "customary law." This transmutation of indigenous mechanisms of social control into the distinct legal mode of control created by the colonial state involved a "crucial shift" (Chanock 1982). This shift required legitimating some representations of a local "legal order" and effacing others.[32] In a similar fashion, in Menggamat "revitalizing *adat*" involving simplifying and selectively presenting *adat* within a process shaped by the predilections of a conservation project. While appearing to take recourse to the past, this process actually involved the selective redefinition and instrumental use of *adat* norms—reinterpreting the *adat* of the past to assert a claim regarding what should occur in the present (cf. Chanock 1978).

In the colonial period in Indonesia, colonial *adatrecht* scholars had collected, organized, and textualized *adat* into a well articulated body of rules that accorded with Western legal thinking (Zerner 1994). In the sites and circumstances where this took place, it led to an *adat* law that was different from that "produced and reproduced in the villages" (Benda-Beckmann and Benda-Beckmann 1985).[33] In some sites, such as the colonial Malukas, this reinterpretation of *adat* served to control the production of valuable commodities (Zerner 1994). Just as the *adatrecht* scholars of colonial times described and appropriated *adat* for specific economic and political needs, the conservationists simultaneously sought to appropriate *adat* for their strategic purposes—to sustain a political movement supporting community-based management of natural resources. Beyond this cultural and legal reading by colonial or conservationist codifiers, however, *adat* arrangements regarding local resource use develop in association with specific conditions and are subject to change, dispute, and renegotiation.

If the "crucial shift" were to take place successfully, it would transform custom in a specific sense. Oral traditions allow for variability in terms of the process of transmission: there may be no unique version or exact codification of an oral tradition, but rather "a structured ensemble which tolerates, and even favours, a form of creativity" (Rouland 2001: 15). In contrast, writing freezes a customary process. While change remains possible, it takes place "only though a chronological accumulation of textual interpretations" that occurs as texts are reorganized and choices are made among them. Subsequently, customs can continue to develop, but "under the influence of lawyers and the powers that they serve, they move away from those that are supposed to observe them." Ironically, however, such processes of institutionalizing custom risk "rendering it obsolete" (Rouland 2001: 15).

If WWF-LP's efforts to restructure *adat* in accordance with state law and conservation norms had succeeded, it would have involved a similar transformation, this time under conservationist hegemony. Because avowedly the intention was to create a more sustainable basis for resource management, from a conservationist perspective this selective reading might be defended against its alternative: the on-going degradation of the local resource base and the watershed forests that provide ecological services to surrounding villages. Nonetheless, it did raise particular problems.

The problem for *adat* interpreters and codifiers was that *adat* is flexible and somewhat ambiguous. Elements of *adat* could be interpreted to allow actors to work in contrary directions: *adat* norms can be used to support and resist exploitative behavior at the same time. In seeking an indigenous logic upon which to found a community conservation initiative, the Leuser CBNRM project conceived *adat* "through a green lens." This version of *adat* necessarily left aside village *adat* practices that failed to fit the need for an *adat* that legitimated the CBNRM agenda (cf. Eghenter 2000). In looking for ecologically sustainable *adat* management practices, the CBNRM analysis and the consequent intervention tended to overlook the history of *adat* allowances for commercial exploitation, the foundations in *adat* for opening agricultural

plots in the "protected" forest and allowing highly exploitative practices such as logging.

In developing a conservationist *adat* that legitimated the CBNRM
agenda and that could be strategically deployed in the service of conservation, project workers and their local collaborators produced and then
promoted a particular form of *adat*. This version of *adat* didn't reveal the
contingent nature of the links—such as there were—between sustainable
village livelihoods and local economic and agro-ecological change. As
in the situation described by Eghenter (2000), this interpretation could
have all sorts of implications for project design and implementation. For
instance, a more historically accurate and broader interpretation of local
institutional arrangements might have revealed the historical precedents
for unsustainable resource use practices and the continuing relevance
of earlier *adat* assumptions for exploitative resource use patterns in the
present. Perhaps if they had read *adat* in this way the CBNRM workers
would have been more cautious about founding a conservation strategy on such a narrow understanding of *adat*; perhaps it could have also
helped project workers develop a more inclusive strategy.

The CBNRM approach made strategic use of the concept of "community" to invert earlier policy models, overturning many of the problems
associated with the state management paradigm (Li 2002). Yet, ironically, a third set of issues emerge as projects attempt to base project interventions on what are perceived to be community values and norms.
In contrast to the idea of a "community" as a group sharing values and
norms, the idea of community as a group living in a particular locale—a
"community of place"—allows for a plurality of diverse interests. These
differences and the conflicts that arise from them can be attributed to a
range of social, political, and economic variables found within a "community of place." Since actors come together around different values,
perceptions, and objectives, they can also form factions, or communities
of interest that vie to influence decision-making processes (Agrawal and
Gibson 1999; Natcher and Hickey 2002). Conservation projects inevitably have particular interests; because these interests are compatible with
those of a particular party within a "community of place," CBNRM
projects find themselves working with some interests and against others. In such an atmosphere of conflict, "one person's helper is another
person's enemy" (Roush 2001). In this fashion, CBNRM projects can be
drawn unwittingly into intra-village disputes.

WWF's initiative in Menggamat gained the support of some village
leaders interested in reinvigorating rules derived from elements of *adat*
and extending them to a newly constituted community forest. When
WWF-LP and their local partners sought to apply the conservationist

interpretation of the local socio-legal order, however, doing so involved imposing a codified version of *adat* in the community conservation forest to regulate resource access in this area. Other actors involved in logging constituted a "community of interest" that contested the CBNRM agenda and its interpretation of *adat* because that agenda worked against their interests. In essence, the implementation of the CCF involved a struggle over the power to control access to local resources.[34] Inevitably, this was also an implicit struggle over the nature and place of "customary law" in the political order, a struggle over "the power of definition"—that is, the power to define the particular version of the essentially negotiable and fluid concepts of *adat* that had gained legal recognition (cf. Oomen 2002). When WWF-LP and its project partners tried to get reality on the ground to work in accordance with this "revitalized" *adat*, they found that while *adat* precepts may be amenable to reformulation on paper, actual understandings within Menggamat were not so easily changed.[35] YPPAMAM and WWF-LP had attempted to found the CCF initiative on community values and norms, but now elements in the villages that supported logging failed to cooperate. Ironically, at one point when the attempt to impose this "revitalized *adat*" met with resistance, advocates of the CCF found themselves trying to invoke state law against those re-calcitrant village elements who refused to accept the CCF initiative.

Community pluralism can be difficult to manage, and this problem can affect project interventions. CBNRM projects like to oversee a multi-stakeholder consultation process leading to a consensus supporting con-servation aims, but if they are drawn into intra-community conflicts, they become just another party in conflict. Narrow interpretations of customary systems and essentialist assumptions regarding community do not provide the conceptual basis to manage "community" differences. Moreover, when a community of interest intent on winning chooses not to participate in a consensus-orientated community-based effort, it is dif-ficult for participatory approaches to proceed (Roush 2001). Further-more, if project interventions elicit only local participation to support pre-set CBNRM agendas, they may include factions whose interests are supported by CBNRM agendas but exclude other actors.

Fourth, CBNRM approaches assume a certain congruence between participatory project methods that help to build village livelihoods and recognize *adat* property with biodiversity conservation. Yet, it is impos-sible to avoid the paradox at the heart of community-based natural re-source management: CBNRM initiatives aim to be "primarily controlled and initiated from within a community" (Lynch and Talbott 1995). In seeking to obtain community support for conservation objectives, they gain much of their legitimacy vis-à-vis other approaches from the claim that they are based on local aspirations. However, CBNRM strategies are usually initiated by outside agencies, who seek to obtain community support for conservation objectives that are exogenously derived (Murphree 1994). In some circumstances villages may come together around CBNRM goals. Alternatively, CBNRMs either end up being exercises in "guided participation" or, alternatively, if community-based approaches are truly democratic, participatory, and equitable, they may lead to decisions that work against conservation (Utting 2000; Roush 2001).

The definition of community problems by conservationist actors according to exogenously derived agendas may conflict with local community perceptions regarding the challenges to local livelihoods. Local problems may be open to local solutions that are not congruent with CBNRM objectives (Li 2002). For example, the aspirations of village actors in the Alas valley to convert natural forests to gardens for cash crop production could not readily be reconciled with CBNRM agendas that sought to retain natural forest.

In supporting the revitalization of customary institutions, CBNRM approaches involve devolving management and decision-making to customary and village authorities. In presenting a positive representation of community and community rights, these interventions aim to further the political and cultural autonomy of local communities and support local struggles to gain local control over resources. Thus, CBWRM approaches face a fifth related issue, that of working with local leaders and customary heads who indeed represent their "communities" in an accountable fashion. In Aceh, *adat* heads gained their legitimacy from their ancestry and reputation, from their knowledge of *adat*, and/or from their ability to offer services, to provide leadership, and to resolve disputes within their villages in a fashion that would win the respect of their peers. There also were social and moral constraints on their behavior: to a certain degree, if they wished to retain respect within the village, they needed to avoid being seen to act against the wider interests of the village. Yet, as this study has also demonstrated, because *adat* heads were not elected, there were few direct means of holding them directly accountable. *Adat* leaders could abuse their positions while engaging in strategies that protected their standing in the village. Operating as compliant clients of wider district networks of power and interest, in some cases local leaders and customary heads became involved in mining local forests for their own benefit, securing their positions by accommodating some groups within the village in logging operations but also undermining the long-term "social-ecological resilience" of the area.

With the increased prominence of *adat* authorities under revitalized *adat* management systems, customary heads can once again become important representatives of communities, but this change alone will not lead to genuine community management and decision-making. Under revitalized *adat* arrangements village governments assert control over community territories and the profit streams to be derived from their exploitation. If these communities lack effective forms of accountability, village elites can divert controls over village resources centralized in their hands to serve their own accumulative strategies (Zerner 1994). The use of customary heads in "participation" can also repeat colonial forms of rural administration. Under the colonial system of indirect rule, colonial authorities administered through compliant local "customary authorities"; similarly, customary leaders can become compliant clients in the hands of state or entrepreneurial patrons. For instance, in Mali and Burkina Faso, as Ribot (1998) has described, participatory forestry initiatives have used customary heads as administrators, intermediaries, or "symbols of the indigenous" to legitimize their projects. These customary heads and chiefs remain unelected and accountable upwards to the state rather than downwards to local populations. Accordingly, timber merchants are still able to obtain lucrative commercial timber licenses while villagers merely obtain labor opportunities and forest privileges rather than secure property rights. If participatory approaches are to have any meaning in these circumstances, Ribot concludes, they need to be supported by new projects and policies that push "for real, generalized and enduring participation." Such initiatives need to encompass empowered, accountable representation, not merely involve customary authorities who it is assumed somehow represent their communities.

Sixth, Indonesian policymakers face the problem of reforming a discredited policy framework for resource tenure inherited from the New Order period. Social and environmental justice in Indonesia requires the recognition of the sovereignty of *adat* orders. In pursuit of village-based notion of justice, in the absence of such policy changes since the end of Suharto's regime in many areas of the archipelago, villagers have taken justice into their own hands, reasserting *adat* claims over local lands and forests (Acciaioli 2002; Li 2001; McCarthy 2004). But the state has failed to develop appropriate official legal rules dealing with *adat* rights over forest and land areas.[36] This failure has significant implications. For instance, in Central Kalimantan, the assertion of *adat* rights can depend on an actor's capacity to negotiate or defend claims in the field rather than on the agrarian law, or on an agreement with mining or timber companies, or on a policy decision from local government agencies. Because they have to defend their rights without effective forms of legal redress under the state law, villages often depend on "people's justice" (*keadilan rakyat*), defending their interests by resorting to intimidation and force (McCarthy 2001a, 2001b).[37] The lack of functioning institutional arrangements and dispute resolution mechanisms governing access to resources by actors from outside the villages means that outsiders may also attempt to impose their will in the field, for instance by deploying thugs (*preman*). This uncertain context means that conflicts can end in violence. Because disputes occur in the shadow of open struggle, in many cases the most powerful local forces prevail (see McCarthy 2004). Consequently, there is a serious need to create not only official legal modes that allow for or recognize *adat* rights, but also effective forms of adjudication and dispute resolution.

In this context a number of questions have emerged concerning the possible role that *adat* could play in sustainable management of natural resources, and how the state should recognize *adat* rights. A detailed examination of the complex legal problems raised by this issue is beyond the scope of this study, but given the pressing need for reform in this area, I will make a few pertinent points.

Those advocating reviving *adat* as a competing legal system have at times fallen into the trap "of romanticising *adat* processes and encouraging a static interpretation of *adat* as a fixed set of customary prescriptions (Campbell 1990: 4)."[38] As Campbell (1999) has observed, "depicting a romanticised version of *adat* as a glorious living tradition of harmony with nature that is fully operative in forest dependent communities, makes it easier for government critics to push their equally simplistic view that *adat* systems (as static self-perpetuating operating systems) have already broken down." The alternative I have taken in this study is to try to understand the complex and shifting ecological, political, and economic challenges facing village actors and how these affect *adat* decision-making processes.

CBNRM strategies that attempt to nest *adat* rights within the state legal regime—such as in Menggamat—have ambiguous implications. Obtaining state recognition for *adat* rights implies a process of mutual adjustment between customary systems and state legal regimes. As we have seen this is not a new phenomenon: in the act of recognizing customary authorities and embedding them within the colonial version of customary government, the colonial state reorganized and (re)constituted them. Subsequently, at times *adat* institutional arrangements and state arrangements have stood in opposition to each other and at the same time existed as mutually adjusting orders. If the state project of recognizing *adat* orders is renewed, once again village actors will be able to use the symbolic capital associated with new official legal formulations to support *adat* property rights. However, if the law is to support *adat* property concepts and village-based notions of justice, positive law needs to provide the requisite legal forms that allow for existing *adat* management systems while developing the mechanisms for accountability, adjudication, and dispute resolution required to provide environmental justice.[39]

As demonstrated in Sama Dua, during the colonial period colonial administrators attempted to understand and then co-opt *adat* concepts. As the colonizing power engaged in a process of learning about and ordering the socio-legal orders of the colonized, it attempted to reinterpret and domesticate these orders to serve its interests (cf. Merry 1992). To an extent, the recognition and emasculation of *adat* occurred simultaneously under the colonial state project.[40]

The creation of a state-accepted customary order involves nesting customary order within the legal mode of control. As we noted earlier, doing this can involve a "crucial shift" in "the mechanisms of social control and the legitimation of authority," turning fluid and negotiable customary forms into rules recognized under official laws (Chanock 1982: 55). Recognizing customary rights and nesting them within the state order can form a part of the process of extending state control into community spaces (Agrawal 2001).[41] If the official legal forms allow for the recognition of customary property rights only under restricted conditions defined by the state, such restrictions may emasculate *adat* orders. If customary rights can become legitimate only if recognized under positive law, after they have been subjected to bureaucratic legality, then they can only be changed through the state. Consequently, protecting customary rights can end up empowering the state (Doolittle 2001). When villagers and NGO projects obtain state recognition of *adat* rights within the confines of official law, villagers can then find themselves relinquishing more pervasive customary rights for the limited rights allowed by the state in accordance with narrow legal and environmental prerequisites (Li 2002; Haverfield 1999; Doolittle 2001).[42]

Furthermore, when an actor gains a privileged position within the state apparatus, she usually also gains discretionary control over the definition and enforcement of property rights. The actor can exert this control through the granting of licenses and certificates and through commanding the application (or nonapplication) of law enforcement measures. Conflicts over resources—that involve struggles over defining the customary order—can turn out to be a part of on-going struggles for power.[43] In some African cases, reforms that have attempted "to rationalize property rights through statutory interventions" have often "served to intensify rather than resolve debate and conflict over land rights, by multiplying the conceptual frameworks and procedures brought to bear on the definitions of rights and the adjudication of disputes" (Berry 1994). Therefore, legal initiatives that change the discretionary powers of state agents in the local domain in these areas need to proceed carefully to avoid inadvertently proliferating the number of competing claims and intensifying conflicts.

Conventional politics favors those with money, contacts, and organizing ability. In South Aceh some decision makers and officials in the district and at the provincial level supported the CBNRM initiative. However, it threatened the material base of district and village interests involved in logging. Local officials, together with wider district networks profiting from timber extraction, held the balance of power: collectively they had a tacit but effective veto over the working of the community forest initiative. Without their support, the community leaders that backed the WWF-LP plan could not effectively implement a new CCF regime. Such actors can readily overpower attempts to apply a participatory community model, and can subvert efforts to look after the ecological systems upon which the welfare of villages depends in the long term.

Legal powers over land and forest tenure can be devolved to *adat* authorities under decentralization initiatives that place the recognition of *adat* rights under the supervision of district governments. If this devolution occurs in the absence of effective forms of representation and accountability under a district government dominated by entrenched regional interests, it would enable those with the money, the contacts, and the organizing ability to exploit newly recognized *adat* rights in ways that work against the interests of marginal villagers. Consequently, unless they are carefully conceived and implemented, legal reforms that attempt to deal with the unresolved problem of *adat* may also unintentionally provide tools for powerful actors to pursue their interests at the expense of marginal villagers.

This problem requires a holistic solution that also aims to provide effective forms of representation, accountability, and dispute resolution.[44] If these problems are faced, more sustainable resource management regimes are more likely to emerge with legitimacy established both by their compatibility with *adat* principles and with their inclusiveness, fairness, and equity established in the eyes of local actors.[45]

Different Approaches, Shared Problems

Although the experiences of the Leuser ICDP contrast with those of the CBNRM project, the dissimilarity is less pronounced than at first seems evident. Both the Leuser ICDP and WWF-LP interventions were driven by donors' concern about biodiversity loss and deforestation. They faced similar problems, shared the same fundamental goals, and could be assessed by equivalent criteria of success. In both cases, the actors driving these interventions wished to change patterns of resource use by creating particular institutional arrangements. Doing so would involve establishing who would be entitled to make decisions and who would benefit, how, when, and under what conditions. Accordingly, each project represented a "project in governance" and advanced a particular "politics of the environment" (Zerner 2000).

The Leuser ICDP intervention aimed to address the managerial and logistical difficulties facing a state forestry agency managing an extensive forest area with few resources and an inability to enforce the law. The project would improve resource management by building or replacing an ineffective state forest regime, making available EU resources, and facilitating a thoroughgoing implementation of state law. In seeking to shape village resource use, the ICDP also set out to provide incentives to encourage local actors to comply with preordained protected area goals. The Leuser ICDP used its access within the New Order regime to effect natural resource decisions and planning. Yet, it neglected the structure of property relations and the power relations that supported them and it was constrained by unworkable state assumptions regarding how resources should be managed. Heavily dependent upon its state patrons, the Leuser ICDP intervention remained reliant on state agencies that were still unable to apply regulatory instruments meant to protect rainforest ecosystems.

In contrast, in accordance with participatory ideas, WWF-LP saw the "community" rather than the state as *the* primary actor in conservation. WWF-LP showed greater acuity at the village level, correctly seeing the need to directly engage village natural resource management practices while it attempted to recognize—as far as the law allowed—the property rights of local populations within the "state forest." WWF-LP faced additional complications, however. Not least of all, this approach ran up against problems derived from narrow assumptions regarding *adat*, and the likelihood that, if there is a link between village livelihoods, community management of resources, and conservation objectives, that link is contingent on a varying array of mediating factors that were difficult to engineer.

Notwithstanding the advantages that the respective projects enjoyed at different political levels and the different philosophies they employed, each project ultimately faced the same problem. District-level networks of power and interest coalesced around logging; various village actors found viable strategies of survival in logging; meanwhile, some also developed income-generating tree crops from pioneer agriculture on the same forest frontier. The concurrence of these parallel but distinct sets of interests among these diverse actors working at different social levels drove the inexorable process of resource depletion. Both the ICDP and the CBNRM project intervention faced the lucrative (if unsustainable) nature of logging for these various actors, and the difficulty of imposing sanctions against these encompassing interests or finding comparatively attractive alternative livelihood options. Neither the WWF-LP nor the Leuser ICDP was able to undermine or defeat entrenched clientelist district networks, nor could they link viable livelihood alternatives to biodiversity conservation, even in the short term. Nor could either of them provide the livelihood alternatives or the exchange options or the disincentives required to rival unsustainable exploitation. Therefore, even if village actors and district government recognized the harmful environmental impact of logging, the accord between parallel sets of interests among different actors would support resource extraction rather than the outcomes advocated by outside interventions.

Consequently, the patterns of unsustainable resource extraction have complex, multidimensional causes that allow for no easy remedy. A wide range of conditions need to change in order to alter the way corresponding sets of interests between actors working at different social levels came together around logging and forest pioneering. These changes would encompass (1) a slackening in the demand for timber; (2) the development of a socio-economic system that enabled farmers to gain sufficient benefits from the crops they produce to provide them with stable livelihoods regardless of fluctuating world markets; (3) the maintenance of resilient agro-ecological systems that offered viable farmers agricultural alternatives to pioneer agriculture; (4) the development of steady revenue streams for local district government budgets and for clientelist politics other than those generated from the timber industry; (5) an increase in local concern for the long-term impact of unsustainable logging; (6) an accountable political system with an active civil society able to control the behavior of local officials and politicians; (7) the renegotiation of customary rules governing resource access and use taking ecological values into account; (8) the reconciliation of those customary rules with legitimate and enforceable official laws allowing for the existence of a parallel *adat* order; and (9) a functioning legal system providing for effective enforcement of state laws. As the Menggamat case suggests, a temporary change in one factor without simultaneous changes in several others would fail to alter the way corresponding interests of different actors come together around logging and forest pioneering in an enduring fashion. Consequently, in the absence of the separatist war that emerged during 1999, it remained likely that the Leuser forests would have been extensively logged before the logging frontier moved on.

The accommodations and exchanges between actors with compatible interests at these various levels—including entrepreneurs and officials, and extending to encompass customary heads and ordinary villagers—constitute the most significant obstruction to biodiversity protection and community forestry interventions alike; this has implications for project interventions. Natural resource management initiatives of whatever hue need to address the clientelist networks of exchange that articulate with *adat* authority systems and formal state structures. This conclusion suggests that, rather than privileging the "community" as *the* primary actor, or the "state" as a key point of intervention, analysis need to take a more encompassing view of the how "state" and "community" actors both interact with emergent clientelist socio-legal orders to create the institutional arrangements shaping resource access.

In retrospect it is easy to draw conclusions regarding these projects. At the time, however, the dedicated and motivated people involved in these projects were trying to make the most of the prospects available under the state policies and the conservationist concepts of the time. Here it is important to note that, to evaluate a project, "it is not enough to say that it engages in simplifications; all social categories simplify even as they bring us to appreciate new complexities" (Tsing 1999). The value of reflecting on these projects lies just here—in the compound lessons they reveal.

The Fourth Circle

In the fourth circle of hell, Dante's Pilgrim sees two groups of angry, shouting souls crashing huge rolling weights against each other with their chests. For hoarding material goods or for wasting them on earth, the souls of the excessively frugal and the profligate face the eternal punishment of this never-ending struggle. In Leuser in the 1996–99 period, like the souls in the fourth circle of Dante's *Inferno*, both these projects seem fated to engage in an eternal and ultimately futile contest with opponents favoring unsustainable extraction.

Transcending this apparently eternal struggle entails avoiding these two extremes—on the one hand attempting to set aside huge areas of pristine, untenanted forests for nature conservation without effective local consent, on the other, allowing precious, highly diverse, steep watershed forests to be degraded. This more sustainable pathway would see sustainable, locally attuned agro-ecological systems cultivated alongside vigorous watershed forests thriving on surrounding steep mountainous areas. But how might this be achieved?

Village aspirations revolve around developing sustainable livelihoods, both immediately and in the long term. The evidence from the three cases discussed in this study show how farmers have worked assiduously to develop gardens planted with trees yielding cash crops for export. Other studies indicate that an agrarian transformation is occurring in forested areas of Sumatra and elsewhere in Indonesia. This transformation involves the conversion of agricultural fallows and forest areas to sedentary agricultural systems involving the cultivation of tree crops (Sunderlin et al. 2000; Angelsen 1995b; Li 2000, 2002). The shift appears to be driven by the appeal of producing export crops for the international market, farmers' need to diversify their agricultural strategies to avoid the risks associated with overdependence on a single crop, and the need for farmers to respond to land pressures by planting trees to secure tenurial rights (Sunderlin et al. 2000; Li 2002).

Dominant policy discourses and the interventions undertaken in their name depend upon standardized, simplified understandings of the causes of environmental degradation and propose solutions on this basis. The populist CBNRM policy narrative describes farmers dependent on natural forests, and seeks to link sustainable management to continuing community management of natural forests. In a parallel fashion, the ICDP approach reflects a global environmental management discourse that aspires to provide technical assistance to improve resource management and the implementation of state policy while providing legal and economic incentives—including the intensification of agriculture in national park buffer zones—to stop villagers from "encroaching" into the forest. Unfortunately, these policy discourses fail to provide adequate frameworks for understanding the complex historical, socio-economic, and agro-ecological contexts in which ecological change is taking place. The logical consequence of these simplified designs for social and environmental management can be seen in their lack of success (cf. Adger et al. 2001).

The evidence suggests that in frontier areas village livelihoods are linked to the continuing transformation of forest ecosystems into productive, cash crop yielding trees (Li 2002). If we accept this linkage, resource policies will have to allow for farmers' aspirations to develop cash crops, if possible in an ecologically and economically sustainable fashion. Project interventions could be important here for two reasons. First, there is a linkage between the economic situation of farmers and the health of these agro-ecological systems. Household well-being is tied to the price of the key products produced by farmers engaging in tree-cropping for export. Where farmers primarily depend upon a single tree crop, price volatility in this commodity is linked to economic insecurity. Where land and capital are available, to avoid falling victim to precarious prices on their main crop (such as during a time of economic instability like that of 1997–99), farmers react to price fluctuations by clearing new land for new crops. In that economic crisis, farmers in the South and Southeast Aceh areas reacted by planting export cash crops both to take advantage of high prices and to increase their short-term and—where they planted tree crops—their long-term security. In this context, project interventions and policies that aim to stabilize farm income and farming systems may assist in decreasing the forest clearing that results from economic insecurity created by price volatility (Sunderlin et al. 2000).

Second, as the Sama Dua and Menggamat cases suggest, tree crop systems are vulnerable to environmental shocks such as pest infestations and the ecological impact of logging nearby areas. Significant disturbances in surrounding watershed forests can lead to fundamental changes to the agro-ecological systems on which villagers depend. If tree crop systems collapse, farmers will be displaced into surrounding forest areas including remnant watershed forests, further adding to the cycle of ecological decline. Such considerations suggest that policy and project interventions need to address these social and ecological problems at the same time.

In other words, ecological and social systems are coupled: shocks to one can upset the other and the resilience of one depends upon the buoyancy of the other. "Social-ecological resilience" is linked to livelihood security, questions of socioeconomic resilience bring up questions of entitlement and access to resources—environmental justice (Berkes et al. 2003). Avoiding the catastrophic economic and ecological shocks that occurred in the 1998–99 period would involve simultaneously attending to the economic, ecological, and social issues that are factors in poverty as well as dealing with associated ecological problems. In the areas studied here, doing so would first entail enhancing the resilience of existing agro-ecological systems upon which villagers depend. Such experiments might aim to help farmers improve existing agro-ecological practices and to increase their capacity to deal with change, for instance by diversifying tree-cropping systems. Policy and project activities in such a context would also need to address questions of environmental justice—tenure, entitlement, access to resources, food security, and maximizing the benefits that farmers obtain from production (Zerner 2000).[46]

Community-based natural resource management (CBNRM) assumes that the market and the state create incentives for communities to use resources unsustainably. Therefore, CBNRM advocates assert that when communities are in control of resource management, the benefits they

receive will create incentives for them to become good stewards of resources (Agrawal and Clark 1999: 633).[47] Not only the complexity of
"community" itself but the interpenetration of community, the market,
and the state undermines this sort of simple analysis. Therefore, rather
than valorizing community-based conservation as an alternative to either state management on the one hand or private management by market actors on the other, analysis (together with interventions) needs to
see clearly how particular constellations of critical interests within "the
state" and "the community" articulate with wider political and economic forces to produce particular outcomes.[48] This examination entails
reconfiguring the place of marginal village actors with respect to markets
and the state in ways that maintain local livelihoods in an ecologically
sustainable manner.

In circumstances where more sustainable systems are either developing or in place, interventions may assist local farmers in pursuing their
livelihood concerns while quietly promoting environmental agendas. Yet,
difficult as it may be to accept, external interventions may not be able to
change evolving agro-ecological systems in desired directions. The ICDP
concept assumes that improving the productivity of agriculture in buffer
zones should lessen forest clearing by pioneer farmers in national parks;
however, research has revealed that the intensification of cash cropping
in frontier forest areas in many cases drives deforestation (Angelsen and
Kaimowitz 1999). As farmers open new areas for cash crop production,
the agricultural systems that emerge do not inevitably aim to provide
food security, sustainability, or crop diversity; rather, farmers may aim to
generate cash quickly, and the cash crop systems that they develop may
be neither ecologically benign nor resilient (Li 2001). Sometimes there
may be no obvious link between protecting biodiversity and improving
local livelihoods. In such instances, it may be difficult to integrate income-generating tree crops into resilient forest-like agro-ecological systems, in parallel with the protection of surrounding frontier forests.[49]

Finally, as I noted earlier, Indonesians refer to the web of political,
economic, and social exchanges surrounding logging at the district level
as a "vicious circle" or a "circle of devils" (*lingkaran setan*). Indeed, like
a chain joined together by many links, each ring in the circle of collusive
accommodations is fixed to its neighbor. In the cases studied in this volume, this circle of devils proved unyielding to outside interventions.

A growing body of literature addresses environmental problems from
an institutionalist perspective, discussing the "design principles" that
favor positive environmental outcomes. These principles encompass
"clearly demarcating boundaries, devising equitable rules for sharing
benefits and costs, establishing effective monitoring arrangements for

imposing graduated sanctions, and creating larger organizations by nesting smaller units within the larger organizations" (Gibson et al. 2000). Such frameworks offer valuable theoretical insights, but they may fail to sufficiently address the way "historically-specific structures of power"—such as the entrenched networks of power and interest found in these districts—account for persisting patterns of unsustainable resource use (Mosse 1997). Breaking out of the circle of devils would entail more than designing institutions; it would involve addressing issues of power and dissolving the close linkages between local patterns of accommodation and the management of local natural resources. It would also entail a more fundamental transformation in the relationship between local government and rural constituencies to improve political representation and participation and, consequently, developing a more transparent and accountable system of governance. In Indonesia, these issues are critical to the country's experiment with "democratic decentralization." In the following section, I briefly discuss this alternative policy formulation, which has emerged since the 1998 *reformasi*, and the implications of this study for its success.

A Decentralized Solution?

Turning to Decentralization

Under the integrated conservation and development approach discussed in the last section, the state transferred areas of responsibilities and authority to a newly created quasi-state agency, the Leuser Management Unit. In contrast, the community based natural resource management project advocated devolution to a community body constituted within state legal norms and according to conservationist principles. A third approach, "democratic decentralization," is now the vogue. This entails devolving administrative powers and resources to local government bodies who are to varying degrees independent of higher levels of government and accountable downwards to local populations (Agrawal and Ribot 1999; Manor 1999).[50] ICDP and CBNRM models can complement democratic decentralization, but they involve distinctive policy approaches. Successful democratic decentralization—were it to occur—would, its theorists suggest, alleviate administrative bottlenecks associated with centralized decision-making processes. Government would be more responsive to local needs, public institutions would be more accountable, service delivery would improve, and diverse groups would find better representation and would more fully participate in decision-making (Frerks and Otto 1996; EPIQ Technical Advisory Group 2001; World Bank 2001).

A body of literature has also argued that decentralization could improve the management of natural resources. State agencies that work with nationwide organizational rules are often unable to accommodate the interests of diverse groups and the variety of variables involved in local settings. Because state agencies impose rules that fail to suit local circumstances, state-imposed governance regimes tend to be poorly designed and to lack legitimacy (Ostrom 1990). In contrast, if locally constituted representative bodies gain the power to make rules, it is argued, decision-making processes will more readily take into account local specificities. Nesting local rule-making processes in larger scales of organization may facilitate the development of appropriate institutional arrangements (Ostrom 1997).

Like the CBNRM approach, this policy model resonates with a political discourse that establishes a seemingly self-evident set of political aims and objectives (Walters 2002). Doing so involves achieving policy aims by mobilizing and enhancing the trust, norms, and horizontal networks of civic engagement—the "social capital"—that are thought to facilitate coordinated social action (Putman 1993). It also entails imagining a social space—with its own processes and dynamics—that can be tapped and harnessed for governmental projects (Walters 2002). In focusing on collective action and self-government as means to address policy issues, democratic decentralization seeks to increasingly deliver development through civil society, relegating the central state more to the role of setting standards, monitoring, and evaluation (Contreras 2000).

Theorists also envisage that democratic decentralization could have distributive effects, and could be associated with environmental justice agendas. If local actors—including disadvantaged ones—can influence decision-making processes undertaken by newly empowered, downwardly accountable local government bodies, they are more likely to benefit from resource policies. Further, just like the proverbial bird wishing to avoid soiling its own nests, local actors may be concerned with the long-term prospects of their region. Accordingly, if local people participate more fully in decision-making, decisions may give increased weight to the environmental future of an area.[51]

However, problems have dogged decentralization programs. Local elites have captured devolved powers and gained advantage from decentralization, proving that empowerment and participation will not necessarily follow from decentralization. Central governments distrustful of lower levels of government have also proved reluctant to grant significant authority and funding to local government bodies. In addition, local government agencies have remained unaccountable, or have lacked the resources and expertise needed for effective planning and implementation. Alternatively, poor planning and coordination across levels of government have hampered decentralization programs (Frerks and Otto 1996; Larson 2003).

Decentralization has also raised specific issues of natural resource management. Studies from Latin America have suggested that the economic importance of forests to revenue and capital accumulation affects policy outcomes. Those who hold power under previous regimes have resisted change, and significant conflicts have emerged during decentralization programs (Larson 2003). Local government actors strengthened by decentralization have allowed forest exploitation to support district budgets and have pursued clientelist agendas (Kaimowitz et al. 1999). Legal frameworks for managing natural resources are particularly complex, involving the nesting of local rules within larger scales. Polycentric organizations involving horizontal and vertical linkages within the state are particularly difficult to manage at the best of times. This complexity and the power struggles that can emerge during decentralization heighten the likelihood of contradictory legal arrangements (Larson 2003).

Decentralizing Indonesia

During 1998 and 1999, Indonesia's political and economic crisis generated a crisis of legitimacy for the centralized form of the state that had existed under the New Order. In areas remote from Jakarta and rich in natural resources, the fervor for *reformasi* also focused on reversing the inequitable distribution of the benefits derived from resource exploitation. In an effort to overturn the corrupt clientelist arrangements of the New Order period, the advocates of change argued that "centralism" had allowed interests close to the regime in Jakarta to siphon off most of the rent generated by exploitation of resources in these regions. This centralism, they argued, also impaired the ability of regional interests to negotiate their own priorities, muzzling the expression of regional aspirations and marginalizing local populations. During 1998, district and provincial actors began to demand a greater role in running their own affairs and a greater proportion of the profits generated from local natural resources. It was widely held that leaders, preferably local-born people (*putra daerah*) representing local interests, should take key positions in local government and in companies active in the regions. Villagers also demanded the return of land and forest resources exploited by outsiders. Seeking to rectify the systematic disregard for *adat* property rights, villagers began taking direct action to redress these grievances, reclaiming land and demanding compensation from companies that had logged or mined their traditional lands (FEER 2/7/98; Warren and McCarthy 2002). As the clamor for reform gathered momentum, some provincial figures called for immediate autonomy. In the disenchanted provinces of Papua, Aceh, Riau, and East Kalimantan, some even discussed ceding from the unitary republic of Indonesia. Press articles began to discuss the fearful prospect of national disintegration (*Suara Pembaruan*, 19/9/98).[52]

Under the pressure of unfolding events, Habibie's transitional government drafted new laws to address the most salient issues, including a new decentralization law. State decrees and policy documents from this period are suggestive of this context and point to the avowed intent of these laws. The State Planning Guidelines (GBHN 1994–2004) and decrees of the supreme national parliament (e.g., TAP MPR IV/2000) repeatedly emphasize the need for a system of regional government that would provide for authority and the allocation of resources to be built upon the main concerns of regional communities.[53]

This shift fitted with a transnational discourse on governance and decentralization that provided ready-made recipes for particular national problems. If policy makers could draft decentralization laws that addressed pressing domestic political issues in accordance with the good governance discourse, they could attract foreign aid for state reform. Being able to articulate the decentralization discourse with a domestic logic of state transformation would provide policy makers a means to address the frustrations generated by decades of overly centralized control in ethnically and geographically distinctive areas of the country, and help the state find a new legitimating narrative.

In crafting this particular decentralization policy model, state planners made strategic political choices. They decided to delegate authority over key areas of government to the districts and municipalities rather than to the provinces. Doing so was seen as a more acceptable alternative than the undesirable options of either instituting federalism or retaining the existing centralized system with slight modifications (Rasyid 2003). Policy makers believed that the districts were too small for separatist or federalist aspirations to take root (Niessen 1999). Moreover, they thought the central government would have more "influence over relatively weak districts compared with strong provinces" (Ahmad 2000: 4). In addition, in their view implementing decentralization at this level would foster competition between districts and provinces and allow the central government to act as an arbitrator. In effect this reform model adopted was that considered to best support the authority of the central government.

Accountability

Theorists argue that accountability is central to the success of democratic decentralization. The argument runs that local authorities with discretionary powers are more likely to want to discern and to respond to local needs and aspirations when local populations can use democratic mechanisms to hold them accountable.[54] Indonesia's framework decentralization law (Law No. 22/1999) sets out a model to make district governments more accountable to their local constituencies. This model relies on a chain of accountability: the executive is held accountable to elected representatives, who are in turn held accountable to the public through elections.[55] Downward accountability is to occur via newly empowered district and municipal legislatives (DPRD). These elected DPRDs have the power to help formulate district laws and budgets, supervise the implementation of district policies and laws, and elect and dismiss the head of the region.

While the new system may appear accountable in form, significant challenges to accountability emerge in practice. The first is the issue of holding the DPRD accountable downwards. Under the existing electoral model, voters chose a party and the party committee then selects representatives from its list of candidates. This model enhances the power of the party at the expense of the electorate. The influence of what Indonesians call "money politics"—where cash is exchanged for favors within the party and the DPRD—further weakens the link between voters and DPRD members.[56]

The second issue is the question of how the DPRD uses the accountability mechanisms provided by the framework decentralization law (Law No. 22/1999). As I mentioned earlier, the legislature is entrusted with powers to elect, supervise, and dismiss the executive. The process of selecting the head is complex and provides amble opportunities for back-door politics and corrupt pressures and practices (Bell 2001). Further, once elected, district heads concentrate their political energies on maintaining the support of the legislature. At key moments district heads dispense favors, projects, and money to members of the regional parliament (DPRD)—for instance, when the *bupati* wants to have his budget and accountability reports accepted. Consequently, political support at these moments can become a commodity, and the accountability mechanisms provided for by the decentralization law end up being put to use within well-established district-level patterns of exchange and accommodation.[57]

A third issue concerns rural constituencies in remote areas. Here a diverse range of geographical, economic, and socio-political factors make the costs of participating in political action beyond the village excessive. Villagers living in places where the media and political and legal information are scarce are isolated and economically disadvantaged, and they also face the entrenched power of public- and private-sector actors who render local organization vulnerable to coercion and co-option (Fox 1990). Such factors undermine the capacity of village actors to use accountability mechanisms as "countervailing powers" in opposition to entrenched networks of exchange and accommodation.[58] Consequently, village actors face significant internal and external obstacles when they attempt to hold the state accountable (Fox 1990). In such situations, villagers may affect outcomes by direct action—by taking the "law into their own hands" (*main hakim sendiri*) (McCarthy forthcoming).

Uncertainty

Policy and legal changes affecting the development of natural resource regimes following decentralization have created uncertainty in Indonesia. The legal framework governing decentralization was contested from the outset. The advocates of decentralization within the transitional Habibie administration needed to pass the decentralization laws quickly before the immanent 1999 election. As a result, the framework laws bear the traces of a political compromise; through complex formulations, legislators managed to accommodate the views of a range of actors. But they did so at the price of precision: the law left many issues outstanding for determination by lower regulations. Implementing regulations that were drafted later might have cleared up these uncertainties, but the disputed politics of this period of regime transition elaborated the many ambiguities. During this period the central government faced difficulties delivering policy guidance and effectively monitoring and supervising the decentralization process, thus compounding the lack of clarity in the legal framework governing regional autonomy.[59]

From the beginning, the Ministry of Forestry worked to maintain its control of the nation's vast forestry estate. The new Basic Forestry Law (Law No. 41/1999) passed during the Habibie period retained the notion of central control of the forestry estate and the assumption of a hierarchical relationship between levels of government. Subsequently, implementation of regulations in the forestry sector—when they eventually appeared—accentuated this trend, with the forestry ministry wanting to leave only operational matters in the hands of the districts and municipalities (Menteri Kehutanan 2002).

In contrast, the decentralization law (Act No. 22/1999) worked on the assumption that districts and municipal governments would attain extensive discretionary powers, with the central government only retaining powers over setting policy, guidelines, and standards. When the government proclaimed the key implementing regulation (PP 25/2000) for the decentralization act, it specified only the areas of responsibility of provincial and central governments. By implication, all remaining responsibilities were left in the hands of district and municipal governments.

During the period of uncertainty, district administrations used the ambiguities presented to advance their own interpretations. Even before the decentralization law formally came into effect, individual regions had moved ahead to quickly establish district regulations (*perda*) according to a decree of the supreme parliament (MPR) that allowed districts to establish their own regulations for matters for which the central government had not yet generated implementing regulations.[60] As a result, individual regions established regulations that ran counter to decisions made by other levels of administration, including those relating to the forestry sector.[61] As a Ministry of Forestry document noted, "many regulations concerning implementing authority regulating forest were overlapping and contradictory, with the result that responsibilities for regulating the forest became unclear" (Menteri Kehutanan 2002).

Laws both reflect and have the power to structure the discourse within which an issue is framed (Merry 1995). The new decentralization laws granted discretionary powers to district governments, including powers to pass regulations governing revenue generation as well as the setting and spending of budgets. Although district powers over natural resources and the environment remained circumscribed, district government actors used their enhanced discretionary authority to directly affect the use of resources in the local domain. District administrations operating at the intersection of an ambiguous set of laws and administrative arrangements could choose which laws on which to base their policies (cf. Agrawal and Ribot 1999). In effect, they chose particular laws that provided the discursive and legal resources to support revenue generation for the district, exchanges and accommodations between key actors, and capital accumulation for district actors. In this way the uncertainty and fragmentation of state arrangements worked to the advantage of entrenched regional elites.[62]

Possibilities

To assess the likely outcomes of Indonesia's decentralization reforms, it is only fair to retain a point of comparison. This comparison can be gained from comparing the emergent situation in Indonesia with the picture offered by the literature surveying the previous timber concession system and the state of affairs in districts prior to decentralization—such as in the two districts discussed in this volume.

Under the New Order, powerful politico-bureaucratic actors at the center had affected policy outcomes in key areas of national life. In the forestry sector decision-makers at the apex of the state had allocated logging concessions to key clients and the families of the politico-bureaucratic actors close to the regime. Under this system, concessionaries accumulated capital from forest rents, the regime amassed the largesse necessary to run its clientelist system, and the state obtained significant surplus for its development programs. In the process, powerful timber tycoons close to Suharto and other key actors in the regime were able to mediate the way the forestry department enforced concession contracts (Barr 2001). Meanwhile, forestry officials both in Jakarta and in the field developed collusive relationships sustained by unofficial fees and kickbacks. In this context, forestry laws and environmental rules were systematically overlooked (Barber et al. 1994). With structural inequalities inherent in legal definitions of tenure and resource use, the *adat* rights of villagers living in areas subject to concession licenses were thoroughly ignored. The system worked against local people, and under the repressive regime villagers could hardly protest (McCarthy 2001a, 2001b). The system also led to the unbridled plunder of the resource base and extensive environmental degradation, leaving Indonesia with a legacy of drought, floods, fires, and resource depletion.

As the case studies in this book have demonstrated, in parallel with this centralized system, during the New Order a highly localized system of resource extraction operated. As long as those who benefited from the system stayed faithful to the regime and made contributions to colleagues and managers within party and state, this system allowed rent seeking by district elites, who operated with a significant degree of independence.[63] As this system worked contrary to state regulations, it also led to outcomes that were antithetical to declared state policy aims and effective environmental governance.

The axis of centralization-decentralization presents a dilemma. The problems created by a regime that concentrated power in the hands of an unrepresentative state bureaucracy provide the reason for moving towards a more locally accountable system of governance. Devolving power to local government bodies presents a particular predicament, however; in the absence of a strong independent civil society that can be counterpoised to the local state, there is a danger that decentralization reforms will be captured by powerful local elites.[64]

Legal sociologists have long criticized the idea that a new law will instrumentally lead to a simple change in behavior. When governments create new rules, all too often the binding power of social arrangements are stronger than the new law. Such arrangements affect how the laws are invoked and the consequences of that invocation (J. Griffiths 1995). Consequently, we would be naïve to expect that a set of decentralization reforms, in some technically rational, instrumental fashion, will lead to a desired set of results.

In Thailand, with decentralization, local elites—particularly town-based entrepreneurs—seeking to gain access to power still found within the state have emerged to monopolize representative positions in decentralized authorities (Arghiros 2001). In the Philippines, decentralization has seen the demise of a centralized state system that amounted to an apparatus that facilitated the extraction of surplus by military and civil elites. Subsequently, the same elites "realized that surplus extraction will be facilitated more by privately initiated and decentralized structures" (Contreras 2000). Thus, according to Contreras (2000: 148), a discourse of "people empowerment" has ironically contributed to the "continuous erosion of state power vis-à-vis the strengthening of local elites in command of decentralized development activities." In a similar fashion, in some areas of Bolivia, local elites—taking advantage of decentralization reforms—have used forest exploitation for district revenue generation and personal enrichment (Kaimowitz et al. 1998).

With regional autonomy in Indonesia, the established arrangements that dominated the district scene have altered along similar lines. After decentralization, clientelist politics—which this study has suggested already shaped environmental outcomes to a significant extent—took a more visible form. Since decentralization offers local elites enhanced opportunities to use district regulations to extract resources, regional elites are able to go about their business with greater "legality." In this sense, rather than establishing a new system, the implementation of regional autonomy may denote a shift towards legitimizing and in some cases extending the well-established *de facto* system discussed in this book. Previously this system was embedded within the wider New Order regime; however, given the transition at the national level, the links between national coalitions and regional networks of exchange and accommodation became contested. Although networks of exchange and accommodation may have crystallized in some districts, in many other areas diffuse and mutually competing networks of patronage continue to vie for dominance.[65] In either case, they work against effective environmental governance.

A key question here is whether the decentralized system—with all its faults—is worse than the highly centralized bureaucratic system that existed in the past. Bearing in mind the problems associated with Indonesia's decentralization reforms, a case can be made that for a number of reasons regional autonomy may yet provide opportunities for advance.

If democratic decentralization is to have substantive meaning for local resource users, it would affect their ability to participate in decision-making. The nature of a decentralization process can then be assessed in terms of how it changes the ability of local groups to advance their interests, to secure local property rights, to control resource access and otherwise benefit from patterns of resource use (Lutz and Caldecott 1996).

In Bolivia, decentralization reforms weakened the influence of national elites controlling concession licenses in some municipalities, opening up opportunities for previously marginalized groups to strengthen their territorial claims and affect negotiations over forest management (Kaimowitz et al. 1998). In Indonesia decentralization has undermined the power of timber concessions that obtained their licenses in Jakarta under Suharto. Previously they could obtain permits from the center and disregard villagers and even local government; now, however, they have to negotiate access to resources at every level. In the process of negotiating in villages, concessionaires and timber brokers have had to allow for the *de facto* existence of *adat* property rights. In the absence of formal recognition of *adat* rights and effective forms of accountability, local people are significantly disadvantaged, with villagers and *adat* authorities vulnerable to coercion and cooption. Nonetheless, compared to the previous situation, villagers may be able to assert *adat* claims and gain at least some benefit from resource extraction (see McCarthy 2004).

Given the ambiguities associated with administrative arrangements and political processes at the national, provincial, and district levels, after the implementation of decentralization laws the emergent socio-legal configurations governing access to resources at the village level remain tentative and subject to dispute.[66] These evolving circumstances provide space for village actors to assert rights over areas subject to *adat* property claims. If local actors can maintain these claims, and if they can gain some official recognition for them, decentralization may yet present the possibility of overcoming some of the structural inequalities inherent in legal definitions of tenure and resource use which have long worked against villagers.

With actors competing for access and control, conditions may be more open than they were under the repressive, bureaucratic New Order regime. Problems such as the corrupt dealings of local politicians are more visible; more frequent exposure of illegal deals may gradually affect patterns of governance. In those situations where political actors such as indigenous groups and NGOs can organize themselves, sometimes with outside support, decentralization and associated reforms offer a new space to carry out advocacy, negotiation, lobbying, and coalition building for positive ends.[67]

Decentralization inevitably broadens the penetration of the state into remote areas. There is always the possibility that local people can capture decision-making power and use the symbolic power of state law for local purposes, slowly turning around the state to serve village interests (Arghiros 2001; Li 2002). Although in many remote areas of Indonesia village actors remain unable to access the accountability mechanisms provided by law, the decentralization reforms could provide a more polycentric form of government. As Ribot has noted, "the balance of powers" is an important structural aspect of accountability, because "a balance of powers in which there are counter powers to the central government can increase accountability by increasing the number of actors with a voice in politics and the ability of non-central actors to scrutinize central institutions" (Ribot 2001: 78). If such a balance of powers is to be achieved, higher levels of government also need to remain locally engaged, for instance in providing technical skills, adjudication, and conflict resolution mechanisms, and in supporting local actors in addressing problems of corruption and gross inefficiency.[68] Paradoxically, to avoid shifting problems along a centralized-decentralized axis, decentralization needs to be balanced with centralization, and doing so necessarily involves the application of political skill locally by a strong central government (Tendler 1997).[69]

In Thailand, decentralization lessened the importance of local bureaucrats, with local politicians increasingly emerging as the new patrons. While these local politicians are integrated into their circles of exchange and accommodation, nevertheless they do mediate the space between the central state and villagers better than the closed bureaucracy of the past. To a limited extent these local patrons are beginning to serve the needs of their rural clients. Decentralization opens the possibility for learning; as the system evolves, villagers may yet find better means of pursuing their interests than that available to them under a closed centralized bureaucratic system (Arghiros 2001).

In Indonesia, clientelist networks of exchange and accommodation described in this study have evolved to exploit the opportunities provided by regional autonomy. Nonetheless, the decentralization era offers villagers greater opportunities to pursue their interests. Finally, conditions remain equivocal: alongside the compound problems decentralization presents, the shift toward such a system also affords some improvements on the authoritarian, bureaucratic governance of the New Order period, when policies were so often pursued that worked against the interests of remote villagers.

Appendix: Fieldwork in Aceh: Research Design, Context, and Experience

Critiques of ethnography have engaged with a significant issue: What is the epistemological status of an account of another culture created by an observer who necessarily bears values and symbols that filter his or her experience? In the face of such questions, these days ethnographers cannot cling to "an ideology claiming transparency of representation and immediacy of experience." Writing ethnographic accounts can no longer be "reduced to method: keeping good field notes, making accurate maps, 'writing up' results" (Clifford 1986: 2). Whether the researcher likes it or not, the construction of an account is caught up in "contexts of power, resistance, institutional constraint and innovation" (Clifford 1986: 3).

While I will not overly dwell on this reflective turn, clearly the account presented in this text was "made or fashioned," and as such it bears the marks of its creation. To that extent it is inherently something "partial" or "incomplete." In order to give an indication of its limits, I will follow the convention of recounting the specific conditions under which the research took place. After doing so, in the following section I will briefly discuss the specific political and socio-economic context of research.

Fieldwork (1996–99)

Carrying out fieldwork in remote areas of Aceh in 1996–99 proved to be taxing. Due to the proximity of Western Australia to Indonesia, however, rather than carrying out the research in one block, I made serial visits. During the three-year period I visited the southern Aceh area five times, spending approximately twelve months in the field. Working in this way had distinct advantages. As I visited the area over three years, I could intersperse periods of reading newspaper clippings, agency reports, and colonial texts with fieldwork. Between periods in the field I reflected on the field notes, refined research questions, and even began writing. This also provided a longitudinal dimension to the research, allowing me to study institutional change over time. On more than one occasion a return trip revealed how, in response to economic or political change, villagers were taking up new strategies, making evident new or previously hidden aspects of the local institutional arrangements. Comparisons across time and field sites carried out through repeated visits over the 1996–99 period were deepened when on two occasions I asked friends from particular villages to accompany me on visits to other areas.

The initial step involved selecting village sites. On preliminary trips during 1996–97, I first traveled to Kutacane (see Map 3). Located on the floor of the Alas valley (Southeast Aceh district) and surrounded by steep mountains on both sides, this small district capital is set in some 10,000 hectares of rice fields. From there I moved north to the Badar area. There, along the banks of the Alas River, I lived for some weeks in villages inhabited by Gayo and Alas people. I then made my way to the picturesque town of Tapaktuan, the capital of South Aceh district. This town lies on the Indian Ocean on the west coast of Sumatra. As a guide-book to Aceh noted without exaggeration, it is perhaps the most picturesque town in all of Sumatra: "the coast line both north of and south of Tapaktuan is extremely beautiful. Great beaches, steep cliffs and big waves" (Bangkaru 1998).[1] In each of these areas I spent time in villages, making contacts, spending long periods of time interacting with local people, and eventually taking up temporary residence in a household.

On the second trip, during the forest fires that ravaged Sumatra during 1997, to deepen my understanding of the area and refine the choice of field sites, I traveled for a month, moving north up the Alas valley, west through the remote Gayo Lues highlands and then south down the coast of South Aceh. I had the good fortune to meet a seventy-five–year-old resident of Badar who wished to return to his home kampung to visit his newly married daughter. In return for paying traveling expenses, he would accompany me. Our destination was Susoh, a town whose well-crafted antique timber houses attest to its history. Susoh had been one of the earliest settlements of Minangkabau colonists (Aneuk Jamee) in the area, and a center of South Aceh's eighteenth-century "pepper coast" (Bulbeck et al. 1998). To reach Susoh we had to travel through the remote highlands of Gayo Lues, a center of Indonesia's clandestine marijuana industry. The road between Terangon and Blangpidie was so neglected that the dilapidated *kijang* that usually served inter-village routes through the mountains could not descend here: only a four-wheel drive could make the perilous trip through the mountains. This journey offered a lesson in fatalism. In many places, the monsoonal torrents had washed away the road or it had fallen away due to landslides, leaving a thin lip of road hanging on to a sheer cliff. Precipitous drops made even well-seasoned local travelers and veterans of the arduous pilgrimage to Mecca fearful of an early death. As we made our way towards Tapaktuan, my companion described how during the 1950s and 1960s he had worked this route as a trader, transporting coffee and tobacco by horseback from Terangon down to the coast. In those days the trip down a forest path took ten days. He lightened up our trip with hair-raising stories of bus accidents he had survived, of working as a member of a slave-labor team during the Japanese occupation, of the depredations and summary executions Indonesian soldiers visited on villagers during the post-independence rebellion, and of being forced to flee into the forest and eking out an existence from forest products during this war. We also discussed the stories I had heard of black magic and poison that still circulate in the most isolated corners of South Aceh. He attempted to reassure me by observing that I was unlikely to get poisoned unless someone really hated me; by avoiding offending or insulting people, I could perhaps avoid this hazard.[2] Besides deepening my understanding of this rich but turbulent place, he also provided insights into local land uses and practices and advice on how to adapt to village life.

During the best part of a month, my companion facilitated interviews with farmers. He enjoyed watching the local reaction to the presence of an "albino" (*bule*) in this remote corner of Sumatra. He was amused when, while visiting one remote community in a district renowned for its *ganja* production, the local police presumed that I could only be in the area to buy marijuana. A local police officer went "undercover," pretending to be a *ganja* salesman and setting up a sting. On a more serious note, he impressed on me the need to find a local collaborator (*pendamping*) who could facilitate my work in remote villages. This proved to be invaluable intelligence.

This area is ethnically and linguistically diverse. Across the three research sites studied in this research there were five languages: Bahasa Aneuk Jamee (a Minang dialect) and Bahasa Aceh in the Sama Dua area, Bahasa Kluet in Menggamat, and Bahasa Gayo and Bahasa Alas in the Alas valley. Because moving several villages in any direction usually entailed a change in language, unless I was to confine my studies to one site, learning a local language did not seem particularly useful, particularly when the national language—based on a Sumatran language similar to Bahasa Aneuk Jamee—was so widely understood.[3] Therefore, insofar as possible interviews were conducted in Bahasa Indonesia, a language in which I am a qualified interpreter. On rare occasions when an older person was less than fluent in the national language, a local collaborator would translate. A village assistant would also explain my presence in terms that made sense to local sensibilities. At times the collaborator would clarify a question or answer in the local language. As he became more familiar with the research, when reviewing an interview afterwards or writing up notes, I would check my interpretation or conclusions with the collaborator, who could clarify a point, offer his interpretation, or suggest alternative informants who might be able to elucidate an issue. On several occasions, my *pendamping* asked additional questions or pushed my line of questioning into areas that I had been cautious about investigating, in this way speeding up the research process. Moreover, the network of contacts provided through extended kinship networks and familiarity with informal leaders ensured that I had access to informants whom I would never have located on my own.

Nevertheless, over-reliance on a local *pendamping* presented its own dangers. For instance, the presence of a collaborator could inadvertently affect the responses elicited from informants. To minimize this problem, I carried out particularly sensitive interviews on my own or varied my use of collaborator. At times, when a principal collaborator was unavailable, I found other people to help me. As I became familiar with the circles to which a particular person had access, I was able to use a specific collaborator to gain access to a particular social milieu. Another problem was that illegal activities—such as clearing land in the national park and logging—were a sensitive subject in some sites, and some villagers clearly wished to conceal information from prying researchers. When informants avoided questions or showed nervousness or even hostility to a line of questioning, I learned to withdraw. I decided that it was more important to maintain access, and leave these problems until later. On at least one occasion a collaborator felt that some issues I was focusing on were too sensitive and tried to explain away issues that had emerged in interviews. Yet, as I stayed longer in the villages, I was able to observe some activities directly. I also sought to minimize this problem by interviewing as many informants as possible over as long a period as possible. Over time I gained the trust of informants. When villagers saw me come and go without any changes to the pattern of law enforcement due to the information I had gained, they were more happy (particularly after the onset of *reformasi*) to talk about conflicts and illegal activities in less guarded tones. By the end of the research I was satisfied that I had an in-depth account.

Several authors have observed that environmentalist discourse has produced overly simplified images or even misrepresented the complexity of forest-dependent communities in Indonesia and elsewhere (Milton 1996; Brosius et al. 1998; Li 1999). As Li has written of the Indonesian context, environmentalists have characterized indigenous or forest-based communities in romantically positive terms:

> They are tribal, or at least backed by centuries-old environmental wisdom; they have long been located in one place, to which they have spiritual as well as pragmatic attachments; they are relatively homogeneous, without class divisions; they are not driven by motives of exploitation and greed; they have limited consumption requirements; and their collective desires focus upon the long-term sustainable management of forest resources for the benefit of future generations. (Li 1999: 23)[4]

Brosius (1997) has argued that in the case of the Penan in Malaysia, environmentalists have tended to transform indigenous "knowledge" into "wisdom, spiritual insight, or some other such quality." In the process this discourse "makes land, resources and people inviolable." The danger with "this meta-commentary of the sacred or ineffable," he warned, was that "it imposes meaning on Penan 'knowledge'" that may be quite imaginary" (Brosius 1997: 65–66).

To move beyond such conceptions, early in the research it was important to gain a better understanding of customary *adat* institutions in this area. Attaining an accurate idea of local institutional arrangements pertaining to the forest in the most accessible and easily studied site would provide a baseline for comparing institutional arrangements in more complex and difficult sites. Sama Dua was the most accessible site and was also free of the conflicts associated with logging networks that made research more difficult in the other sites. Furthermore, I had developed the best contacts there. Therefore, I decided to begin here. In 1997 I took up residence with the family of an unmarried man whom I had met in a local *warang* (coffee shop) during 1996, and who had become a trusted friend. As well as taking me on his motorcycle through the villages of Sama Dua and Menggamat, he proved to be an invaluable *pendamping*. His family were descendants of the founders of Sama Dua and the former *datuk* (primary head) of the community, and he had extensive contacts with formal and informal leaders in Sama Dua. The payments I provided for his assistance also inadvertently helped my research: he used the money to clear and replant his *kebun* in the hills behind Sama Dua. I accompanied him to his *kebun*, and over the course of the project we spent long hours discussing strategies for facing the deteriorating economic situation.

By October 1997, I had established rapport with a network of contacts, and I was able to carry out extended open-ended interviews. It was apparent that the best informants were older villagers, especially the informal leaders, heads of local *adat* institutions and former village heads, people referred to collectively as *pakar adat* or *yang cerdik pandai* (the knowledgeable ones). These village elders officiated at *adat* ceremonies, village councils, and dispute resolution sessions (*sidang adat*) and had the most knowledge of institutional arrangements and institutional change. I supplemented open-ended interviews with these informants with interviews and discussions with other villagers. Many of these interviews and discussions occurred in the daily round of village life—during long hours spent sitting around in my host's house, watching satellite television, sipping tea in local *warung*s, accompanying my friend on a fishing expedition, a trip to his *kebun* or on a visit to friends. I was fortunate to visit South Aceh twice during Ramadan and the Idul Fitri festival that followed. During this period, working hours were reduced, and farmers came back to the village from distant *kebun* and were available for discussions late into the night.

In parallel with my activities in the villages, I was also able to consider the problems faced by two projects working to conserve the tropical rainforest ecosystem. After being encouraged by the Leuser Management Unit (LMU) to consider carrying out the research in the area, from the time of my first field trip in 1996, I regularly visited the headquarters of the joint European Union–Indonesian Government Leuser Development Program's (LDP) LMU office in Medan. LMU generously allowed me access to their extensive collection of newspaper clippings and research reports. On several occasions, informal discussions with LMU project workers in Medan enabled me to check my conclusions. During January 1999, I also interviewed LMU staff regarding the considerable obstacles facing this ambitious and controversial ICDP.

While visiting South Aceh, I encountered field staff from the World Wide Fund for Nature (WWF). Following the termination of the project in 1997, I spent many hours discussing resource management issues, community institutions and the problems WWF faced attempting to revitalize *adat* institutions as a foundation for forest management with local NGO organizers. I then decided to include Menggamat as a third case study, subsequently visiting Menggamat several times.

After a period of prolonged research during 1997, I returned to Australia briefly to collate my notes, locate gaps in my knowledge, and formulate semi-structured interviews. Generally the interviews focused on three specific sets of issues.

1. The nature and history of local agroforestry systems, including how they related to land tenure systems and the conversion of forest to agroforest;

2. The history of local village governance structures, customary governance systems, and *adat* regimes;

3. State agencies and their relationship with *adat* regimes.

Based on general questions about the issues that I needed to answer, I designed a loose set of questions to elicit further information from several key informants. I did not intend to adhere rigidly to these questions in the interview situation. Rather, these questions formed a guide for the information that I was seeking. I selected questions from this list and adjusted them according to the informant and the interview context. Before proceeding, I checked the open-ended questions first with my village collaborator to ensure that the course of the interviews was arranged in a logical order and that questions were generally phrased in terms comprehensible to local informants. During the fieldtrip in early 1998, despite the difficulties mentioned above, I was able to carry out several important interviews.

Then, back in Perth I began writing up the case studies, returning to Aceh again for two months, from December 1998 to January 1999, to check my conclusions with local informants. At this time, I made another trip to Aceh Tenggara, this time with Zulkifli Lubis, an accomplished researcher from Universitas Sumatra Utara. Zul had a great understanding of the importance of local territorial assumptions and helped me investigate their significance in Badar. With his assistance and building on previous research, I was now able to deepen and recheck the conclusions suggested by the earlier work there. By this time, I had noticed the key role of the local agents of the state. Here and later in Aceh Selatan, I interviewed local officials and journalists.

The extended time frame over which the research was carried out offered a further advantage: the last field trip occurred after the fall of Suharto. The political changes that swept Indonesia had changed public culture, and South Aceh now had a new *bupati*. Officials and journalists alike were now prepared to openly discuss what had occurred over the last five years. Moreover, by this time I had established rapport with trusted people who knew key officials well. Visiting these officials in their company facilitated in-depth interviews with several government officials.

In sum, the research utilized a range of sources, including:

Interviews with key informants, including community leaders and local government, forestry agency, and NGO staff;

Interviews with local informants including community leaders, forest farmers, forest squatters, and illegal loggers;

Participant observations and participation in village activities;

Government reports;

Consultancy and NGO reports;

Dutch colonial texts on the area;

Newspaper clippings.

Political and Economic Context

I conducted this research between August 1996 and February 1999, a period of extraordinary crisis. From 1997 an economic crisis struck Indonesia and it acutely affected the research context.[5] Then, during 1998–99, the dormant conflict between the central government and Acehnese separatists erupted.

When I first caught a bus across Sumatra in 1996, at the provincial border of North Sumatra and South Aceh, soldiers in camouflage fatigues flagged down the bus. The soldiers searched the bus and checked the identification papers of the passengers. This signified Aceh's troubled location within the Indonesian Republic. Yet, during 1996–99, in most respects South and Southeast Aceh proved to be more like a sleepy, peaceful idyll.

Long discussions in village coffee shops, by the side of the road, and in people's houses exposed a darker face, however. Older villagers described the various violent conflicts of the century, often coming back to stories of the war against the Dutch, which preoccupied the Acehnese. I was told where the bodies of Dutch soldiers had been dumped, where the Dutch had pursued local guerrillas into the forest, and stories of Dutch planes strafing Acehnese cattle during the revolution. They also told stories of the Japanese occupation and the Darul Islam dan Tentara Islam Indonesia (DI/TII) insurrection against the central government during the Sukarno period.

Aceh is Indonesia's nearest point of contact with the wider Islamic world, and Islam entered the Indonesian archipelago via Aceh. Even today Indonesians refer to Aceh as the "veranda to Mecca" (*serambi Mecca*). My village friends retained animist traditions alongside a folk Islam and ate with me in clandestine restaurants during the fasting month. Nevertheless, the Acehnese cultural identity is closely tied to Islam. In pre-colonial times, Aceh had an organized kingdom, and was one of the areas of Indonesia that had long functioned like a state. For instance, the Aceh sultanate had opened up diplomatic relations with foreign powers, including the British and the Ottoman empires, and had effectively resisted colonial conquest for centuries. Even after the Dutch finally conquered Aceh at the end of the nineteenth century, Acehnese resistance continued for decades. In one of the field sites I visited, villages had waged a guerrilla campaign against Dutch forces as late as 1925–28. Even though the Dutch finally subdued the area for some decades, the people here retained a strong sense of their separateness.

During the 1940s, Aceh played a key role in Indonesia's struggle against the Dutch. At the founding of the Indonesian Republic, some Acehnese intellectuals had wanted an Islamic state, but Indonesia's nationalists opted for a more secular state more accommodative of other religions. During the first decade of the republic, the central government abolished Aceh's separate military command, and incorporated Aceh into the province of North Sumatra, with its capital in Medan. As most commercial activities now occurred at Medan, Banda Aceh lost its historical role as a commercial center for the Acehnese hinterland. Moreover, the Acehnese resented the fact that Batak, Javanese, and other outsiders gained key positions in the provincial government. Consequently, during the early 1950s, significant numbers of Acehnese believed that the republic threatened Aceh's unique cultural and religious traditions and its historical autonomy. They felt that the republic, by overlooking the special role of Aceh during the revolution, had failed Aceh. On 21 September 1953, the Indonesian Islamic State (*Negara Islam Indonesia*) was proclaimed in Aceh, and conflict broke out between the Acehnese and the Indonesian Republic.[6] Over the next few years over four thousand Acehnese were killed in the conflict known as the Darul Islam dan Tentara Islam Indonesia (DI/TII). Finally, in response to local grievances, in 1959 the central government tried to appease the Acehnese by granting the province the status of special region (*Daerah Istimewa Aceh*).[7]

In South Aceh, during a trip through the forest, an older villager showed me where Acehnese guerrilla forces had hidden from the Indonesian army during this period. One day, when we were admiring the picturesque view from the top of a hill, a villager pointed out a coconut tree just down the beach where they had buried the "communists" massacred during 1965–66. Yet, despite these grim stories, Aceh's troubles seemed to be a thing of the past. People would grumble about Suharto and the "*korupsi*" in far-away Jakarta, but no one expressed a wish to see Aceh split from the Indonesian Republic. On several occasions informants spoke proudly of the ethnic diversity of Southern Aceh and several informants told me that they did not want trouble. "You are safe here," a village head reassured me, "and if you have any problems, please come and see me."

If there were problems in Aceh, locals attested that they were relegated to the far north, in Pidie and Aceh Besar, areas reached only after a ten-

or twelve-hour bus trip. A villager with whom I spent many hours told
me of a visit he had made to an uncle, a policeman stationed in North
Aceh. While visiting the uncle, my friend had found work in the small
town where he was stationed. One day he was walking down the street
when a group of men in army fatigues appeared carrying guns. "They
looked just like soldiers," he recalled, "but they were GPK." This was
the New Order acronym for "security-disturbing movements" (*Gerakan
Pengacau Keamanan*), an Orwellian term used to describe those involved
in activities deemed to be endangering "state security." The Free Aceh
Movement (*Gerakan Aceh Merdeka,* or GAM) was held to be GPK. People started shooting at regular Indonesian army soldiers at the other end
of the village. "Many villagers were caught in the cross-fire," he recalled,
"it was very dangerous." The next day my friend fled back to peaceful
South Aceh. A few days later he heard that his uncle had been killed.[8]

Another villager recollected how in the early 1990s he had fled northern Aceh during the "troubles" there. "At this time," he said, "bodies
were turning up on the side of roads and in villages. In some villages
there is nothing left but widows . . . They killed men, women, even children—shot them all. What about social values? What about people's humanity?" "Who?" I asked. "Certain parties [official, i.e. military] on duty
(*oknum tertentu yang bertugas*). In the end I had to flee and come back
here . . . The GPK don't want to be part of Indonesia," he explained,
"while Pak Suharto wants to control everything."[9]

The GAM movement emerged after Tengku Hasan M. Di Tiro proclaimed an independent state of Aceh in 1976. Di Tiro rejected the compromise that had ended the DI/TII rebellion. He believed that Acehnese
aspirations could no longer be met within Indonesia. In the 1980s, GAM
engaged in an episodic guerrilla conflict with Indonesia's security forces.
After GAM guerrillas received training in Libya, the conflict in northern
Aceh intensified. In response, the Indonesian army instituted a severe anti-
insurgency campaign. In 1990 the government declared Aceh a "military
operational area" (*Daerah Operational Militer,* or DOM). During the
subsequent years of military rule, the military suppressed GAM, com-
mitting many human rights abuses in the process. Amnesty International
estimated that there were two thousand extra-judicial killings during the
DOM period from 1990 to 1998 (Sawitri et al. 1999; Lloyd 2000).[10]

At the onset of the East Asian financial crisis in 1997, people in Aceh
appeared to be more worried about the economic crisis than concerned
with resuming Aceh's claim to independence. As the Indonesian rupiah
crashed, the prices of basic commodities rose dramatically. At the same
time, forest fires ravaged Sumatra, enveloping the area with thick smoke
for months.[11] Concerned about getting lost in the haze, farmers were at

times reluctant to visit distant *kebun,* and fishermen feared losing sight of the coast. The crash outside Medan of a Garuda Airbus carrying a number of prominent people—whose bodies were brought back to the village that I was visiting—added to the sense of apocalypse.

When I returned to South Aceh in February–March 1998, the situation had deteriorated further. The fall in the value of the rupiah had pushed up the price of tires and spare parts, and the public mini-buses serving the mountainous route between Tapaktuan and Medan had fallen into neglect. On one occasion a mini-bus I was on blew a tire on a mountain road, almost cascading down a precipice.

Moreover, during the tumultuous period of early 1998, rumors spread of robberies and violence. Villagers became increasingly preoccupied with coping with the economic crisis, and I found it difficult to proceed with interviews. Many became anxious about price fluctuations in basic goods and worried about meeting their essential needs. At this time, virtually all able-bodied men had taken to the hills to open new plots, and there were few people left in the village.

In 1998, following the fall of Suharto, Indonesia entered a new era of openness, and across the archipelago people began to discuss the abuses of recent history. As villagers in northern Aceh spoke of the human rights violations there, NGOs and human rights activists began to help investigate. Investigators unearthed mass graves bequeathed by the previous decade of military operations. At this time, a sense of outrage and injustice mixed with economic grievances against the Jakarta government. In August 1998, General Wiranto, the head of the Indonesian armed forces, cancelled Aceh's status as a military operational area and apologized for the earlier abuses.

Even so, by December 1998 the mood had changed: people were openly discussing how much better off Aceh would be as a separate country. Aceh had long demanded a larger share of the revenues it was sending to the center. By one report, 11 percent of the Indonesian national budget revenue came from the export of Aceh's natural resources. The amount returned in the form of central subsidies, the main source of regional income, was relatively small (Erawan 1999). For instance, a report stated that Jakarta siphoned off about $4 billion a year in natural gas revenues and sent back less than 1 percent of that in development aid (Moreau 1999).

In my interviews in villages, some *adat* heads and Islamic preachers said that Aceh should become independent. During one visit to a government office, the officials there began to discuss this issue, openly expressing their wish for Aceh to leave the republic. Villagers angrily spoke of the abuses perpetrated by the military in northern Aceh. Many also resented the revenue taken from Aceh. "If we were independent," several people told me, "we would be as rich as Brunei." An older villager who had lived through a long litany of conflicts said that things hung in the balance. "It is not clear yet whether Aceh will go for freedom again," he said. "It could go either way. I don't know whether you will be able to come back here by June."[12]

On a visit to the remotest site of Menggamat, we were about to take a boat up the Kluet River with a village head. After a discussion with some villagers, the village head changed his mind, telling me that he could not go that day because of other business. Later I found out that he had discovered that a band of Indonesian special forces (*Kopassus*) had just moved up the river in pursuit of GPK—the so-called "security disturbing movement" GAM.

On my last day in Banda Aceh at the end of January 1999, a taxi driver confirmed that things were heading towards conflict. "We cannot accept it any more," he explained. "Look at Banda Aceh," he said, pointing at the simple buildings found in the center of the provincial capital. "If you compare Banda Aceh to Jakarta or Medan, there is nothing, no development—we have been left behind. Yet how much gas, oil and gold is found in Aceh? Everything has been taken away to Jakarta . . . All of Aceh are solidly together on this . . . " I left South Aceh with a sense of foreboding about another confrontation between the central government and the Acehnese.

The opportunity to avoid conflict that had emerged in 1997–98 proved fleeting. In March 1999 President Habibie went to Aceh and promised an investigation into human rights abuses. Many Acehnese held out hope for change. However, Jakarta's politicians, preoccupied with political intrigues at the center and the East Timor issue, were either reluctant or unable to follow up on this promise. A journalist visiting Aceh concluded that that the military was refusing to withdraw from Aceh or to allow the prosecution of officers responsible "for thousands of murders, rapes and disappearances during the official nine-year campaign against the separatists, which ended in 1998" (Guardian Weekly 1999). In January 1999, demonstrations in northern Aceh were already leading to conflicts—people began dying again. Over the ensuing months, support for separation from Indonesia flourished. In December 1999 an Acehnese activist told an English reporter that if the military had withdrawn at the beginning of 1999 and the government had given Aceh greater autonomy, the government "could have won a referendum on autonomy versus Independence." But by the end of 1999, there was "no chance" (Guardian Weekly 1999). GAM took up arms in the countryside while a student-led movement worked in the cities to secure an East Timor–style referendum on independence. As the military continued its abuses and terror campaign, dissent broadened to a wide cross-section of the population. Students and others who would not have supported GAM in the past began to do so (Sawitri et al. 1999).

By the middle of 1999, the conflict between Acehnese guerrillas and the military began to creep south towards South and Southeast Aceh, and by August I heard from friends in South Aceh that they no longer felt safe. Demonstrations and conflicts with the military spread even to the villages of South Aceh. By late 1999, newspapers reported shootings, disappearances, and murders in the peaceful villages where I had spent so much time in 1996–99.[13] I learned that GAM or the military had beaten up, intimidated, detained, or murdered people I had known well—including villagers I had interviewed during languid tropical days, people I had enjoyed a coffee with, and in one case a man on whose motorcycle I had sat during several trips up the nutmeg–covered hills of South Aceh. Shots were now ringing out over leafy village lanes, and bodies were turning up on the road.

Glossary

Adat	Customary law, custom, customary authority system.
Bambu	A standard bamboo container used to measure volume in Aceh. One *bambu* holds 1.3 kilograms of nutmeg fruit.
Bupati	District (*kebupaten*) head, the chief official in a district.
Camat	Leader of sub-district head (sub-regency, or *kecamaten*).
Cukong	An entrepreneur or financial backer, a businessman who funds logging operations.
Daerah	Region.
Dakut	Headman.
Damar	A resin tapped from *Dipterocarp* trees, especially *Hopea* and *Shorea* species. Damars are used in the manufacture of paint, batik dye, sealing wax, printing ink, varnish, linoleum, and cosmetics.
Desa	An official village, an administrative unit consti-

tuted in accordance with the 1974 village government law.

Dinas	Provincial government office.
Hutan	Forest.
Hutan adat	An *adat* forest, that is, a forest subject to local *adat* authority.
Hutan negara	State forest, area mapped as state forest land.
Jagawana	A forest guard, an employee of the national park or line agency of the forestry department.
Kabupaten	District.
Kampung	Village.
Kebun	Mixed-crop permanent garden, permanent agro-forest garden.
Kecamatan	Sub-district administration.
Kehutanan	Forestry.
Kemiri	Candlenut.
Kepala desa	Village head.
Kepala mukim	Head of village league or *mukim* (known in Kluet as *kemukiman*). See also *mukim*.
Keucik	The village head (*kepala desa*).
Kluet	Ethnic group found around Kluet River in South Aceh.
Ladang	Land area subject to temporary cultivation, swidden plot.
Leuser Ecosystem	The area subject to LDP's conservation concession extending over approximately 2 million hectares.
Minangkabau	Ethnic group from West Sumatra. Migrants from this area moved to South Aceh, where they are known as Aneuk Jamee.
Mukim	In historical Aceh the *mukim* consisted of the villages and hamlets that shared a mosque and over

time had come to consider themselves to be a single community under the leadership of a figure known as the *imam mukim*.

Nilam
Pogostemon cablin, a cabbage-sized, leafy plant that grows to a height of thirty to seventy centimeters. Distilling the dried leaves of *nilam* produces patchouli oil.

Oknum
Literally, "a person acting in a certain capacity." A euphemism for one who abuses his position or otherwise acts contrary to his official responsibilities.

Penebangan liar
Literally, "wild logging." Illegal logging.

Reformasi
A movement for political change in Indonesia that began in 1997.

Retribusi
A tax or fee levied by government usually in return for a service.

Sawah
Wet rice cultivation.

Seuneubok
A collective institution of farmers with gardens (*kebun*) along a forest path. The institution of *seuneubok* controls aspects of farming practice in a specific *seuneubok* territory.

Tauke
A boss in control of logging operations, either independently or as agent of *cukong*.

Uang pembangunan
"Development fee," a fee paid for access to community forest.

Uleebalang
Principal local ruler. In historical Aceh an *uleebalang* ruled over a group of *mukim*.

Notes

CHAPTER I

1. Bromley and Cernea (1989) observed that there is some confusion between institutions and organizations—with banks, hospitals, schools, and government agencies often called "institutions." To avoid this confusion, this study uses the term that they adopted, "institutional arrangements." They argue that institutional arrangements define the property regime over land and related natural resources. For example, they note that "these institutional arrangements define (or locate) one individual vis-à-vis others, both within the group (if there is one), and with individuals outside the group" (Bromley and Cernea 1989).

2. The term *adat* was developed in Dutch colonial legal discourse, because colonial administrators used indigenous institutions as a basis for the colonial system of indirect rule. From the early twentieth century on in particular, a group of legal scholars associated with Van Vollenhoven ("the Leiden school") endeavored to find a way to recognize indigenous rights over land and prevent the alienation of Indonesian land to European interests, especially the rubber and sugar industries (Otto and Pompe 1989). In the post-colonial period, *"adat"* became the generic term for describing the plethora of religious beliefs, customary institutions, and social practices throughout the Indonesian archipelago that have traditionally defined collectivities and established rights and obligations within and between them. According to Bouen, "The term '*adat*,' borrowed from Arabic and taken into general usage throughout archipelagic South-East Asia, encompasses an extensive, perhaps unique, range of meanings. *Adat* may cover single cultures . . . or the archipelago as a whole. In each of these geographic frames the meaning of *adat* varies from a general sense of cultural propriety and social consensus—sometimes explicitly measured against Islam—to specific forms of speech and ritual action, to a set of local sociolegal procedures" (Bowen 1988). During

the Suharto period, the term "*adat*" came to be used in a more narrow sense. As a University of North Sumatra report on *adat* in Southeast Aceh suggested, the term often tends to be used in the narrow sense of the English word "custom," for instance to describe marriage ceremonies and inheritance law (Lembaga Penelitian dan Pengabdian Masyarakat 1978). In a similar way, as Acciaioli observed in Central Sulawesi, in contemporary government discourse, the term had been truncated to include only such things as etiquette, the way to sit before elders, and the way to dress before guests. "Almost by definition beliefs and even procedures for ordering social life are simply not considered *adat* or even culture" (Acciaioli 1985). Clearly, local, government, and NGO usages of the term may be quite different things. However, NGOs and others have continued to use the term *adat* to refer to the customary institutional arrangements of local communities, including those pertaining to the access and use of natural resources. The term has continued to be used (as I use it in this study) as a shorthand for the customary institutional arrangements of local communities. See discussion in Holleman 1981; Warren 1993; Kahn 1993; Li 1997; Burns 1989; Burns 1999; Fitzpatrick 1997.

3. In 1994, Iwabuchi observed that the "ethnic groups in the southern part of the Special Province of Aceh, including the Alas, the Singkil, the Kluet, the Jamee, and the Tamiang, are the least documented in the whole of Sumatra" (Iwabuchi 1994a). In that work Iwabuchi studied the social life and kinship structures of the Alas. Bowen (1991) researched the history and social customs of the Gayo of Central Aceh. However, with the exception of the report by Koesnoe and Soendari (1977), Rijksen and Griffiths' *Leuser Development Programme Masterplan* (1995), and other consultancy reports commissioned by the Leuser Development Programme, little work has been done on this remote area of Sumatra. This volume, a comparative study of customary institutions and natural resource management in a Kluet, a Jamee, and a mixed Alas/Gayo community, contributes to filling this gap in the literature.

4. For an application of this concept to an Indonesian context, see Li 1999.

5. The TGHK mapping exercise referred to here was later revised by the Regional Physical Planning Program for Transmigration (RePPProt). This process is discussed in greater detail in McCarthy 2000.

6. In 1999 a new law was passed to replace UU 11/1967. While making some allowances for *adat* rights in forest areas, the new law (UU 41/1999) also retained the concept of state forest zone (*Kawasan Hutan Negara*). Following the forest fires of 1996–97, the Forestry Department (Dephut) also changed its name to the Department of Forestry and Estate Crops (Dephutbun).

7. So-called "non-cultivation areas" were "designated with the primary function of preservation of the environment" and in turn were divided into protection forests and nature reserves. In contrast, "cultivation areas" were designated for productive purposes and were divided between cultivation, forest production, and transmigration functions (Act No. 24/ 1992 §1).

8. For a more detailed discussion, see McCarthy 2000b.

9. See McCarthy 2000b.

10. McCarthy 2000b.

11. Sunderlin and Resosudarmo (1997) distinguish between the agents involved in changing land use, the immediate causes influencing the decisions of agents, and the underlying causes of changes to forest cover. In this framework, the underlying causes include overarching national, regional, or international forces that influence the "decision parameters" of agents (Sunderlin and Resosudarmo 1997: 15).

12. See KLH and UNDP 1998a and 1998b; WWF 1998a and 1998b.

13. See McCarthy 2000b.

14. For further discussion, see Ostrom 1990; Ostrom 1992a and 1992b; Ostrom 1997; Edwards and Steins 1998; Oakerson 1992; Selsky and Creahan 1996.

15. E. Hoebel, *Anthropology: The Study of Man* (New York: McGraw-Hill, 1996), quoted in Berry (2000: xxii).

16. For recent critiques of this collective action approach, see Mosse 1997; B. Campbell et al. 2001; Johnson 2003.

17. Cf. Moore 1986.

18. Richard Fox, *The Lions of the Punjab: Culture in the Making* (1985), cited by Gupta (1995: 398).

19. As I describe later, the networks involved in timber extraction that are discussed in this study follow this pattern, in a similar fashion to that which Peluso (1992c) found in Java's teak forests.

20. Marc Galanter, "Justice in Many Rooms: Courts, Private Ordering and Indigenous Law," *Journal of Legal Pluralism* 1 (1981), cited in Merry 1988.

21. "Under these conditions," Kahn concluded, "to talk of 'the state' as an institution that is somehow separate from the rest of society is somewhat misleading" (1999: 96).

22. Arghiros has suggested that the term be reserved for "dyadic, multifaceted and asymmetrical relations where there is an evident ongoing personal and reciprocal element to the relationship" (2001: 7).

23. In lowland, irrigated, rice-producing areas, village stratification tends to be based on unequal access to property, particularly land. In contrast, he has suggested, land in upland areas is more equally distributed, and consequently it is difficult to identify a discrete class of landlords.

24. For recent discussions of corruption, see Larmour and Wolanin 2001; Aditjondro 2002; Lindsey and Dick 2002; Sulistoni et al. 2001. For a journalistic account of similar problems in developed contexts, see Palast 2003.

25. For instance, corruption can extend to the provision of sexual services. When a direct approach to a person fails, actors seeking access to resources may also threaten with physical violence those who are initially reluctant to comply.

26. For a compelling discussion of how these sorts of arrangements have persisted over time in the timber sector in East Kalimantan, see Obidzinski 2002. Obidzinski notes that "the lower civil service classes were free to pursue entrepreneurial activities of any kind as long as they displayed political loyalty, cumulatively did not undermine macro-economic conditions in the country and shared proceeds from illegal business by providing 'payments'" (2002: 31).

27. Cf. Berry 1994, 1989, 2001.

28. The timber concession system had already been well studied. See Hurst 1990; FAO 1991; Potter 1991, 1993, 1996; Mubyarto 1992; Ahmad 1993; Ascher 1993; Barber, Johnson, and Hafild 1994; Dauvergne 1994; Peluso, Vandergeest, and Potter 1995; Barber 1998; Barr 1998, 2000, 2001; Latin 1998.

29. See Appendix.

30. For a discussion of multi-sited ethnographic strategies, see Marcus 1998.

31. The research was less open-ended that this suggests, however; from the outset of the project, the pre-existing literature indicated "what phenomena deserve particular attention and what other phenomena can be disregarded or accorded less attention" (Whyte 1984). At the same time, the qualitative approach taken in this study meant that the research was grounded in the social world being studied. Several authors have suggested that, in the course of research and analysis, researchers need to be wary of imposing exogenous categorical frameworks rooted in an *a priori* research agenda upon research data (Patton 1990; Strauss and Corbin 1990; Emerson, Fretz, and Shaw 1995). This implied that *a priori* conceptual schema—suggested by rational choice theory and institutional economics—might do violence to the categories that might emerge in the research process itself. On this note, Ostrom's framework analyzed local institutions from the perspective that seeks to understand local institutional arrangements in terms of conditions that support the emergence of "successful" community institutions for the management of a natural resource. In Ostrom's terms, a successful institution for governing a CPR is one that (at the very least) organizes to produce an outcome—the efficient, sustainable use of a resource (Ostrom 1992b). Yet, this seems to be a rather *a priori* definition of "success." For the communities discussed in this study, sustainable use of a resource is not necessarily the most obvious criterion of success for their institutions. Indeed, as later discussion will demonstrate, the communities living on the edge of the Leuser Ecosystem's forests organize for many reasons. These range from peacefully resolving disputes, supporting village livelihoods, and protecting local property rights, to nourishing the identity of village and farming communities and maintaining indigenous socio-religious traditions. Local institutional arrangements manage the conversion of forest into gardens and, in times of economic need, allow for unsustainable logging of local forest reserves. While the criteria for success built into Ostrom's framework provide a particular perspective on institutional development, these criteria and the set of assumptions that underpin them do not necessarily provide the best interpretative lens for understanding local institutions.

32. This distinction derives from Selsky and Creahan (1996).

33. For overviews of the political ecology approach, see Neumann 1992; Bryant 1992, 1998; Adger et al. 2001; Kirkpatrick 2000.

34. See Escobar 1999.

35. I am grateful to Peter Brosius for this point.

CHAPTER 2

1. Without making extensive use of Dutch colonial archives, this research has relied on published Dutch-language material. Nineteenth-century historical sources concerning South Aceh describe small self-governing coastal principalities

(*negeri*) centered on coastal port towns dealing in pepper. While Acehnese chiefs ruled some of these principalities, others were predominantly Minangkabau. In the course of the Aceh War (1873–1903), in 1899 Netherlands Indies forces under Colijn established colonial control of South Aceh. The Dutch general Van Heutsz chose Lieutenant Colijn to establish Dutch power in the area south of Meulaboh. The colonial forces took Tapaktuan on 1 June 1899. They occupied Tapaktuan itself easily; Acehnese resistance was centered on the Blangpidie area, seventy kilometers to the north (the most southern Acehnese *uleebalang*). Guerrilla war continued in the forest areas for some years, especially in the remote mountainous areas of Gayo Lues behind Blangpidie (Doup and Bakar 1980).

2. Cf. Moore 1986.

3. When Veth visited the South Aceh coast in 1873, just before the outbreak of the Dutch wars against Aceh, he found sizable Minangkabau (Aneuk Jamee) colonies at Bakengon, Tapaktuan, Labuan Haji, Susoh, and Meulaboh; there were also a number of smaller communities, including at Sama Dua just north of Tapaktuan (Veth 1873). According to Veth, at this time Tapaktuan's population consisted of "Malays," primarily from the Minangkabau district of Pasanan, as well as Priaman, Padang, as well as people from Natal (near modern-day Sibolga) and Bengkulu in South Sumatra. He wrote that the people had moved to the area early in the nineteenth century to find a refuge from Paderi dominance or to escape punishment for misdeeds in their places of origin. Although he did not speculate on the issue, it was also likely that they were attracted by the pepper boom. In the seventeenth and eighteenth centuries the most important pepper ports had been in the Minangkabau area, but at the beginning of the nineteenth century the main center of pepper export had moved to the coast of South Aceh (Kreemer 1922). Besides these so-called "Malays," Veth found that the population consisted of a few hundred slaves from the island of Nias as well as the families of Nias people who had gained their freedom. In Tapaktuan itself he found a market and well-built wooden houses clustered around a bay with a small island. Although the bay offered an excellent anchorage, it was completely open and not a safe haven from winds blowing from the south. In an age of warfare, he estimated that the township could raise three hundred soldiers (Veth 1873).

4. Interview with former village head, 12/11/1997. The structure of local *adat* and village government institutions will be discussed later in this chapter.

5. Interview, Sama Dua, 6/1/99.

6. Interview with former *kepala mukim*, Sama Dua, 3/1/99.

7. Interview with *adat* leader, Sama Dua, 1/1/99.

8. This was an underestimate. In addition to the pocket of Aneuk Jamee around Tapaktuan and Sama Dua, there were other Aneuk Jamee pockets at Susoh (near Blangpidie), on the island of Simeulue, and elsewhere along Aceh's southwest coast. The history of Minangkabau and Acehnese settlement is an interesting one. The area is now considered a part of Aceh, but in the seventeenth and eighteenth century South Aceh was a peripheral area located between Banda Aceh and the Minangkabau hinterland of West Sumatra. Minangkabau first settled further north around Susoh during the seventeenth century, but Minangkabau migrants probably did not settle the Tapaktuan and Sama Dua area until

the early years of the nineteenth century. Around the same period or a little later, sizable settlements of Acehnese also developed around the Minangkabau communities. Although the Minangkabau migration into southern Aceh had began as early as the seventeenth century, after the Painan treaty, significant numbers of Acehnese did not settle in South Aceh until the eighteenth and nineteenth centuries (Ahmad 1992). At that time, the Acehnese Sultanate was the predominant political power along the west coast, and the sultanate attempted to extend its long-standing control of the revenue generated by the pepper trade. Merchants, *syahbandar* (port masters), and *uleebalang* tried to open plantations and extend the influence of the Aceh Sultanate to this southern frontier. Farmers from Aceh Besar and Pidie also moved to the South Aceh area in the eighteenth century, not because of food scarcities in their home territories, but rather to take up the agricultural opportunities generated by the high demand for pepper and the boom in pepper prices (Ismail 1991). Large numbers of people from Pidie followed, withdrawing from the bloody Aceh war that waged at the end of the nineteenth century. These people finally founded the Blangpidie principality eighty kilometers north of Tapaktuan, the most southern *uleebalang* potentate (*keuleebalangan*) in Aceh (Ahmad 1992).

9. At the northern end of the valley the pocket of these people ends, however, and the villagers are Acehnese.

10. According to colonial sources there was continuing conflict over whether to apply the matrilineal Minangkabau *adat* or the Acehnese *adat,* which was more heavily influenced by Islam, particularly vis-à-vis inheritance. For instance, when a *datuk* (headman) died, by Minangkabau tradition the title would be transferred to the *datuk*'s nephew or cousin (*kemenakan*) from the female line rather than to his son. Acehnese *adat* applied more strictly Islamic precepts, however; if they were applied the title would be inherited by the son. In a context where both traditions co-existed, following the death of a *datuk*, the parties wishing to attain the position would appeal to either tradition. On the one hand, the son of a deceased *datuk* would reinforce his claim to the title by appealing to support from Acehnese and Islamic leaders for a more thoroughgoing application of Islamic law. On the other, the *datuk*'s family would appeal for a retention of Minangkabau tradition. Colonial sources reveal that there was some of this switching between Minangkabau and Acehnese traditions, but that finally Acehnese traditions prevailed (BKI 1903; 1912).

11. Interview, *adat* leader, Sama Dua, 6/1/99.

12. For a discussion of the problems associated with the construction of discrete, unchanging, but distinctive cultural identities (such as those claimed by individual groups) within a context of continuous cultural variation, ambiguity, and cultural hybridity, see Kahn 1999.

13. The Dutch figures were for a smaller area: at that time the Sedar settlement was not included in Sama Dua.

14. The population censuses of 1980 and 1990 for South Aceh show an average growth rate for the district as a whole of 2.2 percent per year, close to the average population growth for rural Aceh of 2.3 percent per year. However, the rate of population growth has been falling, along with average family sizes. "To-

tal fertility" is the number of children a woman would have if she passed through her childbearing life at the fertility rate that applied in a particular year. In 1970, total fertility in Aceh was 6.3; it dropped to 4.4 in 1990. Census figures show that over this period net migration in South Aceh was close to zero. However, data from a more recent but less accurate inter-censual population survey show a decrease in this growth rate over the period 1990–95 to 1.6 percent, most likely due to falling fertility and some out-migration. See Jordan and Anhar 1997.

15. On *damar*, see the Kluet case study.

16. Interview, Sama Dua, 2/2/98.

17. As I discuss below, a collective institution of farmers having *kebun* on a forest path is known as a *seuneubok*. A *seuneubok* controls aspects of farming practice in a specific (*seuneubok*) territory.

18. Interview, Sama Dua, 2/2/98.

19. For other accounts of this phenomena in Sumatra see Mary and Michon 1987; Michon et al. 2000.

20. Interview, Dinas Kekubunan, Tapaktuan, 12/1/99.

21. Interview, Sama Dua, 13/1/99. Then as now, agroforest patterns did not seem to have been uniform and might have varied with price variations. According to one older informant, in colonial times, villagers planted rubber trees between the clove trees. The clove trees were spaced approximately eight meters apart.

22. This agricultural system involving the cultivation of dryland rice in shifting plots alongside perennial cash crops is somewhat reminiscent of the mixture of smallholder rubber and swidden agriculture that Dove describes for the Kantu' Dayak of West Kalimantan. There, by combining swidden agriculture and perennial cash crops, the Kantu' farmers were able to participate in the cash economy on their own terms. They could meet their subsistence rice needs from swidden fields while avoiding many of the risks involved in exclusive cash cropping (see Dove 1993).

23. See also Dove 1985, 1986; Freeman 1955; Spencer 1966.

24. Interview, Sama Dua, 13/1/99.

25. Interview, Sama Dua, 14/2/98.

26. Genus *Ficus*, family *Moraceae*.

27. Interview, Sama Dua, 18/1/99.

28. Agroforestry has been defined as "a dynamic, ecologically based, natural resource management system that, through the integration of trees on farms and in the agricultural landscape, diversifies and sustains production for increased social, economic and environmental benefits for land users at all levels" (International Council for Research into Agroforestry [http://www.cgiar.org/icraf]). See also B. O. Lundgren, "What Is Agroforestry?" *Agroforestry Systems* (1982) 1 (1): 7–12.

29. Interview, former village head, 12/11/97.

30. Interview with nutmeg wholesaler, Sama Dua, November 1997. A *bambu* is a small bamboo container commonly used to measure volumes. It contains 1.3 kilograms of nutmeg fruit.

31. The road through here was only asphalted in the early 1980s; before

that time it took ten days to reach Medan, a trip that involved passing through extended jungle and fording many rivers. Today this trip takes approximately twelve hours.

32. Interview with nutmeg wholesaler, Sama Dua, 14 November 1997.

33. Interview, Sama Dua, 14/2/98. Poorer farmers from hamlets located farther from the road were isolated from the choices that outside agriculture offered those with networks extending further into the cash economy of Tapaktuan township.

34. As explained below, a *seuneubok* is a specific area consisting of all the forest gardens that lie along a certain forest path.

35. Interview, Sama Dua, 14/2/98.

36. Interview, Sama Dua, 20/1/99.

37. A WWF-IP report noted that birds known to prey on nutmeg pests included Burung Murai Batu (*Copsychus malabaricus*), Murai biasa (*Copsychus saularis*) and Kukang (*Nycticebus coucang*) (WWF n.d.).

38. I have been unable to locate the scientific name of this insect. *Bubuk* means powderlike; therefore it is likely that the name refers to the way the insect produces powder or makes branches powderlike while digging into them.

39. Interview, Dinas Perkebunan, 14/1/99. This theory bears out the concern of Menggamat *adat*, in which *adat* rules attempt to protect the rice fields from pest infestation caused by logging. A village head I interviewed in the course of this study held that, because logging of the forest could disturb insect pests that then infest the rice crops, in Menggamat the *adat* had rules concerning where and when logging could occur (see next case study).

40. Interview, Dinas Perkebunan, 14/1/99. In May 1997 local informants estimated that the nutmeg plants had killed around 20 percent of the nutmeg trees in each *kebun* (Serambi Indonesia 1997). A few months later *Kompas* reported that the local government estimated that over 2,202 trees had been attacked by the pest (Kompas 1997).

41. The US dollar value of the rupiah declined from Rp 2,400 in July 1997 to lows of Rp 16,000 to 17,000 in 1998. However, the rupiah traded in the 8,000–9,000 range during most of the 1997–99 period (Sunderlin et al. 2000).

42. Soaring food prices and a stronger rupiah since October 1998 had gradually made real prices move towards their pre-crisis levels (ibid.).

43. *Nilam* (*Pogostemon cablin*) is a cabbage-sized, leafy plant that grows to a height of 30 to 70 cm. By distilling the dried *nilam* leaves, farmers produce patchouli oil, a product used in cosmetics, perfumes, and aromatherapy.

44. Interview with *nilam* farmers, Manggamat, 6/2/98.

45. This trend was so pronounced that, when I returned to the area in February 1998, few informants were available for interviews. In South Aceh in late 1997, *nilam* seedlings sold for Rp 25 per seedling, and farmers could harvest *nilam* only after six to eight months. Since farmers required capital to move into *nilam* cultivation, poorer farmers had difficulty making the transition.

46. Interview, official from the District Agricultural Office, Tapaktuan, 14/1/99.

47. Interview, Sama Dua, 1/1/99; Interview, Dinas Perkebunan, Tapaktuan, 12/1/99.

48. On sloping land, the *nilam* needs rain, and farmers need to plant before the wet season.

49. Interview with furniture manufacturer, 2/12/97.

50. Interview, Sama Dua, 20/1/99.

51. Interview, Sama Dua, 20/1/99

52. Interview, Sama Dua, 20/1/99.

53. Due to the sensitive nature of this issue in Sama Dua, it was difficult to clarify the exact regimen governing these interactions.

54. Van Dijk has distinguished "tenure" as a different form of control over property from "territoriality." Individuals who invest labor and other assets in intensively developing a resource such as land seek to gain more secure property rights over the benefits derived from that resource. Tenure "may be conceived of as sets of rules and practices specifying whom is to get access to land at which time and in which place" (van Dijk 1996: 18). Tenure tends to relate to land under more permanent and intensive forms of use such as cropping. Yet, the ecology of other resources does not readily lend them to be managed by instituting private property or individual tenurial rights. See Berkes et al. 1991. For instance, in the case of common pool resources (CPR) such as fishing grounds and forests, many actors can extract resource units at the same time. Moreover, CPRs tend to be large, and efforts to exclude potential beneficiaries costly. See Ostrom (1997). In many of these systems the occurrence or "subtractability" of dispersed or non-renewable resources was unpredictable, and resources could be harvested without the constant investment of labour or assets associated with settled agriculture. Consequently the management of many CPR systems has fallen under the more diffuse form of control known as "territoriality." Territoriality is a claim over a territory that is maintained vis-à-vis other groups (van Dijk 1996: 19).

55. Li (1999: 12) has argued that territorial control, "the direct attempt to regulate the relationship between population and resources," was not a feature of tradition bound pre-colonial political systems. However, local groups did have some indigenous concepts concerning the extent of their territory. An example of this is the notion of *pintu rimba* (discussed in Chapter 6).

56. According to colonial records, at first the Minangkabau settlers around Tapaktuan retained the clan (*Suku*) structure of social organization used in their homelands. However, over succeeding generations the memory of the old *Suku* connections was lost and a new Suku arrangement "spread its roots." Henceforth people belonged to a *Suku* named after their place of origin. For instance, there were Suku Priaman and Suku Pasaman. While the *Suku* identity was strong, the Dutch source notes that the Aneuk Jamee did not faithfully follow the original West Sumatran *adat*: for instance, marriage was allowed within the *Suku*. The head of a *Suku* went by the title of *datuk* (BKI 1912).

57. Writing of the situation amongst the Aneuk Jamee of Tapaktuan, a colonial source notes that the *keucik* did not have any authority. One source mentioned their subservient role under the *datuk*, a role typified by the saying "*tangan kaki datuk*" (hands and feet of the *datuk*) (ibid.).

58. For discussions of this process in Indonesia, see Warren 1993; Kahn 1993; F. Benda-Beckmann 1979. For Africa, see Moore 1986; Chanock 1985.

59. For a discussion of the effect of this process on the Minangkabau, see Kahn 1993; for Bali see Warren 1993. For more general accounts see Holleman 1981; Burns 1989.

60. Eventually, in the course of a dispute within the community, the southern settlement established a second *datuk* (BKI 1912).

61. For another account, see BKI 1903.

62. For a discussion of the variable way this worked out across Indonesia, see Li 2000. For a discussion of the process in a neighboring community, see Mc-Carthy 2002.

63. Interview, Sama Dua 1/1/99.

64. Interview, Sama Dua 1/1/99.

65. *Kemenyan* is an incense derived from gum benzoin, obtained from *Styrax* trees, mainly *S. benzoin* (Indonesian Heritage 1996: 71).

66. Interview, Sama Dua 12/1/99.

67. Interview with *Ketua seuneubok*, 15/1/99.

68. A seventy-five-year-old man now living in the Alas valley recalled seeing forest boundary markers throughout South Aceh. As a young man he had ranged widely across the forest, collecting forest products and hunting rhinoceros and other animals. He described how in South Aceh, as elsewhere in the Dutch East Indies, the colonial government distinguished areas under control of local heads (*tanah adat*) from areas classified as state forest, marking these forest boundaries right across South Aceh with posts (*patok*). The colonial authorities could then obtain revenues from areas of state land by giving them over for exploitation. At the same time, by colonial regulation, the *adat* heads collected the *pantjang alas* tax on forest products obtained from land (*tanah adat*) that was still under the control of local heads. This account accords with the colonial forestry regulations for Aceh and elsewhere (Adatrechtbundels 1938a and 1938b). However, he noted that following independence villagers opening plots beyond these boundaries threw away the posts. Subsequently, since technically they had opened plots on state land, it was against their interests to acknowledge the prior existence of these boundaries (interview, Badar, 12/12/98). Whether this was the case for Sama Dua is unclear. In any case, older villagers in Sama Dua failed to mention Dutch forest markers in their area. An official from the regional agricultural office also told me that the Dutch set out forest boundaries but never placed pillars in the field (interview, Dinas Perkebunan, 12/1/99).

69. In Bahasa Indonesia, *petua empat*, or four elders.

70. While the *keucik* has an important role in village life of the communities studied here, the religious leader (*teungku menusasah*) plays a less prominent role than that described for Aceh Besar (see Khariuddin 1992).

71. Under Shafi Islamic law a *mukim* is the area that provides the forty males required for Friday prayer. In the seventeenth century Aceh of Sultan Iskandar Muda, several hamlets or even villages (*gampong*) would share the same mosque. Since the people would come together for Friday prayers and other occasions, this wider community of villages sharing a mosque came to constitute a league of several villages. Usually a religious leader or *imam* officiated over the *mukim*. However, in eighteenth- and nineteenth-century Aceh Besar (north Aceh), the

imam mukim "became secular, hereditary rulers of the *mukim*, distinct from the *imam sembahyang*" leading the prayer (see Reid 1975). At the next level of supravillage organization, an *uleebalang* ruled over a grouping of *mukim*s, which was called a *nanggroe (negeri)*, and a league of *uleebalang* formed a *Sagi*.

72. Previously northern Sama Dua, an area predominantly Acehnese, had formed a part of the administrative area (*Landschap*) of South Lho Pawoh (BKI 1912). But now these hamlets formed the *mukim* of Sedar and joined Sama Dua district.

73. Interview, Sama Dua, 5/1/99.

74. Interview, Sama Dua, 20/1/99.

75. Interview, Sama Dua, 5/1/99.

76. According to the village government law, the *kepala desa* was the head of both the LKMD and the LMD. The LKMD was the village council involved in the routine business of the village, and its members were elected. However, the suitability of villagers for office had to be confirmed by the *kepala desa* and the *camat*. The LMD was a body that met less regularly, for instance to oversee the election of the *kepala desa* and other village officers, to receive reports from the village head, and to approve village decisions and the village budget. Members of the LMD itself were not elected, but rather appointed by the village head, in consultation with other prominent members of the village (Warren 1993). This is also discussed in the Menggamat case study.

77. Interview, Sama Dua, 5/1/99.

78. The centralization of power began during the colonial period with the creation of a single *adat* authority for Sama Dua as discussed earlier. This centralization continued above the village level with the development of the *imam mukim* and then the sub-district head (*camat*), also discussed earlier.

79. Interview, Sama Dua, 5/1/99.

80. I.e., the police could be induced with payments to solve a dispute in favor of one party.

81. Interview, Sama Dua, 20/1/99. To illustrate his point, this villager observed that a granite mine planned for Sama Dua was facing problems. "With the granite mine they went and measured people's *kebun* first [in preparation for buying the land for the granite mine] without consulting or talking to people. So villagers are offended. They went directly to the *camat* rather than via LKMD and village head (*kades*). And so the problem is now stark: people don't want to sell [their land to the mine]."

82. This villager observed that "development projects are not well handled. The money is wasted without seeing any results from the project. For example the problem experienced by *bangdes* (rural development funds): millions of rupiah and we get no signs of progress. The *bangdes* comes down from the [central] government [for village development] and its use should be discussed first by LMD. Only then should the *kepala desa* pass on the decision to LKMD for implementation—after the LMD has discussed what the community most needs. But now the *kepala desa* takes the issue straight to LKMD" (interview, Sama Dua, 20/1/99). Another informant serving on a village LMD also complained about the lack of transparency in village decision making concerning the *bang-*

des: "They rarely organize a discussion (*musyawarah*), for example, how should the *bangdes* be used. Once the money comes down they are silent—so it is not clear how the money is used . . . Money comes down, then they are silent until it comes down the next year . . . Deliberation (*musyawarah*) is rarely held. But how else will we find agreement from so many opinions. The process should be: LMD discusses problem first, then makes a decision, which goes to the *kepala desa*, who then takes it to LKMD, who then implements it. But now the LKMD works directly with the *kepala desa* and the LMD tends to be left behind. It is only called when there is a case that needs to be settled. Otherwise the village head (*kades*) deals directly with LKMD, although this should be based on a decision from the LMD" (interview, Sama Dua, 20/1/99).

83. As we discussed above, the original founders of the *seuneubok* marked the geographical boundaries of this area when the *seuneubok* was founded.

84. Interview, Dr. Mummamad Ismail Gade, February 1998.

85. However, as noted in the later discussion, many of the *seuneubok* norms could not be reduced to a set of regulations.

86. The conditions that gave rise to the writing down of these rules will be discussed later in the chapter.

87. Farmers working in the hills are usually men; women principally tend to go to the hills in groups to collect firewood. Some women also accompany their husbands or fathers to their *kebun* and have considerable knowledge and skill concerning agroforestry. However, women did not generally become the main workers in the *kebun* unless they have been widowed. In common understandings within Sama Dua, *seuneubok* members are almost entirely men.

88. To a degree, this suggests that there is an intersection between a sense of a "community as a social organization" and a "community of shared understanding" (see Agrawal and Gibson 1999). While the two often do not coincide, they are not necessarily separate either.

89. The durian occupies a special place in Sama Dua life. The eating of durian is a social activity that readily lends itself to festivity. As groups of young men camp out under the durian trees, they consume many durians on site.

90. Interview, Sama Dua, 14/2/98. For a discussion of the role of durian in the emergence of a Dayak "ethic of access" in Kalimantan, see Peluso 1996.

91. Interview, Sama Dua, 14/2/98.

92. Interview, Sama Dua, February 1998.

93. For a discussion of the relation of *adat* to law and the notion of *adatrecht*, see Benda-Beckmann and Benda-Beckmann 1985.

94. Cf. Warren 1993. In Habermas' (1987) terms, empirical concepts such as exchange, health, and wealth were interwoven with the scared within an undifferentiated "life-world." This life-world provides "culturally transmitted and linguistically organized stock of interpretative patterns." This provides the background for communication, meaning, and motivation and legitimizes attitudes and social activity.

95. The *Aulia* was known to communicate with people through dreams. He appeared in the form of an Islamic holy man wearing white and was said to show lost villagers the way out of the forest (interview, Sama Dua, 2/2/98). At such

times the *Aulia* had also been known to take the form of a tiger, and it is said that lost villagers have followed tiger tracks that have led back to the villages (interview, Sama Dua, January 1998). According to Islamic tradition, an *Aulia* is a holy man who retreats from human society to become close to God—and then develops mystical powers. *Aulia* were said to be able to help people, and the spirits of dead *Aulia* might appear in dreams (Teuku Reza, personal communication, 12/3/98). According to another informant, "If we are lost we can ask for help from the *Aulia* . . . Ibrahim is a kind of *Aulia* or *kermat*, an *orang alus* [spirit] or something of this kind. If one is lost in the *rimba* [wild jungle], the *Aulia* is the one that invites the person . . . first to his place where he offers food and a place to pray. For those who are not clean enough, if they are lost they should ask for help and he will show the way . . . If they are not clean, they are just shown the way—and it happens that people are shown the way back" (interview, Sama Dua, 15/1/99).

96. For other examples of the mutually supporting and interwoven nature of customary and supernatural orders, see Moore 1986; Zerner 1994.

97. Interview, Sama Dua, 14/1/99.

98. My conversations with older villagers reflect Griffiths' observation that tiger attacks were reasonably rare. See M. Griffiths 1990.

99. Interview, Sama Dua, 18/2/98. As explained below, by tradition each *seuneubok* had an understanding with one or more tigers, known as *seuneubok* tigers, tigers that spent part of each year in the *seuneubok*.

100. Griffiths' (1990) account describes how dreams are understood to work as a medium for awareness of tigers. "Years ago, the lady had a dream in which two orphaned kittens approached her and begged for food. She consented and the kittens expressed their gratitude. The next day while working in her *ladang*, she saw two tigers at the forest's edge. Recognising the significance of her dream, she prepared food and left it at the place where she saw the tigers, whistling as she left. After that she continued to leave food out, and periodically the tigers came to eat—perhaps learning to associate her call and whistle with the opportunity for easy food" (M. Griffiths 1990: 94). For other accounts of beliefs about tigers, see Wessing 1986, 1991; McNeely and Wachtel 1991; Bakels 1994; Boomgaard 2001.

101. Interview, Sama Dua, 14/1/99.

102. "BKSDA Tidak Serius: Harimau di Labuhan Haji Masih Berkeliaran" *Serambi Indonesia*, 10/1/99.

103. "Sumatran Tigers on the Brink of Extinction," *Jakarta Post*, 14/12/99.

104. Interview, Fakmawan, Balai Konservasi dan Sumber Daya Alam, Departmen Kehutanan, Tapaktuan, January 1999.

105. "Sumatran Tiger on the Prowl Toward Extinction," *Jakarta Post*, 21/12/00.

106. Interview, Sama Dua, 14/1/99.

107. Interview, Fakmawan, Balai Konservasi dan Sumber Daya Alam, Departmen Kehutanan, Tapaktuan, January 1999.

108. Former village head, interview, Sama Dua, 13/11/97.

109. Interview, Sama Dua, 3/2/98.

110. Interview, Sama Dua, February 1998.

111. The *kenduri seuneubok* was merely one variety of ritual feast (*kenduri*). *Kenduri* hold a similar place in Sama Dua culture to the *selamatan* in Java: they were held for many reasons, and particularly mark life and harvest cycle events.

112. Interview, Sama Dua, 18/11/97.

113. Interview, Sama Dua, 18/11/97.

114. Interview, Sama Dua, 28/2/98.

115. Interview, Sama Dua, 28/2/98.

116. Ostrom made this statement with respect to institutions governing common pool resources (CPR). It also seemed to be applicable to the *seuneubok,* although it is not stricktly a CPR system.

117. Interview, Sama Dua, 27/2/98.

118. Interview, Sama Dua, 27/2/98.

119. An understanding of these trends emerged only slowly, through repeated interviews with many informants. Comprehending the meaning of a particular *seuneubok* for its members involved understanding the specifics of a particular institutional history. Perceptions regarding the degree of functioning of a *seuneubok* tended to vary between informants. Thus it was difficult to make definitive conclusions about whether a *seuneubok* had ceased to exist in anything more than in name.

120. Interview, Sama Dua, 6/1/99.

121. Interview, Sama Dua, 4/2/98. The prophet Muhammad was an orphan, and Muslims are enjoined to pay special attention to orphans.

122. Interview, Sama Dua, 18/2/98.

123. Interview, Sama Dua, 2/2/98. From interviews with the leader of a neighboring *seuneubok* I learned that several other *seuneubok* also held meetings at this time and had their members sign similar letters.

124. Interview, Sama Dua, 2/2/98.

125. Interview, Sama Dua, 2/2/98.

126. Interview, Sama Dua, 9/2/98.

127. Keebet von Benda-Beckmann has described how in a Minangkabau village, parties in a dispute "forum shop" among the various institutions of dispute settlement. They make a choice between different institutions based on what they hope the outcome of the dispute will be (K. Benda-Beckmann 1981).

128. Interview, Sama Dua, 17/1/99.

129. Wilson (2000) discusses this dynamic in a South African context, calling the continuities between notions of justice developed within different discursive fields "adductive affinities."

130. Interview, Sama Dua, 2/2/98.

131. For a discussion of the impact of the crisis on commodity prices and forest clearance, see Sunderlin et al. 2000.

132. Interview, Sama Dua, 3/2/98.

133. Interview, Sama Dua, 15/1/99. Many of those opening new plots wished to extend the area they were already farming. Moreover, the characteristics of *nilam* meant that farmers preferred to open new forest areas.

134. Interview, Sama Dua, 17/1/99.

135. For a discussion of this process, see Peluso 1995.

136. This is the exercise that was reviewed by RePPProt (1990) to produce the RePPProt maps.

137. Interview, LDP, 16/2/98. Interview with forestry department official, Dinas Kehutanan, Tapaktuan, 15/1/99.

138. Presidential Decree No. 32 1990 and the Act of the Republic of Indonesia No. 24 of 1992 Concerning Spatial Use Management.

139. The Leuser Development Programme Integrated Conservation and Development Programme (ICDP) mapping exercise extended the area to be protected, specifying a Leuser Ecosystem conservation zone that included the limited production forest and some limited areas of the unrestricted state forest around Sama Dua (Rijksen and Griffiths 1995).

140. This seems to be a more accurate description of PT Remaja Timber's concession in South Kluet. The concession behind Sama Dua covers the limited production forest area discussed earlier.

141. Interview with former *bupati*, Medan, 15/12/97.

142. Interview, former village head, 12/11/97.

143. Interview with former *bupati*, Medan, 15/11/97.

144. Interview, Sama Dua, 26/1/98.

145. Given the difficult position of *adat* communities during the New Order, many NGO publications gave this impression. See Arimbi 1994; Moniaga 1994; Rahail 1996.

146. In April 1998, during a dispute in Kualabatee, South Aceh, under the leadership of a village headmen, the community demonstrated against a concession (HGU) for oil palm production that had been issued for over 7,000 hectares of community land. On 13 April the dispute came to a head when the community "ran amuck," destroying plantation buildings and eventually wrecking the local police station ("Kerusuhan: Kemarahan di Aceh Selatan," *Forum Keadilan*, 20/4/98).

147. Sayed Mudhahar, interview, Medan, 5/12/97.

148. For an discussion of the wider history of these competing regimes, see McCarthy 2000b.

149. *Pantjang* (or *pantjung*) *alas* could be translated as "forest marker." In South Sumatra and Jambi, *pantjung alas* was a traditional permitting system under the control of the *Pesirah* (*kepala desa*) for opening a piece of forest (*alas*) for crops (swidden farming). Apparently a farmer would pay a sum of money known as the *pantjung alas* signifying that all the forest cut was under the *adat* control of the community head. *Pantjung* means cut or slash. A *pantjang* (*tiang* meaning pole or fence) was also a marker of land boundaries that was used to mark property rights and to protect property from outside encroachment (Zahari Zen, personal communication). Colonial texts also list the *pantjang alas* as a tax levied by customary heads in Sumatra, suggesting that the colonial government had given the tax a position within the allowable taxation regime governed by customary heads (Goor 1982, no. 563).

150. The notion of a *hak Allah* follows from the Islamic understanding that forest and natural resources ultimately belong to Allah—at least until they be-

come subject to the property claims of a community or other party.

151. Kreemer (1922) noted that in Aceh and its dependencies, the local authorities primarily levied the *pantjang alas* at the point of export from their territory. On the coast, *pantjang alas* was levied in the port; in the interior the fee was imposed at the *pintu rimba* (lit., forest door), the point where the footpath connecting two territories left the inhabited plain and entered the forest (Kreemer 1922).

152. Van de Velde assumed that, whereas in other areas this fee constituted a form of compensation for common property rights (the "right of avail"), in Aceh this probably did not apply anymore. He made this assumption because the *pantjang alas* in Aceh was paid to the head rather than (as he argued) to the community, as in West Sumatra. This error might be based on differences between areas of Aceh (such as the Gayo area described by Bowen, where the primary unit of social organization was the village rather than the clan). Under Gayo and Acehnese *adat*, the "right of avail" was vested in a village head who did not correspond necessarily to the clan head as it does in the Minangkabau area (Bowen 1988). Where the fundamental unit of community organization was the village rather than the clan, it made sense for the fee for access to be paid to a community leader rather than to a clan leader. Nowadays this fee (known as *uang pembangunan*) is paid to the village head, who—under a village agreement—uses the money for the village, for example, to help maintain the mosque. As we will see in the Menggamat case, however, the distinction between what is a fee paid to the community and what is the fee for the (individual) village head is not always clear.

153. *Pantjang alas* taxes collected, 1916–18 (in Dutch guilders)

Area	1916	1917	1918
Gayo Lues	*f* 115	*f* 100	*f* 100
Tapaktuan	8,280	3,115	1,813
Alas land	17	36	-

Source: Kreemer 1923: 143. These are the only figures found in the course of this research. Further archival research in the Netherlands would be required to obtain more definite information.

154. I am grateful to Gitte Heij for pointing out the implications of this text.
155. Interview, Sama Dua, 5/1/99.
156. This system is discussed in the Kluet case study.
157. Interview, Sama Dua, 15/11/98.
158. Interview, Sama Dua, 11/11/98.
159. Outsiders wishing to hunt in the forest behind the *seuneubok* must also ask permission, and they have to make a payment or make a gift of meat (interview, *ketua seuneubok*, 11/11/98).
160. Interview, Sama Dua, 19/11/98.
161. Interview, Sama Dua, 18/1/99.
162. Interview, Sama Dua, 23/1/99.

163. Interview, Sama Dua, 23/1/99.

164. Interview, Sama Dua, 23/1/99.

165. Interview, Sama Dua, 20/1/99.

166. See Warren 1993 for a similar argument with respect to Balinese *adat* communities.

167. For a discussion of the effect of this process on the Minangkabau, see Kahn 1993.

168. Clearly this is not true for other colonial interventions. See, for instance, Zerner 1994.

169. For other examples of this phenomena, see Merry 1991; and Doolittle 2001.

170. For a discussion of this issue, see Li 1999.

CHAPTER 3

1. In the absence of statistics concerning past population growth for Menggamat, it is instructive to note that the average growth rate for South Aceh district was 2.2 percent between 1980 and 1990, which according to Anhar and Jordan "is about the likely rate of natural increase." They conclude that this district has not experienced major gains or losses due to migration (Jordan and Anhar 1997). In interviews with village heads in Menggamat I was told that there are still extensive areas of unused land available for cultivation. Thus, in contrast to the Alas valley, land pioneering associated with population growth and in-migration is not such a significant factor in the dynamics of Menggamat's development. Moreover, due to the area's fearsome reputation for mystical practices and poisons, coastal people are reluctant to take up residence there—with the exception of the hamlet of Sarah Baru, where significant numbers of Aneuk Jamee people from Sama Dua have opened land.

2. When the Dutch colonial government created the Leuser Nature Reserve in 1934, the mountainous area some fifteen kilometers east of Menggamat was included. In 1936 the authorities added the so-called "Kluet reserve," 20,000 hectares of lowland rainforest and coastal swamp in South Kluet some fifteen kilometers south of Menggamat.

3. The connection to the Alas land was more than merely cultural: in the late eighteenth century, the son of a Kluet king appealed to the Alas kingdom in a dynastic dispute that led to a force from the Alas region settling the matter (Ahmad 1992).

4. Stories of poisoning and black magic surround forest-dependent upland groups in other areas. According to Peluso (1992c: 219) the circulation of such stories by villagers "may serve as a form of local resistance to outsiders treading uninvited on village territory." Traditional animist folk beliefs about the forest have remained strong in Menggamat, and a Menggamat *dukun* (herbalist/spiritual adviser) interviewed in January 1999 said that up to the 1970s many villagers had kept poison in their houses. Other informants argued that poison would only be used against enemies in situations where there were few other opportunities for redressing grievances. Nevertheless, they advised against causing unnecessary offense in remote villages such as Menggamat.

5. Colonial control of South Aceh was initiated in 1899. While the Dutch established indirect rule through *kejuren* resident in Kota Fajar (*landschap Kloeet*), Menggamat remained remote from colonial power. During and after the guerrilla war that raged in the Kluet area, the South Aceh region was placed under direct military control (1926–38). At this time Menggamat was a stronghold of guerrilla forces (Doup and Bakar 1980). In 1998 there were also rumors of GAM activity behind Menggamat.

6. In a similar fashion communities living around Java's forests have also been characterized as criminals and outlaws (Peluso 1992c). As Li (1999) has argued, highland peoples have routinely been represented (by the Dutch in the past and by lowland groups and the Indonesian authorities today) as marginal, backward, economically irrational, and unable to respond successfully to development opportunities.

7. Interview, Menggamat, 8/1/99.

8. The limited historical information available indicates that leaders of Acehnese and Minangkabau descent also ruled the Kluet kingdom. As they did elsewhere in South Aceh, migrants from West Sumatra settled on the coast. According to Sayed Mudhahar's account, at one time the head of the Pinim Sikulat clan (*suku*), a clan group of West Sumatran descent—who customarily bore the title of *datuk* Pusako—held the Kluet kingship (Ahmad 1992). Yet another tradition holds that settlers from the Pasai area of Aceh came to the Kluet region under the leadership of a religious leader (*imam*) named *Imam* Gerbung. After discovering the fertility of the area, *Imam* Gerbung decided to settle in Kluet (WWF and ID 0106 Gunung Leuser National Park Conservation Project, n.d.). According to this account, the *datuk* who was head of the Pinim clan group (*suku*) and accordingly held the Kluet kingship was a descendant of *Imam* Gerbung and hence of Acehnese descent.

9. Interview, Menggamat, 8/1/99. A more correct translation might be "to use *gamat* posts."

10. There is also a *marga* in the Alas valley known by this name.

11. Interview, Menggamat, 8/1/99. As the Kluet area has become more integrated with neighboring areas of South Aceh, the *marga* system has fallen into disuse.

12. The Kluet kingdom controlled the plain of the Kluet River, stretching from the mountain pass separating the Kota Fajar area from Tapaktuan just to the north and bordering the Trumon area to the south.

13. In 1923 Kreemer reported that the area (extending beyond the immediate Kluet valley) had a very mixed population. The population was primarily what he called "Malay" (i.e., Aneuk Jamee of Minangkabau descent) in the Bakongan and Kandang area, whereas elsewhere (outside the immediate Kluet area) the inhabitants were primarily Acehnese (Kreemer 1923). Clearly at that time the colonial authorities knew little about the Kluet area, and it is likely that colonial control over this remote settlement was precarious. As well as the "Malay" and Acehnese areas, Kreemer refers to the "original inhabitants," the Kluet people, whom he mistakenly identifies as "Alas or Gayo." Under the *kejuren*, each of these population groups had its own head, creating a "fairly complicated" gov-

ernment structure. In the immediate Kluet area there were fifteen subordinate *uleebalang*, including one for Menggamat Kreemer (1923).

14. Outside of the alluvial river valleys, which have high agricultural productivity, the soils in the mountainous areas are of poor quality and subject to erosion (Rijksen and Griffiths 1995). According to a district report (Pemerintah Kabupaten Daerah Tingkat II Aceh Selatan 1991/1992), the soil of the Menggamat area is classified as "Podsolik Merah-Kunung." The department of agriculture has observed that this variety of soil is suited only for forest, dry land food crops (*palawija*), *alang-alang* grasslands, and rubber (Direketorat Jenderal Perkebunanan 1977).

15. See Michon et al. 2001.

16. Interview, 9/1/99.

17. In Kluet the leader of what was known as a *kemukiman* was called an *uleebalang*. In Acehnese terms, the title *uleebalang* is inappropriate for the head of the *kemukiman Menggamat*. An *uleebalang* in historical Aceh ruled over a group of *mukim* rather than a single *mukim*. Moreover, the position of the *uleebalang* was abolished in Aceh after the Indonesian revolution. In recent years, to bring the terminology in line with that found elsewhere in Aceh, the *uleebalang* is known as the *imam mukim*.

18. Interview, Menggamat, 9/1/99.

19. Interview, Menggamat, 9/1/99.

20. Interview, Menggamat 9/1/99. Interestingly, in some respects village territoriality has remained somewhat undefined until recently. For example, when WWF-LP carried out village surveys in 1995, they found that the boundaries between village territories were unclear—even to the village heads themselves. Subsequently, the villages held discussions (*musyawarah*) to agree on boundaries between territory belonging to each village. Henceforth the boundary of each village territory was clearly marked according to natural features of the landscape—usually along streams running from the steep hills surrounding Menggamat. Each village was then issued a map that showed the area subject to the authority of each respective *keucik*. Consequently, each village territory contained permanently cultivated gardens, less permanent plots, and areas of forest held for future agricultural use. A village head interviewed during 1998 said that residents of a particular village gained use of land belonging to their village subject to approval from the village head. However, villagers from adjacent villages could also use land from another village territory if they got permission from the village head responsible for that area. In this way, the regime worked to specify the boundaries of the resources over which a user had legitimate access as well as to define who had the right to use community resources.

21. Interview with village head, 12/2/98.

22. For any offense of this type against the community forest, there are also material sanctions. The offender also has to plant trees in the watershed, as specified by the *adat* official (*kejuren blang*) who is responsible for regulating the supply of irrigation water and protecting the forest along the river course.

23. Interview, Menggamat, 8/1/99.

24. In 1998 *nilam* prices were so high that there was a danger villagers would

not even cultivate their rice fields. "But they have to plant rice paddy because the *nilam* price can fluctuate so much. Otherwise there may be a time when they need rice and won't be able to afford it" (interview with *camat*, Menggamat, 6/2/98).

25. Dove describes how this system is extensive in terms of land use. However in terms of return on labor, it is much more productive, effectively freeing up labor time for cash crop production (Dove 1985).

26. I am grateful to Dr Susan Moore for pointing out this similarity.

27. This is described at greater length later in this chapter.

28. Interview with *nilam* farmers, Desa Mersak, Menggamat, 6/2/98.

29. Interview, Menggamat, 12/2/98.

30. I write "ideally" because this village head was giving an idealized picture of how the Menggamat *adat* regime operated. As we will see later, there is some variation in application of the *keucik*'s authority.

31. Interview, village head, Menggamat, 8/1/99.

32. Interview, 8/1/99.

33. "Formerly the area under the control of Kemukiman Menggamat had its origin in the prosperous healthy community and extended up to the headwaters and even to Gunung Leuser itself and down to the sub-district at the edge of the road. It shows that the authority followed the course [of streams] down from the peaks of the Menggamat hills. Where the water terminated, there was [the end of] its authority. So automatically this was at the boundary of the kecamatan at the edge of the road" (WWF-LP 1995a). In Dutch times, Menggamat extended to Kampung Tinggi, close to Kota Fajar, but now the boundary is at Pancoran to the south (interview, 8/1/99).

34. Interview, Menggamat, 9/1/99.

35. Interview, Menggamat, 9/1/99.

36. According to Khairuddin, the original ancestors of the Alas and Kluet people settled here after being dispersed by a flood from their *kampung* at "Laut Bankoe," an area further to the south (Khairuddin 1992).

37. Although this territorial claim is inscribed on official maps as far as Sebumbung, originally Kluet territorial claims extended farther up, to the very headwaters of the Kluet River at Gunung Leuser. According to a village elder, to the north five hours up the Kluet River by boat is "the boundary made by our ancestors, at the foot of Gunung Leuser, which is not far beyond this" (interview, Menggamat, 8/1/99.)

38. For further discussion of this issue see Sellato 2001.

39. Interview, Tapaktuan, 9/1/99.

40. See Sellato 2001.

41. Interview, Tapaktuan, 9/1/99.

42. Interview, Tapaktuan, 9/1/99.

43. After the TGHK forest maps under the Regional Physical Planning Program for Transmigration (RePPProt), a team of British and Indonesian scientists reviewed them and prepared a new, revised set of maps.

44. That is, *Keputusan Gubernur Daerah Istimewa Aceh No 188.342/776/94* and *Peraturan Dearah Tingkat II South Aceh No 7 Tahun 1994*.

45. Interview, Menggamat, 9/1/99.

46. Interview, Menggamat, 9/1/99.

47. For instance, the district forestry office (*Dinas Kehutanan*) in South Aceh obtained its first car in November 1998.

48. For a discussion of this problem in Kalimantan, see Gunawan et al. 1998.

49. For a description of this system, see Niessen 1999.

50. Largely for these reasons, as part of the agreement with the IMF the Indonesian government agreed to limit the increase in *retribusi* and to abolish taxes between regions (*Suara Pembaruan* 16/1/98). Subsequently, the number of *retribusi* levied across Indonesia were reduced from 192 varieties to 30, creating a cash crisis for some regional administrations (*Suara Pembaruan* 25/3/98). In South Aceh this led to the abolition of 18 *retribusi* (including the *Retribusi Hasil Bumi dan Industri*), creating a fiscal problem for local government only partly alleviated by hiking the *Sambungan Pihak Ketiga*, a levy on goods exported from the region (interview, Tapaktuan, 5/1/99).

51. Interview, Tapaktuan, 6/1/98.

52. In a similar fashion, HPH would work outside the area that they had agreed to cut that year, only cutting it at the end of the year.

53. Interview, Dinas Kehutanan, Tapaktuan, 15/1/99.

54. In 1996, the regulations were changed, and forestry officials took a large role in the issuing of SAKO.

55. As noted in Chapter 1, significant overlaps exist between the roles of *oknum* (agents of the state indulging in extra-legal activities), *tauke* (organizers of logging teams), and *cukong* (entrepreneurs or large-scale brokers). Some individuals who might be called *oknum* act as *cukong*, and in some cases people who might be termed *cukong* also carry out the function of *tauke*.

56. Interview, Dinas Kehutanan, Tapaktuan, 15/1/99.

57. Interview, Dinas Kehutanan, Tapaktuan, 15/1/99. This pattern is reminiscent of that Peluso (1992) found in Java's teak forests. Here, at times forest guards had faced violent resistance from those appropriating teak from the forests illegally. As a result, forest guards made great efforts to avoid confrontations with illegal loggers.

58. Interview, *jagawana*, South Aceh, 13/1/99.

59. In February 1994, *Waspada* reported that a police official (*oknum Kapolsekda setempat*) coordinated logging in the South Kluet area, using local people as "an instrument" (*Waspada* 1994).

60. For example, informants in South Aceh continued to praise the effectiveness of a former *bupati*, Sayed Mudhahar, in regard to his ability to raise revenue and initiate projects. Sayed Mudhahar had also impressed the *Kompas* reporters with his ability to raise funds (*Kompas* 1991).

61. Interview, *Dinas Kehutanan*, Tapaktuan, 15/1/99.

62. In 1991 *Kompas* had reported that in South Aceh, out of a local government budget of Rp 12.2 billion for 1990/91, the contribution of the forest sector had been only Rp 200 million (*Kompas* 1991). In 1998 the system of logging royalties was changed. Although the Reforestation Fund (DR) was pre-

viously managed by the forestry department outside of the state budget, with large amounts allocated to projects favored by the president and his coterie, from this time the DR became a budget item and was handled by the finance department. At the same time, timber royalties (*Iuran Hasil Hutan,* or IHH)—which were also not registered in tax receipts within the national budget—were abolished and replaced by a resource rent fee known as *Provisi Sumber Daya Hutan* (PSDH) subsequently administered by the Ministry of Finance.

63. Interview, Tapaktuan, 5/1/99.

64. Sayed Mudhahar, interview, Medan, 5/3/98.

65. Interview, Menggamat, 10/1/99.

66. Interview, Menggamat, 9/1/99.

67. Interview, Tapaktuan, 5/1/99. It is difficult to confirm this allegation from official figures because local government figures appearing in reports are not transparent. As Devas noted some years ago, in district budgets *retribusi*—including those levied on forestry operations—are not disaggregated but tend to be grouped together under the category of "other" (Devas 1989).

68. The *Sumbungan Pihak Ketiga* generated some 156 million to district treasuries over 1997/98. However, since the charge was also levied on other goods, it is unclear how much of this amount was generated on timber trucks (Badan Pusat Statistik Kabupaten Aceh Selatan 1997).

69. Interview, Tapaktuan, 5/1/99.

70. Interview, Tapaktuan, 5/1/99.

71. Elsewhere in South Aceh, farmers have described the danger of attack by wild boars short of food and armed with tusks. In the past, tigers kept the number of boars under control, but the reduction in tiger numbers combined with disturbances to the boars' food supply has increased the incidence of boar attacks.

72. Beneath the *Controleur* were the district heads (*landschapshoofd*), each ruling over a territory known as a *landschap.* When indirect rule was extended, the district heads became known as self-governing administrators (*Zelfbestuurder*). In the Kluet territorial area (*landschap Kloeet*), this administrator was the *kejuren* stationed in Kota Fajar.

73. For, as Lev has noted, from the start the Dutch "resolved to respect local law—another way of saying that, by and large, they could not have cared less—except where commercial interests were at stake" (Lev 1985).

74. The LKMD meets regularly to discuss day-to-day village affairs, but the LMD is only called occasionally for discussions of significant issues.

75. As Mubyarto has written of the analogous situation in Jambi's forest villages, under the new law, village government "had the task of arranging development (*pembangunan*) that had already been programmed from above. Although this development program might be considered 'good' by the government, yet the community might not readily accept the program. This was because this program no longer strictly accorded with the local *adat* that remained so strong" (Mubyarto 1992).

76. Interview, Menggamat, 9/1/99.

77. Interview, Tapaktuan, 2/98. The *camat* and the district council (*Muspika*) vetoed several YPPAMAM and *kemukiman* decisions.

78. Interview, Menggamat, 1/99.
79. Interview, Menggamat, 8/2/98.
80. For discussion of a similar *adat* assumption in East Kalimantan, see Eghenter 2000.
81. Interview, Tapaktuan, 5/1/99.
82. According to Ostrom, "the person who engages in corruption receives a disproportionate gain by using his or her power over the allocation of valued resources to extract an illegal payment form someone else." (Ostrom 1992).
83. Interview with *nilam* farmers, Desa Mersak, 6/2/98.
84. Interview, district official, Menggamat, 6/2/98.
85. Interview with villagers, Menggamat, 12/2/98.
86. Interview with former WWF personnel, Tapaktuan, 14/1/99.
87. Interview, Menggamat, 8/2/98.
88. Interview, Menggamat, 8/2/98.
89. Interview, Menggamat, 10/1/99.
90. Interview with loggers, Desa Mersak, Menggamat, 8/2/98.
91. Interview with *camat*, Kluet Utara, 6/2/98.
92. Interview with loggers, Desa Mersak, Menggamat, 8/2/98.
93. The value of the Indonesian rupiah fluctuated widely during the period of this research. In November 1996, one US dollar bought Rp 2,357. Between July and October 1997 it had lost 27 percent of its value, falling from 2,400 per dollar to 3,200. Over the next year it continued to slide, and at the beginning of 1999 one US dollar bought Rp 8,800.
94. See also Obidzinski 2003.
95. Interview, Southeast Aceh, August 1996.
96. Interview, Tapaktuan, January 1999.
97. *Yayasan Perwalian Pelestarian Alam Masyarakat Adat Menggamat.*
98. *Hutan adat* refers to the forest remaining on village lands.
99. *Keputusan Menteri Kehutanan Nomor: 622/Kpts-II/95.*
100. The document bore the title "Cooperative Regulation of Thirteen Villages in the Menggamat *Adat* Community . . . Concerning Community Conservation Forest and Natural Resource Management Rights in the Menggamat Adat Community Area of North Kluet" (*Peraturan Bersama Tiga Belas Desa di Kemukiman Adat Manggamat Kecamatan Kluet Utara Kabupaten Dati II Aceh Selatan Tentang Hutan Kemukiman Konservasi dan Hak Pengelolaan Sumber Daya Alamnya di Wilayah Adat Kemukiman Manggamat Kecamatan Kluet Utara*).
101. Pascal 4, *Peraturan Bersama Tiga Belas Desa di Keumiman Adat Manggamat.*
102. Pascal 9, *Peraturan Bersama Tiga Belas Desa di Keumiman Adat Manggamat.*
103. Pascal 9, *Peraturan Bersama Tiga Belas Desa di Keumiman Adat Manggamat.*
104. Interview with head of YPPAMAM, Menggamat, 8/1/99.
105. *Keputusan Menteri Kehutanan Nomor: 622/Kpts-II/95.*
106. The WWF-LP project finished at the conclusion of the project's funding. Conflict with LMU and local government as well as other issues all contributed

to the mix of problems surrounding the closure of the project. Subsequently, until early 1999, several WWF-LP staff formed Yayasan Bina Alam to continue activities with limited funding from LMU.

107. A WWF-LP worker later noted: "Our understanding was that if only people from Menggamat were doing this, it could be controlled by the *adat* heads or the *kepala mukim*. But if outsiders did it, with people behind them like the *camat*, Koramil, or police, *adat* leaders could not prevent it" (interview, Tapaktuan, January 1999).

108. Cf. Chanock 1982; and Oomen, 2002.

109. The idea that promoting the marketing of non-timber forest products (NTFP) by local communities can save tropical rainforests has been heavily criticized. According to Dove, this strategy rests on the thesis that deforestation occurs because of the poverty of local communities. The argument then is that forest clearance continues because the riches enclosed in the rainforest have been overlooked. In his view, by focusing on the microeconomics of forest dwellers, the approach diverts attention from the broader political-economic causes of deforestation. Thus, it "masks the need for fundamental change in political-economic institutions—a need which outweighs any potential economic benefits that the approach offers to forest dwellers." Moreover, he concludes, the benefits of NTFP may in any case be illusory. If NTFPs were of high value, they would already have been exploited by elites (Dove 1993).

110. Interview, former WWF project worker, Tapaktuan, January 1999.

111. I am grateful to David Edmunds for pointing this out.

112. Interview, former WWF project worker, Tapaktuan, 14/1/99.

113. See Michon et al. 2000: 181.

114. Interviews with WWF-LP and LMU staff in 1996 revealed that relations between the two programs were particularly poor due to disagreements over conversational approaches, rivalry, and personal animosities. To WWF-LP's surprise, they were told to move out of Gunung Leuser National Park by the governor of Aceh at the request of the Leuser Management Unit (LMU), who ran the Leuser Development Programme (LDP). The governor's letter, as well as WWF donors' lack of interest in supporting a project in Leuser, due to the existence of multi-million dollars funding for LDP from the EU, led to the end of WWF-LP's conservation activities in Leuser in 1997. A senior WWF representative later noted, "WWF and LMU met several times to clarify this incident, apologies have been exchanged and accepted—the case is closed." Agus Purnomo, e-mail, 10/5/2001.

115. Interview, former WWF project worker, 14/1/99.

116. Like other commodity prices, the price of patchouli oil fluctuates widely in the world market. For example, over a twelve-month period (1995–96), prior to the extreme impacts of the currency crisis prices in Indonesia, the price dropped by 40 percent.

117. On sloping land *nilam* needs rain, and farmers need to plant before the wet season.

118. According to a CIFOR paper on the economic crisis, towards the middle of 1998 demand began to pick up again (CIFOR 1998).

119. "50 persen kilang kayu di Aceh Barat dan Selatan Tutup," *Waspada* 6/10/98.

120. Interview, 5/1/99.

121. Interview, former WWF project worker, Tapaktuan, 14/1/99.

122. For more about YLI, see Chapter 3, Epilogue.

123. "TPF Hentikan Pengoperasian PT MRT," *Waspada* 15/10/98.

124. According to a report in *Waspada*, the district government's revenues were 45 percent below target for the 1998/99 fiscal year. One reason, the article suggested, was that "the third party tax (*sumbangan pihak ketiga*) that is levied at the border [with North Sumatra] fell as a result of the decrease in the armada of trucks that carry several commodities exported from the region, especially timber." However, the *bupati* associated the fall in revenue with central government regulations (*UU No. 18/1997, Inpres No. 9 dan 10 tahun 1998*) that prohibited the collection of export fees (*retribusi barang ekspor*). These included several taxes on wood exports including the RHBB. ("PAD Aceh Selatan gagal capai target, dispenda-dinas LLAJ saling tuding," *Waspada* 23/3/99.)

125. "Dandim 0107/Aceh Selatan akui ada anggotanya terlibat pencurian kayu," *Waspada* 6/6/98.

126. "Pencurian kayu kembali marak di Aceh Selatan," *Analisa* 25/9/98.

127. "Ekosistem Leuser Terganggu. Puluhan irigasi mulai kurang air," *Serambi Indonesia* 16/5/98.

128. Unpublished LMU survey regarding perceptions of environmental change.

129. "Bencana banjir tak terlepas dari dam KKN," *Waspada*, 21/9/98; "Rapat Tim Koordinasi Leuser: Kaburnya tata batas dan saran peninjauan HPH," *Serambi Indonesia* 25/5/98.

130. These groups included Walhi Aceh, Kesatuan Aksi Reformasi Daerah Aceh Selatan (or KARDAS), and *Rimueng Lamkalut.* ("Desakan pegiat lingkungan hidup: Cabut izin HPH dan IPK." *Waspada* 28/9/98).

131. "Bila izin HPH tak Dicabut, Rimueng Lamkaluet akan Mengamuk" *Serambi Indonesia* 17/9/98; "Seluruh HPH Di Aceh Selatan diancam akan dibumihanguskan," *Waspada* 26/8/98.

132. "Sejumlah HPH/IPK mulai kosongkan camp." *Waspada* 5/10/98.

133. Interview with LMU staff, Medan, 31/1/99. Logging regulations prohibit logging operations on slopes above 40 percent, and for some years LMU had attempted to get HPH on very steep areas cancelled. However, such concessions continued to have legal validity even though they were in the Leuser Ecosystem and enclosed very steep land unsuited for logging. The *bupati* at this time waged an unofficial campaign against these Medan-based concessionaires.

134. It was reported that there were now plans to turn the area into an oil palm plantation (International Campaign for Ecological Justice in Indonesia 1999: 13).

135. See Schmink and Wood 1992.

136. As I observed earlier, this occurred during the colonial period, when, during the *damar* boom, control shifted from village heads to the *uleebalang*. However, as Dutch reports reveal, the colonial state fulfilled this role in other

areas subject to commercial timber exploitation (Kreemer 1923).

137. For a parallel situation in East Kalimantan, see Eghenter 2000.

138. For a discussion of this problem in the context of Malaysian Borneo, see Doolittle 2001. See also Fay and Sirait 2002.

139. Agus Purnomo, e-mail, 10/5/2001.

CHAPTER 4

1. In some studies, population growth caused by in-migration has been linked to deforestation. For instance, Colfer and Dudley (1993) found that in-migration was the principal source of East Kalimantan's population increase, which in turn was one of the principal threats to its forests. For a discussion of the effect of migration on forest management in Kalimantan, see also Potter 1996.

2. To compare levels of prosperity in the different areas of the valley in the absence of statistics, Iwabuchi used the community's own measure of financial achievement—the ability to take the Haj pilgrimage to Mecca. On the one hand, Iwabuchi noted that there were no *hajji* from the rice-cultivating village of Kute Meli, located far from wage-earning opportunities and new lands. On the other hand, in villages near the district capital, where there were wage-earning opportunities, and in the villages in Badar District and other areas on the margins of the valley, there were many *hajji* (Iwabuchi 1994).

3. Interview, Badar sub-district, 15/12/98.

4. Interview, Badar sub-district, 10/12/98.

5. Interview, 15/12/98.

6. Interview, 19/12/98.

7. The total population of these communities was very small. When the Dutch geographer Volz came to the valley in 1905, "he was surprised at the small population—approximately 7,000—compared with the huge area of land" (Iwabuchi 1994a). Before World War II, villagers in Alasland lived in longhouses. Several families occupied each longhouse, and a group of longhouses were clustered together into a village community that was known as a *Kuta* (or *Kute*). The total population of these communities was very small compared with the huge area of forest. The Alas people practiced rice cultivation in irrigated fields along the fertile valley and some shifting agriculture in surrounding dry lands. Forest resources were still abundant and rice production from agricultural lands exceeded local consumption. In addition to agriculture, the Alas people made handicrafts, raised livestock, fished, engaged in petty trade, and collected forest products. In the 1920s Dutch sources reported that the main exports from the area were cattle, tobacco, forest rubber (*getah*), and rattan. For accounts of the history of the area, see Paulus and Stibbe 1921; Koesnoe and Soendari 1977; Kreemer 1922, 1923; Iwabuchi 1994a; Rijksen and Griffith 1995.

8. In the Gayo uplands around Takengen in Central Aceh, according to Bowen, property rules vary depending on what resource has the highest value. In the older core area of the uplands the main resource, *sawah*, was in short supply. An expanding agricultural frontier in the newly settled northern highlands, coffee plantations were being opened. Compared with the core areas, villagers valued wealth created through labor more than inherited land. This was one reason that

in the northern highlands the division of wealth at inheritance and divorce more closely followed Islamic law. Islamic law protected the earned wealth of men and women, who, under certain circumstances, would lose the fruits of their labor under Gayo *adat*. Meanwhile, in the core areas where *sawah* land was highly valued, Gayo *adat* institutions attempted to tightly control use rights over village *sawah* land. The village authorities attempted to stop alienation of land to those not affiliated with the village by residence or marriage (Bowen 1988). For a similar comparison under different circumstances, also see F. Benda-Beckmann and K. Benda-Beckmann 1994.

9. The exception here, as we discussed in the other case studies, was the levying of taxes on the export of valuable forest products.

10. As noted earlier, van Dijk distinguished "tenure" from "territoriality" as different forms of control over property. Tenure "may be conceived of as sets of rules and practices specifying whom is to get access to land at which time and in which place" (van Dijk 1996: 18).

11. Interview, Jongar village, 13/12/98.

12. In the Alas valley the Dutch set about constructing two "self-governing territories" based on the positions of the *kejuren* (lords) that previously existed there. Kreemer reported that in the process of creating these territorial heads, the Dutch also caused fierce enmities between rival families, particularly on the side of the family of the descendants of Raja Kemala (the original overlord), who risked losing their pre-eminent status under the Dutch administration (Kreemer 1922). The primary problem for colonial authorities was that the Dutch concept of a "territorial head" governing an unbroken territory did not accord with the indigenous system of organization. Rather than forming a separate territorial area of several contiguous villages, in Tanah Alas the *kampung*s under each head were interspersed with others according to clan ties. In setting up districts, the Dutch experienced endless disputes between heads concerning real and perceived rights. Eventually, in 1912, a meeting of the indigenous rulers was held and the two districts (*landschappen*) were divided into four territorial units called *mergo*. Despite the name, these *mergo* were constituted by geographically adjacent villages and were not based on clans—except that each area was named after the clan that had been important in the area. However, Kreemer related, the Dutch hoped that eventually each clan group would settle in the area that bore its name (Kreemer 1922). Also, in 1912, the Dutch appointed two *kejuren* as territorial heads (*landschapshoofd*), each ruling over four *mergo*. These heads, along with the underlying territorial chiefs (*pengulu* or *raje*) and tribal village heads (*pengulu suku*), received a salary from the Dutch government in place of *adat* income (Iwabuchi 1994a).

13. Interview, Jongar village, 13/12/98.

14. Interview, Badar, 15/12/98.

15. Subsequently, three hamlets of the composite village (*desa gambungan*) developed on this plantation land.

16. In the early 1920s, travelers making the 109-kilometer journey from Kutacane to Blangkejeren usually walked, sometimes with pack-horses, and the journey took four days. In those days, Kampung Jongar was the most northerly

village (around twelve kilometers from Kutacane); beyond this lay thick lowland rainforest. After Jongar, the first "bivouac" stood (now in Badar sub-district) twenty-one kilometers from Kutacane (Kreemer 1922). Nowadays the gates of the national park lie just before the Ketambe orang-utan research center.

17. To open land here, the founder of the *kampung* obtained written permission from Kejuren Pulo Nas in Kutacane (interview, 17/12/98).

18. See Peluso 1995; Dephut 1992; KLH and UNDP 1997; Ascher 1993; McCarthy 2000b.

19. Since the colonial territorialization process in Badar served the joint purposes of conservation and plantation agriculture, it proceeded more thoroughly than did similar processes in Sama Dua and Menggamat.

20. The ICDP involved a complex territorial strategy that divided the Leuser Ecosystem into zones. As envisaged by the masterplan, LMU would obtain a "conservation concession" (*hak pengelolaan*) over the Leuser Ecosystem area on state forest land. Within this area, LMU would establish distinct zones "on the bases of ecological dictates" (Rijksen and Griffiths 1995). Within these zones, access would be prohibited "except for holders of a specific permit or licence." These zones would be created in accordance with the zoning categories set out in Act No. 5 (1990) concerning conservation of natural resources and their ecosystems. These include "strict nature reserve" (*Cagar Alam*) status and "wildlife sanctuary" (*Suaka margasatwa*) status for the core wilderness area of the ecosystem subject to the strictest protection. The status of "nature conservation area" (*Kawasan Pelestarian Alam*) —"which allows for sustainable utilisation"—and that of "restricted production forest" (*hutan produksi terbatas*) would be given to different categories of bufferzones (Rijksen and Griffiths 1995).

21. For a discussion of this issue, see Koesnoe and Soendari 1977.

22. Although Bowen (1991) studied the history of the *Gayo Deret* and *Gayo Lot* of Central Aceh, to date researchers have neglected the *Gayo Lues* of Southeast Aceh. This was because, as Bowen reported, "the southern and eastern regions [of the Gayo homelands] have remained more isolated from the respective coasts as well as from the Colonial and Indonesian governments. Only in 1988 could the southern region, of 'Spacious Gayo' (*Gayo Lues*), be reached by bus, from Isak to the north or from the Karo Batak area to the south. Many of its approximately forty thousand Gayo inhabitants lived one or two days' further journey from the road" (Bowen 1991).

23. Dove argues that the state takes a dim view of swidden because government programs involve taking over swidden areas, for watershed management or, more commonly, for industrial forestry, export crop production, intensive food crop production, and/or transmigrant settlement (Dove 1985: 176).

24. Because the price of *nilam* was rising when I visited Gayo Lues in late 1997, farmers were planting *nilam* (for its oil and its cosmetic properties) after the first tobacco crop. They planted *sirih* (a product consumed with betel nut) last. Then the land would be left for three to four years for natural forest succession to occur. This secondary forest would then be burned to produce ashes, which would provide nutrients for another round of crops. Farmers continued to hold use rights over land during the fallow period.

25. After independence, Southeast Aceh became a part of Central Aceh Regency. Southeast Aceh, including the five administrative sub-districts of the Alas valley and four *kecamatan* (sub-districts) known as *Gayo Lues*, became an autonomous regency only in 1974.

26. Interview, 14/12/97.

27. Over time the Toba Batak have gained control of large amounts of land in the southern end of the valley. Under the *adat* of the Alas people, individuals had ownership rights and were able to sell their land, and newcomers could obtain rights to unoccupied lands. As *adat* Alas entailed paying a large bride price (*mas kawin*) to marry off one's sons, Alas families often sold land. This practice depended upon the availability of large areas of uncultivated land. Since the colonial era, the Batak have been able to gain access to or to buy large areas of land. The Batak were considered more industrious and commercially oriented than the Alas. According to the *adat* of the Toba Batak, over generations the patrilineal descent group came to own what was known as *golat* land. Only clan members could gain access to this land; normally individuals could not sell the land. Batak in-migrants came to own large sections of land in the southern part of the Alas valley, with two consequences. First, an economic imbalance between the Christian Toba Batak and Muslim Alas population emerged, leading to religious antagonism between the two groups. Second, Alas people displaced by this process have tended to move farther north, opening land on the steep slopes of the conservation area in Badar sub-district (Iwabuchi 1994a, 1994b; Rijksen and Griffiths 1995). Moreover, with the increasing scarcity of free forest land, the traditional Alas pattern of expanding into new forest has fallen into decline (Iwabuchi 1994a).

28. Before that time, each hamlet was an autonomous village with its own head.

29. This area was given protection forest status by the colonial forest authorities in 1940 (Rijksen and Griffiths 1995).

30. Between 1976 and 1982, using USAID funds, the footpath connecting Kutacane and Blangkejeren—a forty-kilometer stretch through the national park—was turned into an asphalt road (Wind 1996). This road facilitated the migration of Gayo people down from Blangkejeren into enclave villages within the reserve. As people began to cultivate candlenut forest gardens, residence in the area became doubly attractive. In many places the state forest lands alongside the road were both unoccupied and undefended. By opening permanent candlenut forest gardens here, along a road that provided easy access to market, Gayo people could live a more prosperous life while avoiding the tedium of the swidden agricultural cycle. Consequently, larger numbers of Gayo people moved south, many moving as far down as Badar sub-district, at the northern end of the Alas valley. They moved in such large numbers that many hamlets were completely Gayo. In addition to the Alas people that moved into the fertile farm land of the former plantation area, other individual Alas families were forced to move north by land shortages caused by Batak migrations into the southern end of the Alas valley.

31. Several hamlets have a majority of Gayo. Those with a majority of Alas live in one hamlet, and a neighboring hamlet has a mixture of Gayo and Alas.

32. Interview, 17/12/98.

33. Interview with former village head, 17/12/98.

34. This growing of coffee and candlenut marks a significant change in agricultural practices—a move away from shifting dry land plots. Rather than abandoning old fields and planting new crops on newly opened plots, forest farmers could establish permanent tree plantations. Candlenut and coffee gardens keep the soil covered with a rather thick layer of vegetation that prevents heavy erosion, whereas small food crops require a fallow period to replenish the soil. Once established, these tree crops continue to produce for many years. For these reasons, farmers found these trees to be much more suitable for steep areas than the small food crops (Strien 1978).

35. Following independence and the extension of special status to Aceh, as in Sama Dua, each sub-district (*kecamatan*) was divided into several village leagues (*kemukiman*) in accordance with the pattern found in northern Aceh. This pattern replaced the territorial areas (*mergo*) of the colonial period. To bring the area into accord with the Aceh hinterland, a chief, or *kepala mukim*, now presided over these *kemukiman* (Iwabuchi 1994a). However, this system had been grafted onto the Alas pattern of customary authority and had very shallow roots in *Tanah Alas*. Consequently, in contrast to Sama Dua and Menggamat, here the *kemukiman* structure virtually disappeared following implementation of the village government law. As a senior village leader in Badar sub-district put it in 1997, "There is still a *kepala mukim*, but he has no seat." By contrast, the *kepala mukim* had much greater significance in South Aceh, where the position has a long history.

36. Interview, 13/12/98.

37. Interview, 16/12/98.

38. Interview, 17/12/98.

39. It is important to note that here, as in other areas of Aceh, the village head plays a key role in both the *dinas* business of the village and the *adat* business.

40. Interview, 17/12/98.

41. Until the early 1970s, nature conservation was the responsibility of Dinas PPA, a small section of the Directorate of Forestry within the Ministry of Agriculture (Rijksen and Griffiths 1995).

42. Interview with village informants, 15/12/98.

43. As Peluso (1992: 229) observed in Java's forests, errant forest guards could expect to be confined to desk work or transferred to another location. Field foresters tended to be jailed "only in the most extreme circumstances."

44. Interview with village informants, 15/12/98.

45. Interview, November 1996.

46. Interview, Kutacane, 21/12/98.

47. In late 1998, the local chapter of the nature lovers' organization *Pecinta Alam* wrote a letter to several officials that illustrates the nature of logging networks. The letter alleged widespread collusion between local officials and illegal loggers in the Bengkung area, southwest of Kutacane. Here, the letter alleged,

fifty chainsaws operated within the national park, with those carrying out the logging making monthly payments of Rp 100,000 per chainsaw to PPA officials, and even to an employee of the Leuser Management Unit (UML/LDP). These officials also passed on cuts to other senior officials working for the national park. In return, the forestry officials and UML employee agreed to give the loggers advance warning of raids. The letter also asserted that when UML, PPA, and Polres personnel who were engaged in anti-logging patrols came across illegal operations, they received large payments. In one case a patrol caught an employee of the logging company PT Wajar red-handed with forty tons of stolen wood and two chainsaws. The PPA official heading the patrol fined the logger 6 million rupiah and returned to Kutacane. Other members of the patrol stayed on, including a police official named Sahlam, who took six tons of the confiscated wood, made a raft, and attached it to the back of the patrol boat, intending to take it down river for illicit sale. On the way down the river, villagers claiming the wood intercepted the raft and a dispute arose. Sahlam threatened the villagers, firing six shots. However, the villagers were able to take the raft from behind the speedboat. In the fighting Sahlam broke his leg, and out of fear, the patrol was then forced to return. (letter from *Pecinta Alam* [Nature Lovers], Southeast Aceh, to head of PPA, Kutacane, 1998). TNGL staff denied the charge, however, arguing that it was a political tactic by *cukong* to make the police (*Polres*) and forest guards (*jagawana*) withdraw from the area ("Surat Kaleng Permbahan TNGL Berumatan Politis," *Waspada* 16/12/98).

48. Drs Martin Desky, the district head of Golkar (the state party), alluded in 1998 that they involve wood theft, widespread gambling, and trafficking in marijuana ("Tiga Permasalahan di Kutacane," *Waspada* 14/10/98).

49. Interview, Tapaktuan, 6/1/99.

50. Interview, confidential informant, Kutacane, 21/12/98.

51. "Format Serahkan Laporan Tertulis," *Waspada* 10/9/98.

52. "Kades Duking bupati Southeast Aceh," *Waspada* 7/8/98: 39.

53. "Buntut Berita Perambahan Hutan: Koresponden *Waspada* Aceh Tenggara Dicari Oknum Tak Jelas," *Waspada* 20/10/98.

54. "LSM Lestari, Keluhan Masyarakat Terhadap Pencurian Kayu di Aceh Tenggara," *Waspada* 18 November 1998.

55. "Buldozer Permabah TNGL Korea Diamanankan," *Waspada* 15/12/98.

56. Interview, Kutacane, 21/12/98.

57. McCarthy forthcoming.

58. *Waspada* 28/1/98, "Petugas Jagawana Tahan 40 Orang Di Hutan TNGL."

59. Interview, 20/12/98.

60. Interview, 17/12/98.

61. "Golkar Minta Kawasan Kappi dan Pasir Luk-luk Dibebaskan," *Analisa* 10/12/98. In Pasir Luk-Luk village heads applied for 30,000 hectares of protection forest (*hutan lindung*) to be excised from the conservation area. Much of the area was flat and suitable for *sawah*. While the request was still in process, but apparently blocked by the central government, the farmers spontaneously occupied 200 hectares, and threatened to openly occupy the whole 30,000 hectares (interview, Kutacane, 21/12/98).

62. For a discussion of policy tools, see Bridgman and Davis 1998; and Schneider and Ingram 1990.

63. An Asian Development Bank report (1997) pointed to the relation of LMU/YLI to the Indonesian state, an issue that drew the criticism of Indonesian NGOs: "LDP has done little to involve local and national NGOs and win their support for the Programme. Unlike the approach of a number of other large conservation and development projects in Indonesia, LDP staff have thus far professed little interest in the opinions or capacities of local or national NGOs. The reaction of the many NGOs in the Aceh Regional NGO Forum is bitter. They reject the claim often made by LDP that 'LIF [YLI] is an NGO' in the sense generally accepted by the Indonesian NGO community. First, they argue it is composed of retired civilian and military officials and is linked more to the power centre in Jakarta than Aceh's rural communities. Second they believe that LIF only came into being when there was a prospect of funding from the EU, in antithesis to the general NGO credo of community-based self-help. The Jakarta-based Leuser Steering Committee has no representation from NGOs or the timber industry, nor does it include the PHPA or officials from the provincial level planning or sectorial agencies. The inter-provincial Leuser Coordinating Committee (LCC), meanwhile, has a large staff—34 members and 15 advisers—but not fully representative membership. The advisers include 7 military representatives and four university rectors, but no one from an NGO or timber concession, and no specialised natural or social science experts" (Barber and Nababan, *Eye of the Tiger: Conservation Policy and Politics on Sumatra's Last Rainforest Frontier* [1996], cited in McNeely 1997). In the wake of this criticism, LMU attempted to involve some local NGOs in its activities.

64. Through ensuring the discussion of the ecosystem and the importance of these decisions in the local press and in workshops, the high (and often controversial) profile that LMU has maintained has successfully raised awareness among decision-makers and the wider public about the unique value of the Leuser ecosystem.

65. Other reports show that local government resents the existence of the park because of its effect on local revenue (Rijksen and Griffiths 1995).

66. The *Leuser Development Programme Masterplan* and the "Financing Memorandum" between the Republic of Indonesia and the Commission of the European Communities state that YLI/LMU would have a "conservation concession" with full management rights over the area. On this point the memorandum reads: "In accordance with Indonesian laws and regulations, the Government will, by decree of the Minister of Forestry, grant a conservation concession for the State Forest Land which encompasses the Leuser Ecosystem. The concession will be in the custody of a non-governmental non-profit-making Foundation of local origin, on the understanding that the implementation of its conservation is delegated to the LMU until the full transition is accomplished . . . The concessionaire (and its permanent successor organisation) will be formally authorised to collect fees, royalties and other revenues derived from sustainable, non-destructive activities in the conservation concession area. These funds would be used to fund on-going project activities" (Republic of Indonesia and Leuser Development

Programme 1995). As set out in the YLI/LMU contract, this would have reduced the Ministry of Forestry agency PHPA to "supporting protection and inspecting whether management is applied to the (legal and contractual) obligations" (Rijksen and Griffiths 1995).

67. Kepmen No. 227/Kpts-II/95, Articles 1 and 6.

68. The decree explicitly stated that YLI/LMU's management mandate "did not extend to rights to have control over or to possess—or anything of that kind in any shape or form—the land and other wealth above or below ground in the Leuser Ecosystem" (Keppres No. 33/1998, paragraph 3). As the Ministry of Forestry retained these rights, the decree represented a victory for that ministry.

69. As I discussed in Chapter 3, LMU/YLI have enjoyed extraordinary support from patrons at the center, and this backing helped LMU/YLI put pressure on local government to enforce regulations and on the HPHs working in the ecosystem.

70. "Temuan Pansus VI DPRD-I di Aceh Tenggara: IPK Tidak Dikeluarkan Tapi Ada 11 Kilang Kayu," *Waspada*, 11/9/98.

71. "TPHT Aceh Tenggara Bongkar Kilang Kayu Illegal," *Waspada*, 11/8/98.

72. "TPHT Aceh Tenggara Bongkar Kilang Kayu Illegal," *Waspada*, 11/8/98. "TPHT Aceh Tenggara Periksa Izin Kilang Kayu," *Waspada* 18/4/98.

73. For instance, LMU's 1997/98 annual report noted that fifty-five microprojects had been carried out in the program area. "Most of these projects were implemented with the involvement of the appropriate dinas and monitored by the local Bappeda offices" (Leuser Management Unit 1997/8).

74. Interview with local NGO, Medan, 19/12/98.

75. Interview, forestry official, Banda Aceh, 29/12/98. Perhaps, from another perspective, insofar as some local figures might gain from their role in implementing projects, this funding might amount to a form of compensation for local options for exploiting the ecosystem.

76. Interview, LMU personnel, January 1999.

77. Interview, 20/12/98.

78. Interview, 5/9/97.

79. Interview, 13/12/98.

80. Interview, 13/12/98. The legal status of the enclave hamlets was also not altogether clear. At the time that the Minister of Agriculture made the formal declaration creating the park, state planning recognized the existence of two communities within the park boundaries—the Geumpang and Marpunge enclaves. These villages were legally recognized because they had existed prior to the declaration of the national park, and even before the Dutch created the first reserve (Department Kehutanan and Universitas Syiah Kuala Darussalam–Banda Aceh 1993). However, the settlements in other sites along the road had a more ambiguous status. Although many villagers had opened gardens here earlier, before a large area was declared a reserve in 1981 (Wind 1996). "I have been here since 1976," a villager I interviewed in 1997 told me, "and it is only because this whole area up to the top of the hill some two kilometers from the road was designated for agriculture that I have held out for all this time. There was a map here that showed this. I saw this back in 1976. But the boundaries have been moved,

and now we are told we have to move" (interview, 8/9/97).

81. See Barber et al. 1995; Li 1999.

82. As Li has argued, coercive removal of small holder farmers categorized as "forest encroachers" has been feature of state territorialization strategies (Li 1999). In Leuser, such removals, as in this case, have aimed to further state territorialization that set aside specific areas for biodiversity conservation.

83. Interview, 8/9/97.

84. Interview, 13/12/98.

85. Interview with *jagawana*, Badar, 16/12/98.

86. Interview, 8/9/97.

87. Interview, 19/12/98.

88. Interview, 17/1298.

89. Interview, 12/12/98. Interview, 13 December 1998.

90. Interview, 12/12/98.

91. LMU was structurally responsible to Bappennas rather than to the Ministry of Forestry.

CHAPTER 5

1. For a discussion of this issue, see Li 1997; Milton 1996; and Brosius 1997.

2. Cf. Warren 1993; Fitzpatrick 1997.

3. I am grateful to Peter Brosius for this point.

4. This suggests the need to avoid a naïve form of rational choice theory that might imply an ahistorical concept of utility maximizing self-interested individual that is without context and fails to account for how subjectivities are constituted through specific social and discursive practices (Agrawal and Gibson 1999).

5. For a comprehensive discussion of how these dynamics have operated historically, see Obidzinski 2003.

6. For discussion of this phenomenon, see Barr 2000; Barr 1998; Robison 200; Robison 1998; MacIntyre 1994.

7. For discussion of how this affected forest management, see Obidzinski 2003.

8. "Lingkaran itu tidak mempunyai ujung (berputar terus) sehingga sulit menentukan mana ujung dan mana buntutnya. Semuanya setan, tapi yang mana yang jadi boss nggak tahu" (interview, Ministry of Environment official, 4/4/2000).

9. Gunarso and Davie (1999) call this phenomenon "pseudo-autonomy."

10. Cf. Mosse 1997; Bourdieu 1990.

11. As I noted earlier, the very shifting, contested, and processual nature of institutional arrangements in these situations diverges considerably from the model of institutional arrangements described by "collective action" scholars. For relevant discussions See Mosse 1997; Campbell et al. 2001; Johnson 2003.

CHAPTER 6

1. Richardson 2000.

2. Incentive tools assume people lack incentives to take the actions needed, and set about creating inducements, charges, sanctions, or even force to encourage compliance. Authority tools—or regulatory instruments—involve statements backed by the authority of government that grant permission, prohibit, or require action under designated circumstances. The use of law enforcement—which is derived from legislative power—is a key policy instrument of this kind. Capacity tools provide information, training, education, and resources to enable individuals, groups, or agencies to take the action needed (Schneider and Ingram 1990).

3. Cf. Brosius, Tsing, and Zerner 1998.

4. *Nederlandsch-Indische Vereeniging tot Natuurbescherming.*

5. "Natuurmonumenten," *Indisch Staatsblad* 1916: 278, 18–3–1916. The *Natuurmonumenten* ordinance was made possible by the guiding assumption of the "domain declaration" (*Domein verklaring*), according to which the colonial regime claimed the right to dispose of all land over which individuals did not exert recognized property rights. This principle facilitated the creation of nature reserves: the government could simply declare certain areas within the forest domain to be nature reserves. The colonial authorities then retained responsibility for management and supervision.

6. In 1925 the foundation of another environmental group, the Netherlands Committee for the International Protection of Nature (*Nederlandse Commissie voor International Natuurbescherming*), added momentum to the push towards creating protected areas in Sumatra. By 1929, this committee, together with the Netherlands Indian Society for the Protection of Nature, entered into negotiations "to select the necessary regions" (Dammerman 1929). It was this committee that first advocated the gazetting of the forest reserves around Gunung Leuser.

7. For a discussion of this history, see Snouck Hurgronje 1906; Reid 1979; Siegel 1969.

8. In the intervening period the committee joined forces with the Society for the Preservation of Nature Reserves (*Vereniging voor Behoud van Natuurmonumenten*) to gain the support of the "Netherlands lobby." Subsequently, the colonial government started to investigate the claim to ensure that the proposal did not jeopardize potential plantations (Rijksen and Griffiths 1995). By 1934, the colonial government had appointed a new governor for Aceh, A. Ph. Van Aken, a man more sympathetic towards nature conservation. He supported the proposal and in February 1934 convened the self-governing territorial heads (*Zelfbestuurders van de Landschappen*) to gain their consent to the first part of what they called the "Gunung Leuser Reserve" (*Wildreservaat Goenoeng Leuser*). As I discussed in the Alas case study, these heads were the primary clients of the colonial state within its system of indirect rule. Therefore it is unclear whether an agreement between the colonial State and these heads meant that the local people gave their assent to the Leuser reserves. For a discussion of this issue, see Koesone and Soendari 1977.

9. The original reserve extended over 416,500 hectares. In 1936, an area of lowland rainforest in the Kluet region of South Aceh (adjacent to the Menggamat

area discussed in Chapter 5) was added, an area of special importance as a corridor for elephant herds. In 1938 an area of the Western Section of the Wilhelmina range in North Sumatra province was also set aside as a reserve (in Langkat and Sekundur). Further areas, such as the Kapi reserve (on the northern end of the Alas valley), were added post-independence. As this complex historical process created several adjoining reserves over a period of forty years, it gave rise to uncertainties about the status and the extent of the various reserve areas. Moreover, at the time when the first reserves were created, the topography was largely unknown, especially of the western mountains. This "resulted in very vague and even erroneous descriptions of the boundaries" (Strien 1978).

10. These approaches were in keeping with the managerial and technocratic assumptions of international environmental discourse and practice at the time. Within this discourse, state bureaucracies have primary responsibility for managing the environment. Representatives of state agencies attend high-level conferences and summits, and developing nations have become parties to international environmental conventions. State environmental agencies develop policies, craft laws and gazette natural areas in line with this international environmental discourse. Multilateral and developing nation donor agencies enter into agreements with recipient States to support the implementation of these laws and policies. As new policy initiatives and project interventions filter down from distant policy-making arenas to the district and village level, they call for people to change their behavior in the hope that changed behavior will lead to desired environmental consequences. This exercise of power is justified by state law. In an instrumental fashion the implementation of the normative model found in the law is supposed to achieve the desired situation (Benda-Beckmann 1989). During the 1970s and 1980s, international conservation agencies often proved themselves to be out of touch with what was appropriate at the local level. Many interventions—such as the creation of an orang-utan rehabilitation center at Bukit Lawang in North Sumatra—had unforeseen but lamentable consequences (McCarthy 1999).

11. This is discussed in more depth in the Alas case study.

12. As Barber argues, "without the active support and participation of farming and fishing communities, indigenous groups, and others who live in or near high-biodiversity areas," progress towards biodiversity conservation "will remain confined to the proliferation of policy statements and meetings by experts based in offices, agencies and cities far from the action" (Barber 1996: 3).

13. See *The Biodiversity Action Plan* (BAP) and *Agenda 21–Indonesia*.

14. However, following several highly publicized critiques of ICDPs in the late 1990s, the term ICDP has fallen from favor.

15. In 1989 a team of ecologists visited the Leuser area as a part of a World Bank–funded technical assistance project, drafting the initial proposal for an ICDP here (Rijksen and Griffiths 1995).

16. At the Global Forest Conference in Bandung in February 1993, developing countries requested support from industrialized nations to conserve tropical biodiversity. Shortly after the conference, the Indonesian planning agency, Bappenas, formally asked for European Union investment in rainforest conservation in Indonesia and the EU responded by granting funds for an ICDP planning project

to produce the *Leuser Development Programme Masterplan*.

17. Ecological studies, such as "Population Viability and Habitat Analysis" (PVHA) of endangered species, served as the main conservation planning criteria. They indicated that the national park (GLNP) covered too small an area to preserve the predominant ecosystem types and the mega-fauna typical of northern Sumatra (Rijksen and Griffiths 1995). To overcome the problems associated with the capacity of the GLNP complex to conserve viable populations of the biodiversity of northern Sumatra, the team of ecologists who designed the project selected a more extensive conservation area known as the "Leuser Ecosystem." The ecologists designed the area to contain the ranges of the major elements of the biological diversity of northern Sumatra. Extending over approximately 1.8 million hectares, the Leuser Ecosystem was said to constitute the largest rainforest reserve in the world (Rijksen and Griffiths 1995). Meanwhile, the customary land uses and patterns of resource management of village actors remained relatively unexplored. Embarking on a territorial strategy to protect the Leuser Ecosystem, LMU later placed boundary markers around this area mapped according to ecological criteria to indicate the extent of the project. This led to many conflicts with villagers who, finding their land staked out with Leuser Ecosystem markers, began to throw away the markers. To assuage village concerns, LMU also found itself engaged in compensation negotiations with villagers whose lands were deemed important to conservation and whose activities did not conform to the Leuser Ecosystem territorial concept (interview with former LMU project staff, August 2001).

18. See McCarthy 2000 for an overview of the state policy regime under the New Order.

19. See note 17 above.

20. For discussion of this problem, see Brechin et al. 2002; and Wilshusen et al. 2002.

21. Cf. Natcher and Hickey 2002.

22. Cf. Li 2002.

23. Ecologists have sharpened the critique of the ICDP approach. Subsequently, though "people orientated approaches to conservation continue to be the norm," the acronym is disappearing from use. See Mogelgaard 2003.

24. Later generations of ICDPs—now under a different name—seem to have taken more sophisticated approaches. See Mogelgaard 2003; Hughes and Flintan 2001.

25. This is not to deny that law enforcement can contribute to forest protection. In Tanah Alas, land clearing and logging inside the conservation area increased due to the weakening of state law enforcement following the political and economic crisis of 1998–99. This same phenomenon also occurred in other areas of Indonesia during the economic and political crisis (see Sunderlin et al. 2000). The key point here is that, if law enforcement lacks legitimacy, and it involves deploying a repressive and often corrupt state apparatus lacking capacity to effectively implement the law, it is usually expensive and inconsistent and leads to ongoing and unresolved conflicts. In these contexts, forest protection through law enforcement remains at best a partially successful strategy. Ironically, in re-

sponse to the perceived failures of the first generation of ICDPs, ecologists have advocated a return to strict protection of protected areas (see Terborgh 1999). This typically involves again turning to top-down strategies of paramilitary law enforcement. However, such approaches bring with them the same legitimacy problems, local resistance, and corruption that undermined the Leuser ICDP's law enforcement efforts. For a critique of this view, see Brechin et al. 2002; Wilshusen et al. 2002. For an earlier analysis of coercive approaches to conservation, see Peluso 1993.

26. Consequently, although the WWF-LP and Leuser ICDP represent contrasting approaches, ICDPs and CBNRM approaches are not necessarily mutually exclusive.

27. For a discussion of CBNRM assumptions and simplifications, see Li 2002; Agrawal and Gibson 1999.

28. At the end of 1993, WWF obtained funding from the British Overseas Development Administration (ODA) and WWF/UK for a project in the vicinity of the Leuser National Park. By this time, the European Union had funded the preparatory stage of the Leuser Development Programme (LDP), a project with similar aims to the proposed WWF-LP project. After consultations with the state conservation agency (PHPA), park management, and local government, WWF-LP redesigned the project, now focusing on "conflicts over the use of natural resources in and adjacent to the park" (Barber 1997). The new WWF Leuser Project (WWF-LP) aimed to resolve human–park conflicts in selected communities on the park's fringes in order to stabilize the park and improve local livelihoods, and in doing so create viable community conservation models for wider application (Barber 1997).

With funding of U.S.$ 571,000, the project would work in the district of South Aceh over a two-year period. Up until the 1990s, in contrast to the Alas area, South Aceh had never been the focus of conservation efforts. This was the first conservation project to concentrate on South Aceh. In 1997, after two years, the WWF-LP project activities were wound back upon conclusion of project funding. As I describe later, this approach ultimately ran up against the shoals and sandbars associated with district networks of exchange and accommodation that supported logging.

29. Of course, these transformations are not necessarily ecologically benign. See Li 2000.

30. See illustrations 11, 12, and 13.

31. Keterapan Majelis Permusyawaratan Rakyat Republik Indonesia NOMOR IX/MPR/2001 Tentang Pembaruan Agraria dan Pengelolaan Sumber Daya Alam.

32. In central Africa the representations that ended up becoming "customary law" emerged victoriously out of intense conflicts between ethnic groups, genders, and generations. Given the control that the colonial state exercised over this process, the representations that were finally accepted as customary law were those that accorded with the ideas and interests of the colonial rulers. Inevitably, the officially accepted "customary law" only partially represented the values and practices of African communities—combining some practices "inherited from the precolonial past" with practices "adapted and reformed in the unfolding present" (Chanock 1985).

33. As Zerner (1994) has shown, in the Malukus this constructed *adatrecht* was used to support particular strategies of resource control.

34. For an parallel situation in East Kalimantan, see Eghenter 2000.

35. Cf. Agrawal and Gibson 1999.

36. Although there have been some legal innovations in this area, these reforms have been constrained by the principle of state control of forests that is enshrined in the constitution and the umbrella forestry act (Down to Earth 2002). Even for areas outside the forestry estate, the basis for *adat* rights within the state remains constrained. Though the regional autonomy law granted authority to the districts and municipalities over land affairs, in 2001 a presidential decree (Keppres 62/2001) re-centralized this authority. In the absence of new legal initiatives to deal with *adat* customary laws, villagers exert *de facto* control according to a resurgent *adat* "legal order" without obtaining *de jure* recognition under state law.

37. "Daerah Dinafikan, Pusat Kian Ngotot Revisi UU Otda," *Jawa Post* 08/04/02.

38. For a discussion of how *adat* has been "constructed" in different contexts, see Li 1999; Kahn 1993.

39. See Fitzpatrick 1997 for a discussion of alternative legal models for recognizing *adat* rights.

40. I am grateful to Amity Doolittle for making this point.

41. As Li (2002) has noted, remote villages often desire such an engagement with the state.

42. For discussion of this dilemma, see also Fay and de Foresta 1998.

43. Cf. Oomen 2002.

44. See Fitzpatrick 1997.

45. Cf. Agrawal 2001.

46. A key problem here is that existing patterns of elite control over access to markets, permits, and transportation networks work to keep villagers poor, affecting the fluctuation in prices, driving down the prices they earn from existing farming systems, and providing incentives for carving out new plots of land in the forest (cf., Zerner 2000). (Even though to a limited extent WWF-LP and Leuser ICDP activities aimed to help agro-ecological systems, this was neither their main point of involvement nor was it effective.)

This problem is illustrated in the case of nutmeg farming in Sama Dua. Farmers everywhere face the uncertainty of commodity prices, but nutmeg farmers are plagued by particularly low margins and unstable prices. After studying the fluctuations experienced by nutmeg farmers in Indonesia, Indrawanto and Wahyidi (n.d.) identified three reasons for the phenomenon. First, due to their poor bargaining position, nutmeg farmers tend to receive very low prices for their product. For instance, pepper farmers receive 80 percent of the pepper export price, but nutmeg farmers receive only 35 percent. Second, the price obtained by the nutmeg farmer does not generally reflect price increases in the international market. Third, government policies do not ensure the optimal functioning of the nutmeg trade for farmers and trading organizations. The lack of a favorable agricultural policy has compounded the other problems (Indrawanto and Wahyidi n.d.).

In South Aceh the rent-seeking behavior of those involved in the nutmeg trade exacerbated these problems. In Sama Dua, nutmeg traders who purchase nutmeg stones and mace directly from farmers achieve large markups. After buying the nutmeg cheaply, village-level nutmeg traders add 250 percent to the value of the nutmeg merely by drying the stones in the sun. A nutmeg wholesaler interviewed in November 1997 revealed that wholesalers were buying nutmeg stones for around 2,000 Rp/kg but selling the dried nutmeg to nutmeg factories for 7,000 Rp/kg. The village nutmeg wholesalers sell nutmeg on to traders who own distilleries, distinguished by the large heap of dark heavy wood standing in front of a furnace. The wood fires a furnace that cooks a tank filled with *pala* stones. Under heat, the nutmeg stones displace the oil, which is then drained via a pipe. These nutmeg traders have ample opportunity to deceive buyers and maximize their own profits. The trick they use is to increase the bulk of the nutmeg oil by mixing it with various other cheaper oils in ways that escape detection by the nutmeg merchants in Medan. Until recently, these merchants lacked the technology necessary to detect the deception; they only discovered that they had been cheated when the oil was tested by buyers in Singapore. Thus Medan-based merchants faced considerable loss, and some have been driven bankrupt.

Local farmers suspect that nutmeg traders from Medan manipulate prices, colluding to keep prices low at the local level (Serambi Indonesia 1997b; Sinar Indonesia Baru, 1989b). However, most likely the "beggar thy neighbor" strategy of nutmeg factories has eroded trust between traders, and nutmeg merchants further up the chain pass on the losses and risks of their trade by demanding lower prices for nutmeg oil at the regional level. At times merchants cost in the low value of impure oils, and the deception contributes to fluctuations in nutmeg prices.

However, opportunistic behavior is not restricted to traders. For instance, in the 1970s, the *Bupati* of South Aceh, Drs Sukardi, passed a provincial regulation giving the company PT Adi a monopoly on buying nutmeg. Because farmers had no choice but to accept the low prices offered by PT Adi, "this period was a disaster that farmers were forced to endure with determination." Eventually the next *Bupati*, Drs Ridwansyah, abolished this monopoly (Ahmad 1992).

As Indonesia produces 70–75 percent of world nutmeg, in the 1980s some argued that Indonesia should create a "sellers' market." In this view, competition between Indonesian nutmeg producers in the international market had meant that nutmeg prices were too low. To this end, in 1983, the Indonesian Nutmeg Association (ASPIN) was formed. By uniting Indonesian nutmeg exporters, ASPIN aimed to raise the export price of nutmeg. In 1986 the Minister of Industry specified that ASPIN members alone could export nutmeg. The minister also established a Coordination Board for the Marketing and Export of Nutmeg and Mace (BKPB) to coordinate nutmeg export prices; henceforth, only ASPIN members could export nutmeg and mace, and only through BKPB. Initially this policy helped control export prices of nutmeg, and the nutmeg price obtained by farmers in many parts of Indonesia increased from below 1,000 Rp/kg to above 2,000 Rp/kg. However, just as clove farmers had experienced when clove marketing was monopolized, producers did not benefit from the new arrangement; farmers had a weak bargaining position vis-à-vis ASPIN and enjoyed only a small percentage of the rise in export prices. However, in 1990 this policy was abolished, and the nutmeg prices obtained by farmers fell from above 2,000 Rp/kg to around 800 Rp/kg (Indrawanto and Wahyidi n.d.).

These price rises do not seem to have reached South Aceh, however. As of 1989, three exporters had been chosen to export nutmeg from Aceh. However, newspaper reports from this time complained that farmers in Aceh Selatan were not enjoying the price rises expected to follow from the government's agricultural initiative concerning nutmeg (Sinar Indonesia Baru, 1989d). Once again, in 1989 the newspaper reported that nutmeg farmers in Sama Dua were abandoning their nutmeg gardens. Over a three-month period the price of rice rose from Rp 750 to Rp 1,000 per *bambu*. In the same time the price of *pala* dropped from Rp 1,250 to Rp 800 per *bambu* [Sinar Indonesia Baru, 1989c].

47. As Agrawal (1999) has pointed out, "community" is complex: it is easy to conflate a sense of a community existing as a social organization with a notion of a community as group of people tied together by shared understandings and collective interests. Although the first sense of "community" is more apparent, the second is problematic.

48. Cf. Agrawal 1999; Li 2001.

49. Angelsen and Kaimowitz (1999: 20) have noted: "How improvements in agricultural technology affect forest clearing cannot be determined a priori, without information regarding the type of technology and the output and factor market elasticises . . . Intensification programs targeted at farmers living near the forest frontier make farming more profitable and may shift resources to forest clearing and attract new migrants, although the effect may be at least partially outweighed by the resulting downward pressure on agricultural prices and upward push on wages." They conclude that, in the absence of countervailing measures or effects (such as supply of capital or labor or price effects), there are "trade-offs between poverty reduction and forest conservation" (CIFOR 1999).

50. This issue of accountability is stressed because "administrative theorists have always emphasized that bureaucratic performance is to a large extent related to the effectiveness of mechanisms for encouraging or sanctioning organisational commitment and role performance" (Crook and Manor 1994).

51. For a summary of these arguments, see Blair 2000; Ribot 2001; Kaimowitz et al. 1998; and Larson 2003.

52. For accounts of this period, see Hull 1999; Forrester 2001.

53. State Planning Guidelines (GBHN 1994–2004); decrees of the supreme national parliament (TAP MPR IV/2000). See also Hull 1999; and Booth 1999.

54. For discussion, see Ribot 2001.

55. For discussion, see Turner 2000; Lederman 2001.

56. See *Gatra* 23/11/00; *Tajuk* 30/3/00. A report in the national magazine *Tajuk* describes how during the previous election money politics had colored virtually all district, municipal, and provincial elections. A candidate for chief executive required hundreds of millions or even trillions of rupiah. Though the intricacies of this process varied somewhat across regions, according to *Tajuk* the system typically involved three steps. Candidates would use the money to pay off party colleagues to ensure the candidature within the party, then make payments to ensure that a second faction within the parliament supported them, and finally make payments across the legislature on election day. A candidate who cannot outbid rivals may fail at any of these points. Besides money, this system of ex-

change also involves promises. It involves extensive collusion between business and politicians: "In many cases, financial contributors for politicians were local entrepreneurs—or national entrepreneurs that had local interests. They made political investments: financing potential candidates in return for future business concessions" (*Tajuk* 30/3/00).

57. See Turner 2001; "Suap-menyuap di Mana-mana," *Tajuk* 30/3/00.

58. Cf. Agrawal and Ribot 1999.

59. See "Pemerintah Kabupaten Siap Koreksi Otonomi Daerah," *Kompas* 31/5/2002; Andi A. Mallarangeng, "Paradigma Sentralistik," *Tempo Agustus* 2/9/2001; and McCarthy 2004.

60. Ketetapan Majelis Permusyawaratan Rakyat Republik Indonesia Nomor IV/MPR/2000 Tentang Rekomendasi Kebijakan Dalam Penyelenggaraan Otonomi Daerah.

61. "Regional Autonomy Policy May End in Chaos," *Jakarta Post* 21/12/00.

62. See McCarthy 2004.

63. Cf. Obidzinski 2002.

64. Cf. Arghiros 2001.

65. Cf. Hadiz 2003.

66. Cf. Tsing 1999.

67. For some positive examples among many from other areas of Indonesia, see Fauzi and Zakaria 2002.

68. See Low et al. 2003.

69. Young (2000) has argued that given the coexistence and density of institutional arrangements, rather than selecting the proper level of social organization at which to respond to particular problems, solving environmental problems requires dealing with "institutional interplay," the way institutional arrangements interact horizontally and vertically, politically, and functionally. For a similar conclusion from a development administration perspective, see Frerks and Otto 1996.

Appendix

1. The characteristics of the people living in both these areas are considerably different from those commonly associated with the residents of North Aceh area. As Bangkaru notes with respect to South Aceh, "the South Acehnese are different from the North Acehnese. They use two local languages, Acehnese and Jamu, the latter a dialect of the Minangkabau language" (Bangkaru 1998).

2. In a further piece of advice, he recommended that I patronize only one coffee shop (*warung*) in this particular village, as jealous proprietors had been known to poison disloyal clients. Other villagers suggested that I purchase a special stone with mystical powers. If I placed this stone in my shirt pocket, it would warn me of danger by stirring when I entered a place where there was poison. A traditional healer (*dukun*) in a remote village explained that, even though some households retained stocks of poison into the 1970s, he had helped to clear out the village using just such a stone. Villagers in Sama Dua maintained that in this isolated area people were still poisoned, warning me not to eat or drink anything when visiting there. (Stories of black magic were also prolific in these villages.) In

1999 I returned to the area with a friend from Sama Dua, who, despite my embarassment, turned down the food and drink offered according to local traditions of hospitality. His refusal proved particularly awkward when we were invited to a ritual feast and found ourselves placed next to the *adat* heads. I later asked a senior forestry official in Banda Aceh about these rumors. He told me how he stayed with a Dutch forester in a village here some time earlier. After leaving the area, the Dutchman fell ill, clutching his stomach in violent pain. Suspecting that villagers had poisoned the Dutchman, the official turned the car around and took the unfortunate man back to the house where they had stayed. According to custom, a poisoner always has the antidote for his poison, and so he too demanded the remedy. The Dutchman, he assured me, recovered soon after taking it.

3. As Yin (1994) has noted, the evidence offered by comparative or multiple case studies is often considered more compelling. For this reason, from the outset I attempted to include more than one case study. The main advantage proved to be that, by making comparisons across cases, I could understand what was unique about each case and what were the common patterns determining outcomes across more than one site.

4. For several examples of these characterizations, see Li 1999: 22–23.

5. As I have discussed elsewhere, this period produced significant challenges for local communities and researchers working in Indonesia (McCarthy 1998a, 1998b, 1999a).

6. Initially Darul Beureuch, the leader of the revolt aligned with The Darul Islam struggle for an Islamic State. Later Darul Beureuch declared an Aceh Islamic Republic and realigned with the anti-communist PRRI/Permesta rebellions.

7. "Aljumhuriyah Alindunisiah di Simpang Jalan," *Tempo*, 21/11/99. This special status recognized the special role for Islam in Acehnese affairs and gave special autonomy to the province with respect to education and customary law.

8. Interview, South Aceh, 4/11/97.

9. Interview with villager, South Aceh, 2/12/96.

10. See also Robinson 1998; Siegel 2000; Human Rights Watch 2001; Aspinall 2002; ICG 2003; PMB-LIPI and LASEMA-CNRS n.d.

11. The impact of the forest fires on the Leuser Ecosystem was insubstantial. An LMU report notes: "Widespread forest fires in Kalimantan and the southern half of Sumatra created a dense smoke haze over the southern half of the programme area that threatened to close down programme activities. All flights between Jakarta, Medan and Banda Aceh were closed for days and restrictions on land travel to Aceh Tenggara lasted for a week" (Leuser Management Unit 1997/98).

12. Interview, Southeast Aceh, 2/1/99.

13. From this time on, an often-shadowy conflict has continued. With various factions of GAM and those claiming to be GAM carrying out actions, the responsibility for a variety of human rights abuses has often remained obscure (Djalal 2000; Kamaruzzaman 2000).

Bibliography

Abdullah, A. 1994. "Konsepsi Masyarakat Aceh tentang Tata Ruang." In *Bunga Rampai Temu Budaya Nusantara PKA-3*, 272–97. Banda Aceh: Syiah Kuala University Press.

Abrams, P. 1988. "Notes on the Difficulty of Studying the State 1977." *Journal of Historical Sociology* 1 (1): 58–89.

Acciaioli, G. 1985. "Culture as Art: From Practice to Spectacle in Indonesia." *Canberra Anthropology* 8 (1/2), 148–72.

———. 2000. "The Re-emergence of Customary Claims to Land Among the To Lindu of Central Sulawesi: The Revitalisation of Adat in the Era of Reformasi in Indonesia." Paper presented at the annual conference of the Australian Anthropological Society, University of Western Australia, Perth, 19–23 September.

Acheson, J. M. 1989. "Where Have All the Exploiters Gone? Co-Management of the Marine Lobster Industry." In *Common Property Resources: Ecology and Community-Based Sustainable Development*, ed. F. Berkes, 199–217. London: Belhaven Press.

———. 1994. "Welcome to Nobel Country: A Review of Institutional Economics." In *Anthropology and Institutional Economics*, ed. J. M. Acheson, 3–42. Lanham, Md.: University Press of America.

Achmadi, H. 1988. "Forest Fires: Logging Is to Blame." *Inside Indonesia* (14): 16–17.

Adatrechtbundels. 1938a. "De pantjang-alas in Atjeh (1931)." In *Adatrechtbundels; bezorgd door de commissie voor het adatrecht en uitgegeven door het koninklijk instituut voor de taal-, land-en volken-kunde van Nederlandsch-Indie,* 39: 136–38. The Hague: Martinus Nijhoff.

————. 1938b. "Matriarchaat in Tapatoean (1899)." *Adatrechtbundels,* 27: 30–31.

Adger, W. N., T. A. Benjaminsen, K. Brown, and H. Svarstad. 2001. "Advancing a Political Ecology of Global Environmental Discourses." *Development and Change* 32(4).

Aditjondro, G. J. 2002. *Kembar Siam Penguasa Politik dan Ekonomi Indonesia: Investigasi Korupsi Sistemik Bagi Aktivis dan Wartawan.* Jakarta: Lembaga Studi Pers dan Pembangunan.

Agrawal, A. 2001. "State Formation in Community Spaces? Decentralization of Control over Forests in the Kumaon Himalaya, India." *Journal of Asian Studies* 60: 9–40.

Agrawal, A., and C. C. Gibson. 1999. "Enchantment and Disenchantment: The Role of Community in Natural Resource Conservation." *World Development* 27 (4): 629–49.

Agrawal, A., and J. C. Ribot. 1999. "Accountability in Decentralization: A Framework with South Asian and African Cases." *Journal of Developing Areas* 33 (Summer 1999): 473–502.

Ahmad, S. M. 1992. *Ketika Pala Mulai Berbunga (Seraut Wajah Aceh Selatan).* Pemerintah Daerah Tingkat II Aceh Selatan.

Alwy, M. 1998. "Palm-Oil Plantation and the World Bank/IMF: Its Relation and Impacts in Indonesia." http://www.latin.or.id/palm_oil.htm 15/2/99.

Analisa. 1996. "Pengawasan Ketat Diperlukan agar tidak Terjadi Kebocoran di Dispenda." *Analisa* (13 Mar.).

————. 1997. "Karena Harga Minyak Nilam Terus Melonjak Penduduk Tinggalkan Usaha Kayu." *Analisa* (17 Nov.).

Angelsen, A. 1995a. "The Emergence of Private Property Rights in Traditional Agriculture: Theories and a Study from Sumatra." Paper presented to the Fifth Common Property Conference: Reinventing the Commons, Bobo, Norway, 24–28 May.

————. 1995b. "Shifting Cultivation and 'Deforestation': A Study from Indonesia." *World Development* 23 (10): 1713–29.

Angelsen, A., and D. Kaimowitz. 1999. "Rethinking the Causes of Deforestation: Lessons from Economic Models." *World Bank Research Observer* 14 (1): 73–98.

Anon. 1998. "Local Conflicts over Forest Resource Access (Opposition to Indonesian Government's Forest Resource Activities)." *Environment* 40 (4): 32–34.

Appanah, S. 1994. "Natural Forest Management." In *Dipterocarps State of Knowledge and Priorities and Needs for Future Research,* ed. S. Appanah and C. Cossalter. Bogor: CIFOR.

Arghiros, D. 2001. *Democracy, Development and Decentralization in Provincial Thailand.* London: Curzon.

Arimbi, H. P., ed. 1994. *Seminar on the Human Dimensions of Environmentally Sound Development.* Jakarta: WALHI.

Arora, D. 1994. "From State Regulation to People's Participation Case of Forest Management in India." *Economic and Political Weekly* (19 Mar.): 691–98.

Ascher, W. 1993. *Political Economy and Problematic Forestry Policies in Indonesia: Obstacles to Incorporating Sound Economics and Science.* Durham, N.C.: Center for Tropical Conservation, Duke University.

Asian Economic News. 1998. "Indonesia Mulls Indigenous Rights to Forest Resources." *Asian Economic News* (7 Sept.).

Aspinall, E. 2002. "Sovereignty, the Successor State, and Universal Human Rights: History and the International Structuring of Acehnese Nationalism." *Indonesia* (April).

Badan Pusat Statistik Kabupaten Aceh Selatan. 1992. *Aceh Selatan Dalam Angka.*

———. 1997. *Aceh Selatan Dalam Angka.*

Bakels, J. 1994. "But His Stripes Remain: On the Symbolism of the Tiger in the Oral Traditions of Kerenci, Sumatra." In *Text and Tales: Studies in Oral Tradition*, ed. J. G. Ooslen, vol. 22, pp. 33–51. Leiden: Netherlands Research School.

Balai Penelitian Tanaman Rempah dan Obat. 1991. "Tanaman Kemiri." *Edisi Khusus Littro* 7 (2).

Bangkaru, M. 1998. *A Hand Book to Aceh.* Banda Aceh: CV Penerbit Baolhan Haloban.

Bappeda. 1992. *Profil Kabupaten Daerah Tingkat II Aceh Selatan.* Badan Perencanaan Pembangunan Daerah Kabupaten Daerah Tingkat II Aceh Selatan.

Bappenas. 1993. *Biodiversity Action Plan for Indonesia.* Jakarta: Ministry of National Development Planning/National Development Planning Agency.

Bappenas/World Bank. 1993. "Integrated Conservation and Development Project (ICDP) Kerinci National Park Biodiversity Conservation and Park Management." Background Report no. 2.

Barber, C. V. 1996. "Community-Based Biodiversity Conservation: Challenges for Policymakers and Managers in Southeast Asia." Paper presented to the DANCED International Meeting on Biodiversity, Chiang Rai, Thailand, 14–19 January.

———. 1997. "Conservation and Development at Gunung Leuser National Park: An Evaluation of the WWF ID 0106 Gunung Leuser National Park Conservation Project." World Resources Institute Report to the Worldwide Fund for Nature Indonesia Program.

———. 1998. "Forest Resource Scarcity and Social Conflict in Indonesia." *Environment* 40 (4): 4–20.

Barber, C. V., N. C. Johnson, and E. Hafild. 1994. *Breaking the Logjam: Obstacles to Forest Policy Reform in Indonesia and the United States.* Washington, D.C.: World Resources Institute.

Barber, C. V., S. Afiff, and A. Purnomo. 1995. *Tiger by the Tail? Reorienting Biodiversity Conservation and Development in Indonesia.* Washington, D.C.: World Resources Institute.

Barr, C. 1998. "Bob Hasan, The Rise of Apkindo, and the Shifting Dynamics of Control in Indonesia's Timber Sector." *Indonesia* 65.

———. 2000. "Will HPH Reform Lead to Sustainable Forest Management?: Questioning the Assumptions of the "Sustainable Logging" Paradigm in Indonesia." In *Banking on Sustainability: A Critical Assessment of Structural Adjustment in Indonesia's Forest and Estate Crop Industries*. CIFOR and WWF-International's Macroeconomics Program Office. Unpublished manuscript.

———. 2001. "Banking on Sustainability: Structural Adjustment and Forestry Reform in Post-Suharto Indonesia." Bogor: WWF Macroeconomics for Sustainable Development Program Office, Centre for International Forestry Research.

Baskoro, L. R., and L. F. Sihotang. 1996. "Memburu Kayu Siluman di Bumi Khatulistiwa." *Forum Keadilan* 4 (12 Feb.): 50–55.

Bell, G. F. 2001. "The New Indonesian Laws Relating to Regional Autonomy: Good Intensions, Confusing Laws." *Asian-Pacific Law and Policy Journal* 2 (1): 1–44.

Benda-Beckmann, F. von. 1979. Property in *Social Continuity*: Continuity and *Change* in the *Maintenance* of *Property Relationships Through Time* in Minangkabau, West Sumatra. The Hague: M. Nijhoff.

———. 1989. "Scape-Goat and Magic Charm, Law in Development Theory and Practice." *Journal of Legal Pluralism* 28: 129–48.

Benda-Beckmann, F. von, and K. von Benda-Beckmann. 1985. "Transformation and Change in Minangkabau." In Change and *Continuity* in Minangkabau: *Local, Regional,* and *Historical Perspectives* on West Sumatra, ed. L. L. Thomas and F. von. Benda-Beckmann, 235–78. Athens, Ohio: Ohio University Center for International Studies.

———. 1994. "Property, Politics, and Conflict: Ambon and Minangkabu Compared." *Law and Society Review* 28 (3): 589–607.

———. 1999. Legal *Complexity, Ecological Sustainability* and *Social (In)security* in Indonesia. Legal complexity, ecological sustainability and social (in)security in the management and exploitation of land and water resources in Indonesia, Padang, Indonesia.

———. 2002. "Recreating the Nagari: Decentralisation in West Sumatra." Unpublished paper.

Benda-Beckmann, F. von, K. von Benda-Beckmann, and A. Brouwer. 1995. "Changing 'Indigenous Environmental Law' in the Central Moluccas: Communal Regulation and Privatization of Sasi." *Ekonesia* 2: 1–38.

Benda-Beckmann, K. von. 1981. "Forum Shopping and Shipping Forums: Dispute Processing in a Minangkabau Village in West Sumatra." *Journal of Legal Pluralism* 19" 117–59.

Benda-Beckmann, K. von, and F. von Benda-Beckmann, eds. Forthcoming. *Legal Complexity, Ecological Sustainability and Social (In)Security in Indonesia.*

Berge, L. A. van den. 1934. "Memorie van overgave van de onderafdeling Alaslanden, Koetatjane 22 November 1934." TS, 15 pp. Algemeen Rijksarchief. The Hague.

Berkes, F., J. Colding, and C. Folke. 2003. "Introduction." In *Navigating Social-Economic Systems: Building Resilience for Complexity and Change*, ed. F. Berkes and C. Folke. New York: Cambridge University Press.

Berkes, F., P. George, and R. J. Preston. 1991. "Co-management: The Evolution in Theory and Practice of the Joint Administration of Living Resources." *Alternatives* 18 (2): 12–18.

Berry, S. 1989. "Social Institutions and Access to Resources." *Africa* 59: 41–55.

———. 1994. "Resource Access and Management as Historical Processes: Access, Control and Management of Natural Resources." In Access, Control and Management of *Resources in Sub-Saharan Africa—Methodological Considerations*, ed. C. Lund and H. S. Marcussen, 23–44. Occasional Paper no. 13. Roskilde, Denmark: International Development Studies, Roskilde University.

———. 1997. "Tomatoes, Land and Hearsay: Property and History in Asante in the Time of Structural Adjustment." World Development 25(8): 1225–41.

———. 2001. Chiefs Know Their Boundaries: Essays *on* Property, Power, *and the* Past *in* Asante, 1896–1996. Portsmouth, Oxford, Cape Town: Heinemann, James Currey, David Philip.

Bisnis Indonesia. 2000. "Dephutbun bekukan pungutan levy and grant." *Bisnis Indonesia* 28 March.

BKI. 1903. "Mededeelingen Betreffende de Atjehsche Onderhoorigheden." *Bijdragen tot de taal-, land- en volkenkunde van Nederlandsch-Indie* 55: 241–44.

———. 1912. "Mededeelingen Betreffende Atjehsche Onderhoorigheden." *Bijdragen tot de taal-, land- en volkenkunde van Nederlandsch-Indie* 55: 405–40.

Blackwood, E. 2001. "Representing Women: The Politics of Minangkabau Adat Writings." *Journal of Asian Studies* 60 (1): 125–50.

Blaikie, P. M., and H. Brookfield. 1987. *Land Degradation and Society*. London: Methuen and Co.

Blair, H. 2000. "Participation and Accountability at the Periphery: Democratic Local Governance in Six Countries." *World Development* 28 (1): 21–39.

Blower, J. H. 1980. *Management Problems in the Gunung Leuser National Park*. Bogor: World Wildlife Fund Indonesia Programme.

Boomgaard, P. 2001. *Frontiers of Fear: Tigers and People in the Malay World, 1600–1950*. New Haven: Yale University Press.

Booth, A. 1999. "Survey of Recent Developments." Bulletin of Indonesian Economic Studies (35): 3–33.

Bourdieu, P. 1990. "From Rules to Strategies." In his *Other Words: Essays Towards a Reflexive Sociology*, trans. Matthew Adamson, 59–75. Stanford: Stanford University Press.

Bowen, J. R. 1988. "The Transformation of an Indonesian Property System: *Adat*, Islam, and Social Change in the Gayo Highlands." *American Ethnologist* 15: 274–93.

———. 1991. *Sumatran Politics and Poetics: Gayo History, 1900–1989*. New Haven: Yale University Press.

Bowles, I. A., R. E. Rice, R. A. Mittermeirer, and G.A.B. da Fonseca. 1998. "Logging and Tropical Forest Conservation." *Science* 280 (19 June): 1899.

Brandon, K. E., and M. Wells. 1992. "Planning for People and Parks: Design Dilemmas." *World Development* 20 (4): 557–70.

Brechin, S. R., S. R. Brechin, P. R. Wilshusen, C. L. Fortwangler, and P. C. West. 2002. "Beyond the Square Wheel: Toward a More Comprehensive Understanding of Biodiversity as Social and Political Process." *Society and Natural Resources* 15: 41–64.

Bridgman, P., and G. Davis. 1998. *Australian Policy Handbook*. St. Leonards, N.S.W: Allen and Unwin.

Bromley, D. W. 1989. "Property Relations and Economic Development: The Other Land Reform." *World Development* 17 (6): 867–77.

———. 1991. *Environment and Economy Property Rights and Public Policy*. Cambridge, Mass.: Blackwell.

———. 1997. "Environmental Problems in Southeast Asia: Property Regimes as Cause and Solution." EEPSEA Special Papers. Singapore: IDRC Research Programs.

Bromley, D. W., and M. M. Cernea. 1989. "The Management of Common Property Natural Resources: Some Conceptual and Operational Fallacies." Discussion Papers no. 57. Washington, D.C.: World Bank.

Brosius, J. P. 1997. "Endangered Forest, Endangered People: Environmental Representations of Indigenous Knowledge." *Human Ecology* 25 (1): 47–69.

———. 1999. "Comments on A. Escobar, 'After Nature.'" *Current Anthropology* 40 (1).

Brosius, J. P., A. L. Tsing, and C. Zerner. 1998. "Representing Communities: Histories and Politics of Community-Based Natural Resource Management." *Society and Natural Resources* 11 (2): 157–68.

Brown, D. 1994. "Indonesia: Neo-Patrimonialism and Integration." In his *The State and Ethnic Politics in Southeast Asia*. London and New York: Routledge.

Brown, K. 2002. "Innovations for Conservation and Development." *Geographical Journal* 168 (1): 6–18.

Bruce, J. W., and L. Fortmann. 1991. "Property and Forestry." *Journal of Business Administration* 20 (1–2): 471–98.

Bruce, J. W., L. Fortmann, and C. Nhira. (1993) "Tenure in Transition, Tenures in Conflict: Examples from the Zimbabwe Social Forest." *Rural Sociology* 58 (4): 626–42.

Bryant, R. L. 1992. "Political Ecology: An Emerging Research Agenda in Third World Studies." *Political Geography* 11 (1): 12–36.

———. 1998. "Power, Knowledge and Political Ecology in the Third World: A Review." *Progress in Physical Geography* 22 (1): 79–94.

Bulbeck, D., A. Reid, L. C. Tan, and Y. Q. Wu. 1998. *Southeast Asian Exports Since the 14th Century: Cloves, Pepper, Coffee, and Sugar*. Leiden: KITLV Press.

Burns, P. 1989. "The Myth of Adat." *Journal of Legal Pluralism* 28: 1–127.

———. 1999. *The Leiden Legacy: Concepts of Law in Indonesia*. Jakarta: Pradnya Paramita.

Callicott, J. B. 1991. "Conservation of Biological Resources: Responsibility to Nature and Future Generations." In *Challenges in the Conservation of Biological Resources: A Practitioner's Guide,* ed. D. Decker, M. E. Krasny, G. R. Goff, C. R. Smith, and D. W. Cross. Boulder, Colo.: Westview Press.

Campbell, B., A. Mondondo, et al. 2001. "Challenges to Proponents of Common Property Resource Systems: Despairing Voices from the Social Forests of Zimbabwe." *World Development* 29 (4): 589–600.

Campbell, J. Y. 1999. *Hutan Untuk Rakyat, Masyarakat Adat, atau Kooperasi? Plural Perspectives in the Policy Debate for Community Forestry in Indonesia.* Seminar on Legal Complexity, Natural Resource Management, and Social (In)Security in Indonesia, Padang.

Chambers, R. 1983. *Rural Development: Putting the Last First.* London: Longman.

Chanock, M. 1978. "Neo-traditionalism and the Customary Law in Malawi." *African Law Studies* 16: 80–91.

———. 1982. "Making Customary Law: Men, Women and Courts in Colonial Northern Rhodesia." In *African Women and the Law: Historical Perspectives,* ed. M. J. Hay and M. Wright, 53–67. Boston, Boston University Press.

———. 1998. *Law, Custom and Social Order: The Colonial Experience in Malawi and Zambia.* Cambridge: Cambridge University Press.

CIFOR. 1998. "The Economic Crisis and Indonesia's Forest Sector." Bogor: CIFOR.

———. 1999. "CIFOR Researchers Challenge Common Assumptions That More Intensive Farming Will Save Tropical Forests." Bogor: CIFOR.

Clifford, J. 1986. "Introduction: Partial Truths." In *Writing Culture: The Poetics and Politics of Ethnography, a School of American Research Advanced Seminar,* ed. J. Clifford and G. E. Marcus, 1–26. Berkeley: University of California Press.

Cohen, M. 1998. "Tackling a Bitter Legacy Reformasi Advocates in Provinces Focus on Corrupt Officials." *Far Eastern Economic Review* (2 July).

Coleman, J. S. 1990. "Commentary: Social Institutions and Social Theory." *American Sociological Review* 55: 333–39.

Colfer, C. J. P. 1987. "Change and Indigenous Agroforestry in East Kalimantan." In *Whose Trees? Proprietary Dimensions of Forestry,* ed. L. Fortmann and J. W. Bruce, 306–9. Boulder, Colo.: Westview Press.

Colfer, C. J. P., with Richard G. Dudley. 1993. "Shifting Cultivators of Indonesia: Marauders or Managers of the Forest? Food and Agriculture Organization of the United Nations." Community Forestry Case Study Series 6. Rome, Italy: FAO.

Colijn, H. 1938. "Adatheffingen in TapaToean (1901)." In *Adatrechtbundels,* 22–29.

Collier, J. F. 1976. "Political Leadership and Legal Change in Zinacantan." *Law and Society* 11 (Fall 1976): 131–63.

Conklin, H. C. 1975. *Hanunoo Agriculture: A Report on an Integral System of Shifting Cultivation in the Philippines.* Northford, Conn.: Elliot's Books.

Contreras, A. P. 2000. "Rethinking Participation and Empowerment in the Uplands." In *Forest Policy and Politics in the Philippines,* ed. P. Utting. Manila: Ateneo de Manila University Press.

Crook, R., and J. Manor. 1994. *Democracy and Decentralization in South Asia and West Africa.* Cambridge: Cambridge University Press.

Crotty, M. 1998. *The Foundations of Social Research : Meaning and Perspective in the Research Process.* St. Leonards, N.S.W.: Allen and Unwin.

Crouch, H. 1998. "Indonesia's 'Strong' State." In *Weak and Strong States in*

Asia-Pacific Societies, ed. P. Dauvergne, 93–113. St. Leonards, N.S.W.: Allan and Unwin.

Dali, J., and A. N. Guntings. 1981. *Cara Penanaman Kemiri*. Departemen Pertanian, Badan penelitian dan Pengembangan Pertanian, Lembaga Penelitian Hutan.

Dammerman, K. W. 1929. *Preservation of Wild Life and Nature Reserves in the Netherlands Indies*. Fourth Pacific Science Congress, Java, 1929. Weltevreden: Emmink.

Dauvergne, P. 1994. "The Politics of Deforestation in Indonesia." Pacific Affairs 66 (4) Winter 1993–94): 497–518.

———. 1997a. "Globalisation and Deforestation in the Asia-Pacific." Australian National University, Dept. of International Relations Working Paper no. 1997/7.

———. 1997b. *Shadows in the Forest: Japan and the Politics of Timber in Southeast Asia*. Cambridge, Mass., and London: MIT Press.

———. 1997c. "Weak States and the Environment in Indonesia and the Solomon Islands." Australian National University, Dept. of International Relations Working Paper no. 10.

———. 1998. "Environmental Insecurity, Forest Management and State Responses in Southeast Asia." Australian National University. Dept. of International Relations Working Paper no. 1998/2.

Department Kehutanan and Universitas Syiah Kuala Darussalam–Banda Aceh. 1993. "Rancangan Pemindahan Perambah Hutan dari dalam Kawasan Taman Nasional Gunung Leuser." Department Kehutanan dengan Universitas Syiah Kuala Darussalam–Banda Aceh Draft Laporan Akhir.

Dephut. 1992. "Sekilas tentang sejarah dan perkembangan Tata Guna Hutan Kesepakatan: makalah Menteri Kehutanan Republic Indonesia pada Rapat kerja Gubernur Kepala Daerah Tingkat I dan Bupati Kepala Daerah Tingkat II Seluruh Indonesia tahun 1992." Departemen Kehutanan, Direktorat Jenderal Inventarisasi dan Tata Guna Hutan.

Dephutbun. 1988. "Keputusan Kepala Kantor Wilayah Departmen Kehutanan dan Perbeunan Kantor Wilayah Propinsi Daerah Istimewa Aceh Nomor: 445/ KPTS/KWL-4/1988." Departmen Kehutanan dan Perkebunan Kantor Wilayah Propinsi Daerah Istimewa Aceh.

Detektip. 1996. "Aparat Mandul Hutan Gunduk." *Detektip* (20 Dec.).

Devas, N., with B. Binder, K. Davey, and R. Kelly. 1989. *Financing Local Government in Indonesia*. Athens: Ohio University Center for International Studies.

DFID and WWF Indonesia. 1998. "Laporan Perkembangan Sawmill Wilayah Selatan Taman Nasional Bukit Tigapuluh dan di Sekitar Areal KPHP Pasir Mayang." Department for International Development (Jambi), World Wide Fund for Nature (1D 0117) Laporan PFM/KPHP/98/7.

Dijk, H. van. 1996. "Land Tenure, Territoriality, and Ecological Instability: A Sahelian Case Study." In *The Role of Law in Natural Resource Management*, ed. J. Spiertz and M. G. Wiber, 17–45. The Hague: VUGA.

Direketorat Jenderal Perkebunanan. 1977. "Vademekum Perkebunan." Direketorat Jenderal Perkebunanan. Department Pertanian.

Direktorat Jenderal Perkebunan. 1994/1995. "Petunjuk Teknis Penyediaan Setek Nilam." Direktorat Jenderal Perkebunan Direktorat Bina Perbenihan.

Djalal, D. 2000. "A Bloody Truce." *Far Eastern Economic Review* (5 Oct.): 16–18.

Doolittle, A. 2001. "From Village Land to Native Reserve Changes in Property Rights in Sabah 1950–1996." *Human Ecology* (March).

Doup, M. A., and A. B. A. Bakar. 1980. *Ikhtisar sejarah pendudukan Belanda di Tapak Tuan dan daerah-daerah kenegerian di bahagian Selatan daerah Aceh. [Judul asli: Beknopt overzicht van de Krijgsgeschiedenis van Tapa Toean en de Zuidelijke Atjehsche landscahppen ("Korps Marechaussee Atjeh, 2 April 1890–1940)].* Sei Informasi Aceh Th IV No. 2. Banda Aceh: Pusat Dokumentasi dan Informasi Aceh.

Dove, M. R. 1983. "Theories of Swidden Agriculture, and the Political Economy of Ignorance." *Agroforestry Systems* 1: 85–99.

———. 1985. "Government Perceptions of Traditional Social Forestry in Indonesia: The History, Causes and Implications of State Policy on Swidden Agriculture. In *Community Forestry: Socio-economic Aspects*, ed. Y. S. Rao, N. T. Vergara and G. W. Lovelace. Bangkok: Regional Office for Asia and the Pacific, FAO.

———. 1986. "Peasant Versus Government Perception and Use of the Environment: A Case-Study of Banjarese Ecology and River Basin Development in South Kalimantan." *Journal of Southeast Asian Studies* 17 (1): 113–36.

———. 1993a. "A Revisionist View of Tropical Deforestation and Development." *Environmental Conservation* 20 (1): 17–24.

———. 1993b. "Smallholder Rubber and Swidden Agriculture in Borneo: A Sustainable Adaptation to the Ecology and Economy of the Tropical Forest." *Economic Botany* 47 (2): 136–47.

———. 1996. "So Far From Power, So Near to the Forest: A Structural Analysis of Gain and Blame in Tropical Forest Development." In *Borneo in Transition People, Forests, Conservation and Development,* ed. C. Padoch and N. L. Peluso. Kuala Lumpur: Oxford University Press.

———. 1997a. "Dayak Anger Ignored: Inequities in State Development in Kalimantan." *Inside Indonesia* (51).

———. 1997b. "The Political Ecology of Pepper in the Hikayat Bandjar: The Historiography of Commodity Production in a Bornean Kingdom." In *Paper Landscapes: Explorations in the Environmental History of Indonesia,* ed. P. Boomgaard, F. Colombijn, and D. Henley, 341–78. Leiden: KITLV Press.

Down to Earth. 1999. "Reforms Bring Few Gains." *Down to Earth* (quarterly newsletter of the International Campaign for Ecological Justice in Indonesia) 40.

———. 2000. "Regional Autonomy and the IFIs." *Down to Earth* 46.

Echols, J. M., and H. Shadily. 1989. *Kamus Indonesia-Inggris: An Indonesian-English Dictionary.* Jakarta: PT Gramedia.

Eckersley, R. 1995. "Disciplining the Market, Calling in the State: The Politics of Economy-Environment Integration." In *Ecological Modernisation,* ed. R.

Eckersley, S. Young, and J. van der Straaten. London: Routledge.

Ecologist, The. 1995. "'Policy Failure': Protecting Against Blame." *The Ecologist* 25 (6): 218.

Edwards, V. M., and N. A. Steins. 1998. "Developing an Analytical Framework for Multiple-Use Commons." *Journal of Theoretical Politics* 10 (3): 347–83.

Effendi, E. 1998. "An Initial Resource Economic Valuation of the Leuser Ecosystem." Unpublished report.

Eghenter, C. 2000. "What Is *Tuna Ulen* Good For? Considerations on Indigenous Forest Management, Conservaation and Research in the Interior of Indonesian Borneo." *Human Ecology* 28 (3): 331–57.

Elliott, J., A. Khan, and Z. Saad. 1993. "Developing Partnerships: A Study on NGO-Donor Linkages in Kerinci-Seblat and Lore Lindu National Parks." Program Keterkaitan Lingkungan Hidup and Ekonomi Masyarakat (LELI) and PACT.

Ely, M. 1991. *Doing Qualitative Research: Circles Within Circles*. London: Falmer Press.

Emerson, R. M., R. I. Fretz, and L. L. Shaw. 1995. *Writing Ethnographic Fieldnotes*. Chicago: University of Chicago Press.

EPIQ Technical Advisory Group. 2001. "Environmental Policy Implementation: Lessons Learned II." USAID Global Environment Center.

Erawan, I. K. P. 1999. "Political Reform and Regional Politics in Indonesia." *Asian Survey* 39 (4): 588.

Escobar, A. 1999. "After Nature." *Current Anthropology* 40 (1).

Evers, P. J. 1995. "Preliminary Policy and Legal Questions About Recognizing Traditional Land Rights in Indonesia." *Ekonesia* 3: 1–24.

FAO. 1991. *Situation and Outlook of the Forestry Sector in Indonesia*. Volume 1: *Issues, Findings and Opportunities*. Jakarta: Ministry of Forestry, Government of Indonesia; Food and Agriculture Organization of the United Nations.

Fauzi, N., and R. Y. Zakaria. 2002. "Democratizing Decentralization: Local Initiatives from Indonesia." International Association for the Study of Common Property, 9th Biennial Conference, Victoria Falls, Zimbabwe.

Fay, C., and H. de Foresta. 1998. *Progress Towards Recognising the Rights and Management Potentials of Local Communities in Indonesian State-Defined Forest Areas*. Workshop on Participatory Natural Resource Management in Developing Countries, Mansfield College, Oxford University.

Feeny, D., F. Berkes, B. J. McCay, and J. M. Acheson. 1990. "The Tragedy of the Commons: Twenty-two Years Later." *Human Ecology* 18 (1): 1–19.

Ferretti, E., ed. 1997. *Cutting Across the Lands: An Annotated Bibliography on Natural Resource Management and Community Development in Indonesia, the Philippines, and Malaysia*. Ithaca, N.Y.: Southeast Asia Program, Cornell University.

Fitzpatrick, D. 1997. "Disputes and Pluralism in Modern Indonesian Land Law." *Yale Journal of International Law* 22 (1): 172–212.

FKKM. 1998. "New Era for Indonesian Forestry, Forest Resource Management Reformation." Forum Komunikasi Kehutanan Masyarakat, 22 Sept.

Forster, R. R. 1973. *Planning for Man and Nature in National Parks Reconciling*

Perpetuation and Use. Morges, Switzerland: IUCN Publications.

Forrester, G. 2001. "Decentralization in Indonesia: Options for Australian Aid." Unpublished report.

Fortmann, B. 1981. "You Cannot Develop by Act of Parliament: Rethinking Development from the Legal Viewpoint." Paper presented to the Symposium Recht en Ontwikkeling, Bureau Buitenland, RijksuniversiteitLeiden, Leiden, 19–21 May.

Fortmann, L. 1990. "Locality and Custom: Non-aboriginal Claims to Customary Usufructuary Rights as a Source of Rural Protest." *Journal of Rural Studies* 6 (2): 195–208.

Fox, J. 1990. "The Challenges of Rural Democratisation in Latin American and the Phillipines." *Journal of Development Studies* 26 (4): 1–18.

———. 1993. "The Tragedy of Open Access." In *South-East Asia's Environmental Future: The Search for Sustainability,* ed. H. Brookfield and Y. Byron, 302–15. Kuala Lumpur and New York: Oxford University Press.

Fox, J., and K. Atok. 1997. "Forest-dweller Demographics in West Kalimantan, Indonesia." *Environmental Conservation* 24 (1): 31–37.

Freeman, J. D. 1955. *Iban Agriculture: A Report on the Shifting Cultivation of Hill Rice by the Iban of Sarawak.* London: Her Majesty's Stationery Office.

Freeman, M. 1993. *Pemerintahan Indonesia: The Indonesian Government System.* Canberra: Australian International Development Assistance Bureau Country Programs Division, Commonwealth of Australia.

Frerks, G., and J. M. Otto. 1996. "Decentralization and Development: A Review of Development Administration Literature." Van Vollenhoven Institute for Law and Administration in Non-Western Countries Research Report 96/2, Leiden University.

Geertz, C. 1963. *Agricultural Involution: The Process of Ecological Change in Indonesia.* Berkeley: University of California Press for the Association of Asian Studies.

Gellert, P. K. 1998. "A Brief History and Analysis of Indonesia's Forest Fire Crisis." *Indonesia* 65 (April): 63–85.

Gibson, C. C., M. A. McKean, and E. Ostrom. 2000. "Forests, People, and Governance: Some Initial Theoretical Lessons." In *People and Forests: Communities, Institutions, and Governance,* ed. C. C. Gibson, M. A. McKean, and E. Ostrom, 227–42. Cambridge, Mass.: MIT Press.

Gilley, B. 1999. "Sticker Shock." *Far Eastern Economic Review* (14 Jan.).

Goenner, C. 1998. "Conflicts and Fire Causes in a Sub-District of Kutai, East-Kalimantan, Indonesia." http://www.iffm.or.id/Fire_Causes.html 12/02/99.

Goor, C. P. van. 1982. *Indonesian Forestry Abstracts: Dutch Literature Until About 1960.* Wageningen: Pudoc.

Gray, K. 1995. "The Ambivalence of Property." In *The Earthscan Reader in Sustainable Development,* ed. J. Kirby, P. O'Keefe, and L. Timberlake, 223–36. London: Earthscan Publications Limited.

Griffiths, J. 1979. "Is Law Important?" *New York University Law Review* 54: 339–74.

———. 1995. "Legal Pluralism and the Theory of Legislation—With Special

Reference to the Regulation of Euthanasia." In *Legal Polycentricity: Consequences of Pluralism in Law,* ed. H. Peterson and H. Zahle, 201–34. Aldershot: Dartmouth.

Griffiths, M. 1990. *Indonesian Eden: Aceh's Rainforest.* Baton Rouge: Louisiana State University Press.

———. 1992. *Leuser.* Jakarta: The Directorate General of Forest Protection and Nature Conservation and the World Wide Fund for Nature Indonesia Programme.

Groves, R. H. 1996. *Green Imperialism: Colonial Expansion, Tropical Island Edens and the Origins of Environmentalism, 1600–1869.* Cambridge: Cambridge University Press.

Guardian Weekly. 1999. "Aceh Threatens to Destabilise Indonesia." *Guardian Weekly* (9–15 Dec.).

Guha, R. 1994. "Fighting for the Forest: State Forestry and Social Change in Tribal India." In *The Rights of Subordinated Peoples,* ed. O. Mendelsohn and U. Baxi. Delhi: Oxford University Press.

Gunarso, P., and J. Davie. 1999. "How Decentralization Can Improve Accountablity of Forest Resources Management in Indonesia." School of Natural and Rural Systems Management, University of Queensland. Unpublished draft.

———. N.d. "The Management of Wildlife Reserves in Production Forests in Indonesia." Unpublished manuscript.

Gunawan, R., J. Thamrin, and E. Suhendar. 1998. *Industrialisasi Kehutanan dan Dampaknya Terhadap Hasyarakat Adat. Kasus Kalimantan Timur.* Bandung: Yayasan AKATIGA.

Gupta, A. 1995. "Blurred Boundaries: The Discourse of Corruption, the Culture of Politics, and the Imagined State." *American Ethnologist* 22 (2): 375–402.

Habermas, J. 1987. *The Theory of Communicative Action.* Vol. 2 of *The Critique of Functionalist Reason.* Oxford: Polity Press.

Hadad, M., E. A. Syakir, and M. Syakir. 1992. "Pengadaan Bahan Tananman Pala." *Littro* 8 (1): 1–7.

Hadiz, V. 2003. "Local Power: Decentralisation and Political Reorganisation in Indonesia." Paper presented at the conference on Globalisation, Conflict, and Political Regimes in East and Southeast Asia, Fremantle, August.

———. 2004. "Indonesian Local Party Politics: A Site of Resistance to Neoliberal Reform." Critical Asian Studies 36 (4): 615–36.

Hafild, E. 1999. "A New Paradigm for Rights Based Advocacy Strengthening NGO Alliances in the Post-Suharto Era." Paper presented at the Australian Council for Overseas Aid Annual Council Meeting, Canberra, 11 September.

Haryanto, ed. 1998. *Kehutanan Indonesia Pasca Soeharto: Reformasi Tanpa Perubahan.* Bogor: Pustaka Latin.

Haverfield, R. 1999. "*Hak Ulayat* and the State: Land Reform in Indonesia." In *Indonesia Law and Society,* ed. T. Lindsey. Leichhardt, N.S.W.: Federation Press.

Hayes, A. C. 1997. "Local, National and International Conceptions of Justice: The Case of Swidden Farmers in the Contexts of National and Regional Developments in Southeast Asia." Resource Management Working Paper Series,

Australian National University Working Paper no. 1997/14.

Hecht, S. B., A. B. Anderson, and P. May. 1988. "The Subsidy from Nature: Shifting Cultivation, Successional Palm Forests, and Rural Development." *Human Organization* 47 (1): 25–35.

Hess, C. 1996. "Workshop Research Library Bibliography of IAD Framework." http://www.indiana.edu/workshop/wsl/iadbib.htm 2/7/97.

Hidayah, Z. 1997. *Ensiklopedi suku bangsa di Indonesia*. Jakarta: LP3ES.

Holleman, J. F. 1981. *Van Vollenhoven on Indonesian Adat Law*. The Hague: Martinus Nijhoff.

Hooker, M. B. 1978. *Adat Law in Modern Indonesia*. Kuala Lumpur: Oxford University Press.

Howlett, M., and M. Ramesh. 1995. "Policy Instruments." In their *Studying Public Policy: Policy Cycles and Subsystems*, 80–101. Ontario: Oxford University Press.

Hughes, R., and R. Flintan. 2001. *Integrating Conservation and Development Experience: A Review and Bibliography of the ICDP Literature*. London: IIED.

Hull, T. H. 1999. "Striking a Most Delicate Balance: The Implications of *Otonomi Daerah* for the Planning and Implementation of Development Cooperation Projects." Demography Program, RSSS, ANU. demography.anu.edu.au/EPI/Striking.pdf/.

Human Rights Watch Asia. 1997. "The Horror in Kalimantan." *Inside Indonesia* (51): 9–12.

Human Rights Watch. 2001. "Indonesia: The War in Aceh." http://www.hrw.org/reports/2001/aceh/.

Hurst, P. 1990. *Rainforest Politics: Ecological Destruction in South-East Asia*. London: Zed Books.

IASCP. 2000. "CPR Virtual Library of Common Property and Common Pool Resources." http://www.indiana.edu/iascp/library.html. International Association for the Study of Common Property, Indiana University.

ICG. 2003. "Aceh: How Not to Win Hearts and Minds Briefing." (23 July). http://www.crisisgroup.org /home/index.cfm?id=1778&l=1.

IFFM/GTZ. 1998. "Fire in East-Kalimantan in 1998." http://www.iffm.or.id/FiresinEast2.html: IFFM/GTZ.

Ihromi, T. O. 1994. "Inheritance and Equal Rights for Toba Batak Daughters." 28 (3): 525–37.

Indonesian Heritage. 1996. *Indonesian Heritage*, ed. T. Whitten and J. Whitten, vol. 4. Jakarta: Buku Antar Bangsa for Grolier International.

Indrawanto, C., and A. Wahyidi. N.d. "Analisis Tataniaga Pala di Indonesia." Balittro, Bogor. Unpublished report.

Ismail, M. R. 1990. *Sistem Morfologi Verba Bahasa Kluet*. Jakarta: Pusat Pembinaan dan Pengembangan Bahasa, Departemen Pendidikan dan Kebudayaan.

———. 1991. "Seuneubok Lada, Uleebalang, dan Kumpeni Perkembangan Sosial Ekonomi di Daerah Batas Aceh Timur, 1840–1942." Ph.D. diss., Rijksuniversiteit te Leiden.

———. 1996. "Budaya Ekonomi Pengembangan Pertanian Rakyat di Aceh pada

Abad Ke-19." Jurusan Pendidikan Ilmu Sosial Fakultas Keguruan dan Ilmu Pendidikan Universitas Syaiah Kuala. Unpublished paper.

ITFMP. 1999. "A Draft Position Paper on Threats to Sustainable Forest Management in Indonesia: Roundwood Supply and Demand and Illegal Logging." Indonesia–UK Tropical Forest Management Programme PFM/EC/99/01.

Iwabuchi, A. 1994a. *The People of the Alas Valley: A Study of an Ethnic Group of Northern Sumatra.* Oxford and New York: Clarendon Press.

———. 1994b. "The Toba Batak Migration to the Southeastern Part of the Special Province of Aceh, Indonesia." *Southeast Asian Studies* 34: 120–37.

Jakarta Post. 1998a. "Forestry Bill Does Not Support Small Firms." *Jakarta Post* (10 Dec.).

———. 1998b. "Habibie Seeks Review of Forestry Policies." (18 Dec.).

———. 1999. "Forestry Concessions to Be Auctioned." (15 Jan.).

Johnson, C. 2003. "Uncommon Properties: The 'Poverty of History' in Common Property Discourse." Paper presented at RCSD Conference on Politics of the Commons: Articulating Development and Strengthening Local Practices, Chiang Mai.

Jordan, D., and A. Anhar. 1997. "Population and Economy in Aceh Tenggara and Aceh Selatan." Leuser Development Program. Unpublished report.

Kahn, J. S. 1993. *Constituting the Minangkabau: Peasants, Culture, and Modernity in Colonial Indonesia.* Providence, R.I.: Berg.

———. 1999. "Culturalising the Indonesian Uplands." In *Transforming the Indonesian Uplands : Marginality, Power and Production,* ed. T. M. Li, 79–104. Amsterdam: Harwood Academic.

Kaimowitz, D., C. Vallejos, P. B. Pancheco, and R. Lopez. 1998. "Municipal Governments and Forest Management in Lowland Bolivia." *Journal of Environment and Development* 7 (1): 45–60.

Kaimowitz, D., P. B. Pancheco, and J. Johnson. 1999. *Local Governments and Forests in the Bolivian Lowlands.* London: Rural Development Forestry Network, Overseas Development Institute.

Kamaruzzaman, S. 2000. "Women and the War in Aceh." *Inside Indonesia* 64 (Oct.–Dec.).

Kano, H. 1996. "Land and Tax, Property Rights and Agrarian Conflict: A View from Comparative History." Paper presented to the Tenth INFID Conference on Land and Development, Canberra, 26–28 April.

Kantor Statistik Kabupaten Aceh Selatan. 1995. *Aceh Selatan Dalam Angka 1995.*

Kato, J. 1996. "Review Article: Institutions and Rationality in Politics—Three Varieties of Neo-Institutionalists. *British Journal of Political Science* (26): 553–82.

Keating, J. 2000. "New Hope for Indonesia's Forests." *Jakarta Post* (11 Feb.).

Kelompok Kerja WWF ID 0106. N.d. "Konflik Kepentingan Dalam Pengelolaan Sumber Daya Alam Pedesaan, Studi Kasus: Kemumiman Manggamat, Kec. Kluet Utara–Kab. Aceh Selatan." Unpublished report.

Kemp, J. H. 1984. "Strategies and Structures in Thai Society." In *Strategies and Structures in Thai Society,* ed. H. t. Brummelhuis and J. H. Kemp. Antropologisch-sociologisch Centrum universiteit van Amsterdam, no. 31. Amsterdam: Publikatieserie Zuiden Zuidoost-Azie.

Khairuddin. 1992. *Dampak Perkembangan Sarana Transportasi Darat Terhadap Perubahan Hubungan Sosial Masyarakat Kluet*. Banda Aceh: Pusat Pengembangan Penelitian Ilmu-Ilmu Sosial, Universitas Syaiah Kuala.

Khanna, V. 1998. "Flaws in People's Economy.' *Business Times* (15 Dec.).

Kirkpatrick, J. 2000. "The Political Ecology of Biogeography." *Journal of Biogeography* 27 (1): 45–48.

KLH and UNDP. 1997. *Agenda-21 Indonesia: A National Strategy for Sustainable Development*. State Ministry for Environment Republic of Indonesia, United Nations Development Programme.

———. 1998a. "Executive Summary Forest and Land Fires in Indonesia." Kantor Menteri Negara Lingkungan Hidup, United Nations Development Programme.

———. 1998b. "Kebakaran Hutan dan Lahan di Indonesia." Dampak, Faktor dan Evaluasi. State Ministry for Environment Republic of Indonesia, United Nations Development Programme.

Klinken, G. van. 1998. "Taking on the Timber Tycoons." *Inside Indonesia* 53: 25.

Koesnoe, M., and S. Soendari. 1977. "The People and Their Adat in Tanah Alas (South-East Aceh Regency)." Austerlitz: World Wildlife Fund, Netherlands Gunung Leuser Committee.

Kompas. 1991. "Drs Sayed Mudhahar Akhmad Bupati Pencinta Lingkungan Hidup." *Kompas* (26 Feb.).

———. 1996. "Gangguan Hutan Sangat Serius." *Kompas* (12 April).

———. 1997a. "Curi Kayu di Taman Nasional." *Kompas* (11 Feb.).

———. 1997b. "Ribuan Hektar Pala Terserang Hama." *Kompas* (14 Oct.).

———. 1998. "65 Persen Kebakaran Hutan Kaltim karena HTI." *Kompas* (3 April).

Kreemer, J. 1922. *Atjeh: algemeen samenvattend overzicht van land en volk van Atjeh en onderhoorigheden*, vol. 1. Leiden: E. J. Brill.

———. 1923. *Atjeh: algemeen samenvattend overzicht van land en volk van Atjeh en onderhoorigheden*, vol. 2. Leiden: E. J. Brill.

Langenberg, M. van. 1990. "The New Order State: Language, Ideology, Hegemony." In *State and Civil Society in Indonesia*, ed. A. Budiman, 121–50. Clayton, Victoria: Centre of Southeast Asian Studies, Monash University.

Larmour, P., and N. Wolanin, eds. 2001. *Corruption and Anti-corruption*. Canberra: Asia Pacific Press.

Larson, A. M. 2003. "Decentralization and Forest Management in Latin America: Toward a Working Model." *Public Administration and Development* 23 (2).

Latin. 1998. "Reorientasi Sektor Kehutanan Untuk Menduking Pemberdayaan Ekonomi Rakyat." http://www.latin.or.id/berita_biotrop.htm 15/2/99.

Lay, C. 1999. "The Management of Natural Resources for the Strengthening of the Local Base, the Social and Political Perspective: Introductory Notes." Paper presented to the Seminar-Workshop Towards the Management of Natural

Resources for Strengthening of the Local Base, YLBHI-Publicity and International Relations Division (PIRD), Jakarta.

LDP. 1997. "Leuser Development Programme Overall Work Plan." Leuser Management Unit. Unpublished report.

Leach, M., R. Mearns, and I. Scoones. 1999. "Environmental Entitlements: Dynamics and Institutions in Community-Based Natural Resource Management." *World Development* 27 (2): 225–47.

Lederman, D. 2001. "Accountability and Corruption: Political Institutions Matter." World Bank.

Lembaga Penelitian dan Pengabdian Masyarakat. 1978. "Laporan penelitian, beberapa lembaga hukum adat yang masih hidup di masyarakat propinsi Riau dan Aceh (Aceh Tenggara)." Lembaga Penelitian dan Pengabdian Masyarakat, Fakultas Hukum Universitas Sumatra Utara.

Leuser Management Unit. 1996/97. "Leuser Development Programme, Annual Report 1996/7." Government of the Republic of Indonesia, European Union.

———. 1997/98. "Leuser Development Programme, Annual Report 1997/8." Government of the Republic of Indonesia, European Union.

Leuser Management Unit and Leuser International Foundation. N.d. "Leuser Development Programme: Cooperation Between the Government of Indonesia and the European Union." Pamphlet.

Lev, D. S. 1985. "Colonial Law and the Genesis of the Indonesian State." *Indonesia* 40 (Oct.): 57–75.

Li, T. M. 1997. "Constituting Tribal Space: Indigenous Identity and Resource Politics in Indonesia." Unpublished paper.

———. 1999. "Marginality, Power and Production: Analysing Upland Transformations." In *Transforming the Indonesian Uplands: Marginality, Power and Production*, ed. T. M. Li, 1–45. Singapore: Harwood Academic Publishers.

———. 2000. "Agrarian Differentiation and the Limits of Natural Resource Management in Upland Southeast Asia. Institutions and Uncertainty: New Perspectives on Natural Resource Management." Paper given at Institute of Development Studies Workshop, University of Sussex. 6–8 November.

———. 2001. "Local Histories, Global Markets: Cocoa and Class in Upland Sulawesi." *Development and Change* 33 (3): 415–37.

———. 2002. "Engaging Simplifications: Community-Based Resource Management, Market Processes and State Agendas in Upland Southeast Asia." *World Development* 30 (2): 265–83.

Li, T. M., ed. 1999. *Transforming the Indonesian Uplands: Marginality, Power and Production*. Amsterdam: Harwood Academic.

Lindsey, T., and H. Dick, eds. 2002. *Corruption in Asia: Rethinking the Governance Paradigm*. Annandale, N.S.W.: Federation Press.

Litfin, K. 1993. "Ecoregimes: Playing Tug of War with the Nation-State." In *The State and Social Power in Global Environmental Politics*, ed. R. D. Lipschutz and K. Conca, 94–117. New York: Columbia University Press.

Little, P. D. 1994. "The Link Between Local Participation and Improved Conservation: A Review of Issues and Experiences." In *Natural Connections: Perspectives in Community-Based Conservation*, ed. D. Western, R. M. Wright, and S. C. Strum, 347–72. Washington, D.C., and Covelo, Calif.: Island Press.

Lloyd, G. 2000. "Indonesia's Future Prospects: Separatism, Decentralisation and the Survival of the Unitary State." Parliament of Australia Parliamentary Library Current Issues Brief 17.

L.P.T.I. Bogor Sub Bagian Publikasi/Dokumentasi. 1970. "Pedoman Bertjotjok Tanam Nilam (Patchouly)." Departemen Pertanian Direktorat Djenderal Perkebunan Lembaga Penelitian Tanaman Industri Bogor.

Low, B., E. Ostrom, C. Simon, and J. Watson. 2003. "Redundancy and Diversity: Do They Influence Optimal Management?" In *Navigating Social-Economic Systems: Building Resilence for Complexity and Change,* ed. F. Berkes, J. Colding, and c. Folke., 83–114. New York: Cambridge University Press.

Lowndes, V. 1996. "Varieties of New Institutionalism: A Critical Appraisal." *Public Administration* 74 (Summer): 181–97.

Lubis, M. Y. 1992. "Budidaya Tanaman Pala." *Littro* 8 (1): 8–23.

Lubis, Z. 1996. "Repong Damar: Kajian tentang Pengambilan Keputusan dalam pengelolaan Lahan Hutan pada Dua Kommunitas Desa di Daerah Krui, Lampung Barat." *Ekonesia* 4: 62–126.

Lutz, E., and J. Caldecott, eds. 1996. *Decentralization and Biodiversity Conservation.* Washington, D.C.: World Bank.

Lynch, O. J., and J. B. Alcorn. 1994. "Tenurial Rights and Community-based Conservation." In *Natural Connections. Perspectives in Community-based Conservation,* ed. D. Western, R. M. Wright, and S. C. Strum. Washington, D.C., and Covelo, Calif.: Island Press.

Lynch, O. J., and K. Talbott. 1995. *Balancing Acts: Community-Based Forest Management and National Law in Asia and the Pacific.* Washington, D.C.: World Resources Institute.

MacAndrews, C. 1998. "Improving the Management of Indonesia's National Parks: Lessons from Two Case Studies." *Bulletin of Indonesian Economic Studies* 34 (1): 121–37.

MacAndrews, C., ed. 1986. *Central Government and Local Development in Indonesia.* Singapore: Oxford University Press.

MacIntyre, A. 1990. *Business and Politics in Indonesia.* Sydney: Allen and Unwin.

Maksum, M. N. 1983. "Meunasah di Dua Desa. Studi Perbandingan Aktivitas di Meunsasah Gampong Cacang, Kecamatan Labuhan Haji dan Meunasah kelurahan Padang, Kecamatan Tapak Tuan, Kabupaten Aceh Selatan." Pusat Latihan Penelitian Ilmu-Ilmu Sosial, Aceh. Laporan Hasil Penelitian.

Malinowski, B. 1922. *Argonauts of the Western Pacific: An Account of Native Enterprise and Adventure in the Archipelagoes of Melanesian New Guinea.* London: Routledge and Kegan Paul; New York: E. P. Hutton.

Manor, J. 1999. *The Political Economy of Democratic Decentralization.* Washington, D.C.: World Bank.

March, J. G., and J. P. Olsen. 1996. "Institutional Perspectives on Political Institutions." *Governance: An International Journal of Policy and Administration* 9 (3): 247–64.

Marcus, G. E. 1998. Ethnography Through Thick and Thin. Princeton, N.J.: Princeton University Press.

Mary, F., and G. Michon. 1987. "When Agroforests Drive Back Natural Forests: A Socio-economic Analysis of a Rice–Agro-forest System in Sumatra." *Agroforestry Systems* 5: 27–55.

Mattugengkeng. 1987. "Agama dan adat dalam konflik birokrasi pemerintahan tingkat pedesaan di Aceh: suatu studi dari segi anthropologis." Balai Penelitian Lektur Keagamaan Ujung Pandang, Badan Penelitian dan Pengembangan Agama, Departemen Agama R.I.

Mayer, J. 1996. "Environmental Organizing in Indonesia: The Search for a Newer Order." In *Global Civil Society and Global Environmental Governance, The Politics of Nature from Place to Planet,* ed. R. D. Lipschutz. Albany: State University of New York Press.

McCarthy, J. F. 1998a. "Letter from Medan." *Inside Indonesia* (April/June): 27.

———. 1998b. "Villagers Take to the Hills." *The Australian* (1 May).

———. 1999a. "In Perspective." *Eureka Street* 9 (9): 12–14.

———. 1999b. "Nature Based Tourism Case Study: Gunung Leuser, Indonesia" http://www.science.murdoch.edu.au:80/teach /n279/n279content/casestudies/g-leuser/leuser.html.

———. 2000a. "'Wild Logging': The Rise and Fall of Logging Networks and Biodiversity Conservation Projects on Sumatra's Rainforest Frontier." Centre for International Forestry Research, Occasional Paper 31. *http://www.cifor.cgiar.org/publications/Occpaper.htm.*

———. 2000b. "The Changing Regime: Forest Property and Reformasi in Indonesia." *Development and Change* 31 (1): 91–129.

———. 2001a. "Decentralization and Forest Management in Kapuas District." Centre for International Forestry Research. http://www.cifor.cgiar.org/publications.htm.

———. 2001b. "Decentralization, Local Communities and Forest Management in Barito Selatan." Centre for International Forestry Research. http://www.cifor.cgiar.org/publications.htm.

———. 2004. "Changing to Gray: Decentralization and the Emergence of Volatile Socio-legal Configurations in Central Kalimantan, Indonesia." World Development 32 (7): 1199–1223.

———. Forthcoming (a). "Decentralization Blues: Dominance, Resistance and Acquiescence in a Forest Dispute in Central Kalimantan." In Paper Tiger: Enforcement, Law and Environmental Disputes in Indonesia, ed. A. Bedner and J. McCarthy.

———. Forthcoming (b). "'Main Hakim Sendiri': Environmental Governance and Forest Disputes in Central Kalimantan." Paper prepared for INSELA seminar on Environmental Disputes and Enforcement of Environmental Law: Indonesia in a Comparative Perspective, Leiden University.

McCawley, T. 1998. "A People's Economy." *Asiaweek* (18 Dec.): 62–66.

McLean, J., and S. Straede. 2003. "Conservation, Relocation, and the Paradigms of Park and People Management—A Case Study of Padampur Villages and the Royal Chitwan National Park, Nepal." *Society and Natural Resources* 16: 509–26.

McNeely, J. A. 1997. "Mobilizing Broader Support for Asia's Biodiversity: How

Civil Society Can Contribute to Protected Area Management." Asian Development Bank Draft Report.

McNeely, J. A., and P. S. Wachtel. 1991. *Soul of the Tiger: Searching for Nature's Answers in Southeast Asia.* Singapore: Oxford University Press.

Menteri Kehutanan. 1995. "Keputusan Menteri Kehutanan Nomor: 622/Kpts-II/95 tentang Pedoman Hutan Kemasyarakatan."

———. 2002. "Revitalisasi kebijakan Pembangunan Kehutanan untuk Mensukseskan Pelaksanaan Otonomi Daerah." Arahan Menteri Kehutanan dalam rapat kerja kepala daerah seluruh Indonesia Tahun 2002 Department Dalam Negeri, 29 January.

Merry, S. E. 1988. "Legal Pluralism." *Law and Society Review* 22 (5): 869–96.

———. 1991. "Law and Colonialism." Law and Society Review 25 (4): 890–922.

———. 1992. "Anthropology, Law and Transnational Processes." *Annual Review of Anthropology* 21: 357–79.

———. 1995. "Resistance and the Cultural Power of Law." *Law and Society Review* 29 (1): 11–26.

Michon, G., de Foresta, H., Levang, P., and Kusworo. 2000. "The Damar Agroforests of Krui, Indonesia: Justice for Forest Farmers." In People, Plants, and Justice, ed. C. Zerner, 2–20. New York: Columbia University Press.

Migdal, J. S. 1988. *Strong Societies and Weak States: State-Society Relations and State Capabilities in the Third World.* Princeton, N.J.: Princeton University Press.

———. 1994. "The State in Society: An Approach to Struggles for Domination." In *State Power and Social Forces: Domination and Transformation in the Third World*, ed. J. S. Migdal, A. Kohli, and V. Shue, 7–34. Cambridge: Cambridge University Press.

Milton, K. 1996. *Environmentalism and Cultural Theory: Exploring the Role of Anthropology in Environmental Discourse.* London and New York: Routledge.

Mogelgaard, K. 2003. "Helping People, Saving Biodiversity: An Overview of Integrated Approaches to Conservation and Development." Population Action International Occasional Paper, March.

Momberg, F., K. Atok, and M. Sirait. 1996. *Drawing on Local Knowledge: A Community Mapping Training Manual: Case Studies from Indonesia*: Ford Foundation–Yayasan Karya Sosial Pancur Kasih–WWF Indonesia Programme.

Moniaga, S. 1993. "Toward Community-based Forestry and Recognition of *Adat* Property Rights in the Outer Islands of Indonesia." In *Legal Frameworks for Forest Management in Asia: Case Studies of Community/State Relations*, ed. J. Fox, 131–50. Occasional Papers no. 16. Honolulu: East-West Center, Program on Environment.

———. 1994. "The Systematic Destruction of the Indigenous System of Various Adat Communities Throughout Indonesia." In *Seminar on the Human Dimensions of Environmentally Sound Development*, ed. H. P. Arimbi. Jakarta: WALHI.

Moore, S. F. 1973. "Law and Social Change: The Semi-Autonomous Social Field as an Appropriate Subject of Study." *Law and Society Review* 7 (4): 719–46.

———. 1986. *Social Facts and Fabrication: Customary Law in Kilimanjaro 1880–1980.* Cambridge: Cambridge University Press.

———. 1990. "State Theory and Sociolegal Research: An Example from Tanzania." *Studies in Law, Politics and Society* 10: 77–84.

Moreau, R. 1999. "The Promise of Trouble: Wahid Offers a Vote on Independence, and Aceh Erupts." *Newsweek International* (22 Nov.): 52.

Mosse, D. 1997. "The Symbolic Making of a Common Property Resource: History, Ecology and Locality in a Tank-Irrigated Landscape in South India." *Development and Change* 28 (3): 467–504.

Mubyarto. 1992. *Desa dan perhutanan sosial: kajian sosial-antropologis di Prop. Jambi.* Yogyakarta: Aditya Media.

Murphree, M. W. 1994. "The Role of Institutions in Community-based Conservation." In *Natural Connections: Perspectives in Community-Based Conservation*, ed. D. Western, R. M. Wright, and S. C. Strum, 403–27. Washington, D.C., and Covelo, Calif.: Island Press.

Muryadi, W., W. Panggabean, and M. R. Amady. 1997. "Kabut Mengambang, Tetangga Berang." *Forum Keadilan* (25 Aug.).

Muryono, S. 1998. "Kebakaran Hutan dan Elegi Lingkungan Hidup 1997." *Analisa* (10 Jan.).

Myers, N. 1980. "Role of Forest Farmers in Conversion of Tropical Moist Forests." http://www.ciesin.org/docs/002-106/002-106a.html.

Nababan, A. 1996. "Pemerintahan Desa and Pengelolaan Sumberdaya Alam: Kasus Hutan Adat Kluet-Menggamat di Aceh Selatan." Paper presented to the Analyisis Dampak Implementasi Undang-Undang No. 5 Tahun 1979 tentang Pemerintahan Desa terdadap Masyarakat Adat: Upaya Penyysybab Kebijakan Pemerintahan Desa Berbasis Masyarakat Adat, Wisma Lembah Nyiur—Cisarua.

Najiyati, S., and Danarti. 1992. *Budidaya dan Penanganan Pasca Panen Cengkih.* Jakarta: Penebar Swadaya.

Natcher, D. C., and C. G. Hickey. 2002. "Putting the Community Back Into Community-based Resource Management: A Criteria and Indicators Approach to Sustainability." *Human Organization* 61 (4): 350–63.

Neumann, R. P. 1992. "Political Ecology of Wildlife Conservation in the Mt. Meru Area of Northeast Tanzania." *Land Degradation and Rehabilitation* 3: 85–98.

———. 2000. "Land, Justice, and the Politics of Conservation in Tanzania." In *People, Plants, and Justice*, ed. C. Zerner, 117–33. New York: Columbia University Press.

Newman, H., A. Ruwindrijarto, D. Currey, and H. Hapsoro. 1999. *The Final Cut, Illegal Logging in Indonesia's Orangutan Parks.* Environmental Investigation Agency, Telapak Indonesia.

Newmark, W. D., and J. L. Hough. 2000. "Conserving Wildlife in Africa: Integrated Conservation and Development Projects and Beyond." *Bioscience* 50 (7): 585–92.

Niessen, N. 1999. *Municipal Government in Indonesia: Policy, Law and Practice* of Decentralization and Urban Spatial Planning. Leiden: CNWS Publications, Leiden University.

North, D. C. 1990. Institutions, Institutional Change and Economic Performance. Cambridge: Cambridge University Press.

———. 1993. "Economic Performance Through Time." The American Economic Review 84 (3).

NRMP and Bappenas. 1994. Policy Towards Protected Areas in Indonesia: Final Report. Natural Resources Management Project, BAPPENAS–Ministry of Forestry assisted by USAID, Associates in Rural Development for Office of Agro–Enterprise and Environment USAID–Jakarta 38.

Oakerson, R. J. 1992. "Analyzing the Commons: A Framework." In Making the Commons Work: Theory, Practice and Policy, ed. Daniel W. Bromley, 41–59. San Francisco, Calif.: Institute for Contemporary Studies Press.

Obidzinski, K. 2002. Logging in East Kalimantan, Indonesia: The Historical Expedience of Illegality. Amsterdam: University of Amsterdam.

Oi, J. C. 1989. "Peasant Politics in a Communist Economy." In State and Peasant in Contemporary China: The Political Economy of Village Government. Berkeley and Los Angeles: University of California Press.

Oomen, B. 2002. Chiefs! Law, Power and Culture in Contemporary South Africa. Leiden: Universiteid Leiden.

Ostrom, E. 1990. Governing the Commons: The Evolution of Institutions for Collective Action. Cambridge: Cambridge University Press.

———. 1992a. Crafting Institutions for Self-Governing Irrigation Systems. San Francisco, Calif.: Institute for Contemporary Studies Press.

———. 1992b. "The Rudiments of a Theory of the Origins, Survival and Performance of Common-Property Institutions." In Making the Commons Work: Theory, Practice and Policy, ed. D. W. Bromley, 293–318. San Francisco, Calif.: Institute for Contemporary Studies Press.

———. 1997. "What Makes for Successful Institutions to Govern Common-Pool Resources." Paper presented at the "Local Institutions for Forest Management: How Can Research Make a Difference" conference, Bogor.

Otsuka, M. 1998a. "Challenges to Land Use Intensification under Commercialization of Local Farming Systems: A Case from West Sumatra, Indonesia." Unpublished paper.

———. 1998b. "Impacts of Changing Farmers' Land Control Patterns on Deforestation: A Case Study from Minangkabau Villages, West Sumatra, Indonesia." Tropics 7 (3/4): 257–69.

Otto, J. M., and S. Pompe. 1989. "The Legal Oriental Connection." In Leiden Oriental Connections 1850–1940, ed. Willem Otterspeer, 230–49. Studies in the History of Leiden University no. 5. Leiden: J. Brill.

Palast, G. 2003. The Best Democracy Money Can Buy: The Truth About Corporate Cons, Globalization and High-Finance Fraudsters. New York: Plume.

Pannell, S. 1996. "'Homo nullius' or Where Have All the People Gone? Refiguring Marine Management and Conservation Approaches." Australian Journal of Anthropology 7 (1): 21–42.

———. 1997. "Managing the Discourse of Resource Management: The Case of Sasi from 'Southeast' Maluku, Indonesia." Oceania 67 (4): 289–308.

Parlindungan, A. P. 1997. "Hukum Pertanahan Dalam Hubungannya Dengan Pembangunan Perhutanan." In *Hutan Rakyat: Hutan untuk Masa Depan,* ed. D. V. Barus, 199–220. Jakarta: Penebar Swadaya.

Patton, M. Q. 1990. *Qualitative Evaluation and Research Methods.* Newbury Park, Calif.: Sage Publications.

Paulus, J., and G. D. Stibbe, eds. 1921. *Encyclopaedie van Nederlandsch-Indie,* vol. 4. The Hague: Nijhoff.

Peluso, N. L. 1990. "A History of State Forest Management in Java." In *Keepers of the Forests: Land Management Alternatives in Southeast Asia,* ed. M. Poffenberger, 27–55. West Hartford, Conn.: Kumarian Press.

———. 1992a. "The Ironwood Problem: (Mis)Management and Development of an Extractive Rainforest Product." *Conservation Biology* 6 (2): 210–19.

———. 1992b. "The Political Ecology of Extraction and Extractive Reserves in East Kalimantan, Indonesia." *Development and Change* 23 (4): 49–74.

———. 1992c. *Rich Forests, Poor People: Resource Control and Resistance in Java.* Berkeley: University of California Press.

———. 1993. "Coercing Conservation: The Politics of State Resource Control." In *The State and Social Power in Global Environmental Politics,* ed. R. Lipschutz and K. Conca, 46–70. New York: Columbia University Press.

———. 1995. "Whose Woods Are These? Counter-Mapping Forest Territories in Kalimantan, Indonesia." *Antipode* 27 (4): 383.

———. 1996. "Fruit Trees and Family Trees in an Anthropogenic Forest: Ethics of Access, Property Zones, and Environment Change in Indonesia." *Comparative Studies in Society and History* 38 (3): 510–48.

Peluso, N. L., P. Vandergeest, and L. Potter. 1995. "Social Aspects of Forestry in Southeast Asia: A Review of Postwar Trends in the Scholarly Literature." *Journal of Southeast Asian Studies* 26 (1): 196–218.

Pemerintah Kabupaten Daerah Tingkat II Aceh Selatan. 1991/92. "Rencana Umum Tata Ruang Daerah (RUTRD) Kabupaten Daerah Tingkat II Aceh Selatan Propinsi Daerah Istimewa Aceh Tahun 1991–2001."

Pemerintahan Desa di Kemukiman Manggamat. 1995. "Peraturan Bersama Tiga Belas Desa di Kemukiman Adat Manggamat Kecamatan Kluet Utara Kabupaten Dati II Aceh Selatan Tentang Hutan Kemukiman Konservasi dan Hak Pengelolaan Sumber Daya Alamnya di Wilayah Adat Kemukiman Manggamat Kecamatan Kluet Utara." Pemerintah Daerah Tingkat II Kabupaten Aceh Selatan Kecamatan Kluet Utara Pemerintahan Desa di Kemukiman Manggamat.

Perbatakusuma, E., A. Elfian, S. Lusli, K. Rauf, and L. Lubis. 1997. "Stabilizing People-Park Interaction Areas Through Community-Based Sustainable Rural Resources Management: Lessons-Learned from WWF-DB, FPNC Leuser Conservation Project in Gunung Leuser National Park-Indonesia." Paper presented to the Regional Workshop on Participation of Local Communities in Protected Area Management UNESCO-ROTSEA–KEHATI FOUNDATION–LIPI, Medan, 17–20 March.

Peristiwa. 1990. "Sayed Mudhahar Akhmad yang Diperbincangkan Orang" *Peristiwa* (11 July).

Phillips, J. D. 1949. *Pepper and Pirates: Adventures in the Sumatra Pepper Trade of Salem.* Cambridge, Mass.: Riverside Press.

PMB-LIPI and LASEMA-CNRS. N.d. "Aceh, Pusat Penelitian Kemasyarakatan dan Kebudayaan (PMB-LIPI)." Social Sciences for the Study of Conflict in Indonesia, Laboratoire Asie du Sud-Est et Monde Insulindien (LASEMA-CNRS). *http://www.communalconflict.com/aceh.htm.*

Poffenberger, M., ed. 1990. *Keepers of the Forest: Land Management Alternatives in Southeast Asia.* Manila: Ateneo de Manila University Press.

Potter, L. 1991. "Environmental and Social Aspects of Timber Exploitation in Kalimantan, 1967–89." In *Indonesia: Resources, Ecology and Environment,* ed. J. Hardjono, 121–77. Singapore and New York: Oxford University Press.

———. 1993. "The Onslaught on the Forests in South-east Asia." In *South-east Asia's Environmental Future,* ed. H. Brookfield and Y. Byron, 103–23. Kuala Lumpur: Oxford University Press,

———. 1996. "Forest Degradation, Deforestation, and Reforestation in Kalimantan: Towards a Sustainable Land Use?" In *Borneo in Transition: People, Forests, Conservation and Development,* ed. C. Padoch and N. L. Peluso, 13–40. Kuala Lumpur: Oxford University Press.

———. 1997. "A Forest Product Out of Control: Gutta Percha in Indonesia and the Wider Malay World, 1845–1915." In *Paper Landscapes: Explorations in the Environmental History of Indonesia,* ed. P. Boomgaard, F. Colombijn, and D. Henley, 281–308, Leiden: KITLV Press.

Putman, R., with R. Leonardi and R. Nanetti. 1993. *Making Democracy Work: Civic Traditions in Modern Italy.* Princeton, N.J.: Princeton University Press.

Quigg, P. W. 1978. "Protecting Natural Areas." National Aubudon Society International Series no. 3.

Rahail, J. P. 1996. "Masyarakat Adat yang dipinggirkan and terpinggirkan kasus masyarakat di wilayah adat maur ohoi-wut." Paper presented to the Analyisis Dampak Implementasi Undang-Undang No. 5 Tahun 1979 tentang Pemerintahan Desa terdadap Masyarakat Adat.

Rahmadi, T. 2000. "Pengelolaan Sumberdaya Hutan: Pembagian kewenangan antara Pemerintah Pusat dan Pemerintah Daerah." Paper presented to the Seminar Insela, Jakarta.

Rasyid, M. R. 2003. "Regional Autonomy and Local Politics in Indonesia: Local Power and Politics in Indonesia." In *Decentralisation and Democratisation,* ed. E. Aspinall and G. Fealy, 63–71. Singapore: ISAS.

Razzaz, O. M. 1994. "Contestation and Mutual Adjustment: The Process of Controlling Land in Yajouz, Jordan." *Law and Society Review* 28 (1): 7–39.

Reid, A. 1975. "Trade and the Problem of Royal Power in Aceh." In *Pre-colonial State Systems in Southeast Asia: The Malay Peninsula, Sumatra, Bali-Lombok, South Celebes,* ed. A. Reid and L. Castles. Kuala Lumpur: Royal Asiatic Society, Malaysian Branch.

———. 1979. *The Blood of the People: Revolution and the End of Traditional Rule in Northern Sumatra.* Kuala Lumpur: Oxford University Press.

RePPProt. 1988. "Sumatra 0519 Tapaktuan Present Land Use and Forest Status." Direktorat Bina Program, Direktorat Jenderal Penviapan Pemukiman, Departemen Transmigrasi.

———. 1990. "The Land Resources of Indonesia: A National Overview."

Republika. 1998. "Lima Keppres Bidang Kehutanan dan Perkebunan Dicabut." *Republika* (30 Dec.).

———. 1998. "Enam Juta Hektare HPH akan segera Dilelang."

Republic of Indonesia and Leuser Development Programme, The. 1995. "Financing Memorandum." Memorandum no. ALA/94/26.

Reuters. 1998. "Indonesia Fires Devastate 3 Million Hectares of Forest." *Reuters* (25 Nov.).

Ribot, J. C. 1998. "Theorizing Access: Forest Profits Along Senegal's Charcoal Commodity Chain." Development and Change (29): 307–41.

———. 1999. "Decentralization and Participation in Sahelian Forestry: Legal Instruments of Central Political-Administrative Control." *Africa 69*.

———. 2000. "Rebellion, Representation, and Enfranchisement in the Forest Villages of Makacoulibantang, Eastern Senegal." In *People, Plants, and Justice*, ed. C. Zerner, 134–58. New York: Columbia University Press.

———. 2001. "Local Actors, Powers and Accountability in African Decentralizations: A Review of Issues." Paper prepared for International Development Research Centre of Canada Assessment of Social Policy Reforms Initiative. United Nations Research Institute for Social Development (UNRISD).

Richards, M. 1997. "Common Property Resource Institutions and Forest Management in Latin America." *Development and Change* 28: 95–117.

Richardson, B. J. 2000. "Environmental Law in Postcolonial Societies: Straddling the Local-Global Institutional Spectrum." *Colorado Journal of International Environmental Law and Policy* 11(1): 1.

Rijksen, D. H. D., and M. Griffiths. 1995. *Leuser Development Programme Masterplan*. Supported by the European Union, prepared by the Integrated Conservation and Development Project for Lowland Rainforest in Aceh.

Robinson, G. 1998. "Rawan Is as Rawan Does: The Origins of Disorder in New Order Aceh." *Indonesia* (Oct.): 127–99.

Robison, R., and A. Rosser. 1998. "Contesting Reform: Indonesia's New Order and the IMF." *World Development* 26 (8): 1593–1609.

Ross, M. 1996. "Conditionality and Logging Reform in the Tropics." In *Institutions for Environmental Aid: Pitfalls and Promise*, ed. R. O. Keohane and M. A. Levy, 167–97. Cambridge, Mass.: MIT Press.

Rouland, N. 2001. "Custom and the Law." In *Custom and the Law*, ed. P. de Deckker and J.-Y. Faberon. Canberra: Asia Pacific Press at the Australian National University.

Roush, J. 2001. "Community-Based Conservation." In *Solutions for an Environment in Peril*, ed. A. B. Wolbarst. Baltimore, Md., and London: Johns Hopkins University Press.

Rudel, T. K., and B. Horowitz. 1993. Tropical *Deforestation: Small Farmers* and *Land Clearing in the Ecuadorian Amazon*. New York: Columbia University Press.

Sack, R. D. 1986. *Human Territoriality: Its Theory and History*. Cambridge: Cambridge University Press.

Salafsky, N., B. L. Dugelby, and J. W. Terborgh. 1992. *Can Extractive Reserves Save the Rain Forest? An Ecological and Socio-economic Comparison of Non-*

Timber Forest Product Extraction Systems in Peten, Guatemala and West Kalimantan, Indonesia. Durham, N. C.: Center for Tropical Conservation, Duke University.

Sawitri, I., A. Zamzami, and B. Wiwoho. 1999. *Simak dan Selamatkan Aceh.* Jakarta: PT Bina Rena Pariwara.

Sayer, J., N. Ishwaran, and J. Thorsell. Forthcoming. "Tropical Forest Biodiversity and the World Heritage Convention." *Ambio* 29 (6): 302–9.

Schindler, L. 1998. "The Indonesian Fires and SE Asean Haze 1997/98: Review, Damages, Causes and Necessary Steps." Paper presented at the Asian-Pacific Regional Workshop on Transboundary Atmospheric Pollution, Singapore, 27–28 May.

Schmidt, S. W., ed. 1977. *Friends, Followers, and Factions: A Reader in Political Clientelism.* Berkeley: University of California.

Schmink, M., and C. H. Wood. 1992. *Contested Frontiers in Amazonia.* New York: Columbia University Press.

Schneider, A., and H. Ingram. 1990. "Behavioural Assumptions of Policy Tools." *Journal of Politics* 52: 510–29.

Schwarz, A., and J. Fredland. 1992. "Green Fingers: Indonesia's Prajogo Proves That Money Grows on Trees." *Far Eastern Economic Review* (23 July): 36.

SCMP. 1997. "Plantations Reject Blame for Blazes." *South China Morning Post* (30 Sept.).

Scott, J. C. 1977. "Patron-Client Politics and Political Change in Southeast Asia." In *Friends, Followers, and Factions: A Reader in Political Clientelism*, ed. S. W. Schmidt, L. Guasti, C. H. Lande, and J. C. Scott, 123–46. Berkeley: University of California Press.

———. 1985. *Weapons of the Weak: Everyday Forms of Peasant Resistance.* New Haven and London: Yale University Press.

Sellato, B. 2001. *Forest, Resources and People in Bulungan: Elements of a History of Settlement, Trade, and Social Dynamics in Borneo, 1880–2000.* Jakarta: Center for International Forestry Research.

Selsky, H. W., and S. Creahan. 1996. "The Exploitation of Common Property Natural Resources: A Social Ecology Perspective." *Industrial and Environmental Crisis Quarterly* 9 (3): 346–475.

Serambi Indonesia. 1990. "Menhut Belum Terima Laproan Polemik HPH di Aceh Selatan." *Serambi Indonesia* (5 July).

———. 1994a. "Dari Rakor IDT di Kutacane TNGL Rahmat atau Laknat." *Serambi Indonesia* (8 Dec.).

———. 1994b. "Menyingkap Aksi Pencurian Kayu (1) Cukong = Curi Kayu Kong Kalikong." *Serambi Indonesia* (28 Nov.).

———. 1994c. "Puluhan Kilang Kayu Liar Masih Beroperasi di Aceh Selatan." *Serambi Indonesia* (27 Oct.).

———. 1994d. "Tambang Dolar di Gunung Leuser." *Serambi Indonesia* (26 July).

———. 1995a. "Cerita Tentang Pencurian Kayu Di Aceh Selatan (2) Kalau Mau Jujur, Seluruh Kilang Kayu Harus Ditutup." *Serambi Indonesia* (7 Sept.).

———. 1995b. "Mengatasi Pencurian Kayu." *Serambi Indonesia* (13 Nov.).

———. 1995c. "Oknum Aparat Diduga Terlibat Rusak Hutan, Pemerintah Lumpuh." *Serambi Indonesia* (7 Aug.).

———. 1997a. "Irigasi Macet, Masyarakat Rambah Hutan." *Serambi Indonesia* (14 July).

———. 1997b. "Kelompok Tani Romas Ekspor Pala ke Amerika Serikat." *Serambi Indonesia* (29 May).

———. 1997c. "Operasi TNGL Sukses Besar Rubuan Batang Balok Ditangkap." *Serambi Indonesia* (28 Jan.).

———. 1997d. "Penebangan Liar Beralih ke HKKM." *Serambi Indonesia* (20 Jan.).

———. 1998. "Bila Izin HPH tak Dicabut, Rimueng Lamkaluet akan Mengamuk. *Serambi Indonesia* (17 Sept.).

Siegel, J. T. 2000. "Jafar Siddiq Hamzah." Indonesia. Oct.: 171–75.

Simon, H. 1998. "Indonesian Government Must Immediately Implement Just and Democratic Forest Resource Management." http://www.latin.or.id/diskusi_fkkm_jogja_english.htm 15/2/99.

Sinar Indonesia Baru. 1997. "Ini Tamparan Luar Biasa." *Sinar Indonesia Baru* (4 Oct.).

———. 1989a. "Bupati Aceh Selatan Kembangkan Industri Lokal." *Sinar Indonesia Baru* (12 June).

———. 1989b. "Calon Jemaah Haji Aceh Selatan 62 Orang." *Sinar Indonesia Baru* (12 June).

———. 1989c. "Harga Pala di Aceh Selatan Turun Drastis, Harga Gula Melonjak." *Sinar Indonesia Baru* (21 Nov.).

———. 1989d. "Petani Aceh Selatan Belum Rasakan Nikmat Produksi Pala." *Sinar Indonesia Baru* (11 July).

———. 1989e. "Tiga Lokasi Lahan Transmigrasi di Aceh Selatan Disurvey." *Sinar Indonesia Baru* (14 July).

Slaats, H., and K. Portier. 1992. *Traditional Decision-making and Law: Institutions and Processes in an Indonesian Context.* Yogyakarta: Gadjah Mada University Press.

Soedjiatono, B. 1999. "Kalau Koboi Swasta Mengadili Rakyat." *Tempo* (19 Dec.).

Soekirman, S. 1994. "Bandingan Terhadap Makalah Konsepsi Masyarakat Aceh Tentang Tata Ruang." In *Bunga Rampai Temu Budaya Nusantara PKA-3.*

Sonius, H. W. J. 1981. "Introduction." In *Van Vollenhoven on Indonesian Adat Law,* ed. J. F. Holleman. The Hague: Nijhoff.

Soule, M. E., and M. A. Sanjayan. 1998. "Conservation Targets: Do They Help?" *Science* 279 (27 Mar.): 2060–61.

Spencer, J. E. 1966. *Shifting Cultivation in Southeast Asia.* Berkeley: University of California Press.

Spiertz, J., and M. G. Wiber. 1998. "The Bull in the China Shop: Regulation, Property Rights and Natural Resource Management: An Introduction." In *The Role of Law in Natural Resource Management,* eds. J. Spiertz and M. G. Wiber, 1–16. The Hague: VUGA.

Spiertz, J., and M. G. Wiber, eds. 1996. *The Role of Law in Natural Resource Management.* The Hague: VUGA.

Steering Committee. 1998. "Mengembalikan Kalimantan Timur Kepada Rakyat sebagi Bentuk Pendaulatan Rakyat." Paper presented to the Lokakarya Usulan Kaltim untuk Reformasi Bidang Kehutanan dan Perkebunan, Samarinda, 28–29 July.

Strauss, A., and J. Corbin. 1990. *Basics of Qualitative Research: Grounded Theory Procedures and Techniques*. Newbury Park, Calif: Sage Publications.

Strien, N. J. van. 1978. "Draft Proposed Gunung Leuser National Park Management Plan 1978/79–1982/83." International Union for Conservation of Nature and Natural Resources World Wildlife Fund Project 1514–Gunung Leuser Reserves, Sumatra, Indonesia–Management Programme.

Suara Pembaruan. 1998a. "Dephutbun Kekurangan Petugas Jagawana." *Suara Pembaruan* (10 April).

———. 1998b. "Masyarakat Sanggau Menunggu SK HPHKM." *Suara Pembaruan* (26 Dec.).

———. 1998c. "Pasokan Kayu Bulat Tahun 1999 Turun 25 Persen." *Suara Pembaruan* (27 Nov.).

———. 1998d. "Pencurian Kayu Di Jambi Masih Merajalela." *Suara Pembaruan* (27 Nov.).

———. 1998e. "Pengusaha HTI Lebih Suka Ekspor Kayu." *Suara Pembaruan* (12 Jan.).

Sudaryani, T., and E. Sugiharti. 1998. *Budidaya dan penyulingan nilam*. Jakarta: PT Penebar Swadaya.

Suhendar, E. 1994. *Pemetaan Pola-pola Sengketa Tanah di Jawa Barat*. Bandung: Yayasan AKATIGA.

Sulardi. 2000. "Bencana Otonomi Daerah." *Kompas* (28 April).

Sulistoni, G., E. Kaffah, and Syahrul. 2001. *Mencabut Akar Korupsi*. Jakarta: SOMASI NTB.

Sunderlin, W. D. 1998. "Between Danger and Opportunity: Indonesia's Forests in an Era of Economic Crisis and Political Change." http://www.cgiar.org/cifor (11 Sept.).

Sunderlin, W. D., and I. A. P. Resosudarmo. 1997. "Rate and Causes of Deforestaion in Indonesia: Towards a Resolution of the Ambiguities." CIFOR Occasional Paper 9. http://www.cifor.cgiar.org/publications/Occpaper.htm: CIFOR.

Sunderlin, W. D., I. A. P. Resosudarmo, E. Rianto, and A. Angelsen. 2000. "The Effect of Indonesia's Economic Crisis on Small Farmers and Natural Forest Cover in the Outer Islands." CIFOR Occasional Paper no. 28. http://www.cifor.cgiar.org/publications/Occpaper.htm

Siegel, J. T. 1969. *The Rope of God*. Berkeley: University of California Press.

Snouck Hurgronje, C. 1906. *The Achehnese*. Leyden: Brill.

Surya Karya. 1990. "SK HPH Didasarkan Rekomendasi Pemda." *Surya Karya* (3 Aug.).

Suryohadikoesumo, I. D. 1997. "Pembangunan Hutan Nasional." In *Hutan Rakyat Hutan untuk Masa Depan*, ed. D. V. Barus, 3–11. Penebar Swadaya.

Sylva Indonesia. 1998. "Pe
termuan Menhutbun dengan Mahasiswa Fakultas Kehutanan UGM Yogyakarta." http://www.latin.or.id/info_bagi_kudeta.htm

(accessed 15/2/99).

Taale, T. 1995. "The Problem of Forest-Protection at the Local Level Case Study 1: Cuyabeno." In *The Role of Law in the Protection of the Tropical Forest in Ecuador's Amazon Region: Final Report*, ed. T. Taale and J. Griffiths, 95–150.

Taale, T., and J. Griffiths, eds. 1995. *The Role of Law in the Protection of the Tropical Forest in Ecuador's Amazon Region: Final Report*. Tropenbos Foundation, University of Groningen, Corporacion lationamericana para El Dessarrollo, Institutito Ecuatoriano Forestal y de Areas Naturales.

Tendler, J. 1997. *Good Governance in the Tropics*. Baltimore and London: John Hopkins University Press.

Terborgh, J. 1999. *Requiem for Nature*. Washington, D.C: Island Press.

Thelen, K., and S. Steinmo. 1992. "Historical Institutionialism in Comparative Politics." In *Structuring Politics: Historical Institutionalism in Comparative Analysis*, eds, K. Thelen and S. Steinmo, 1–32. Cambridge: Cambridge University Press.

Timber and Wood Products. 1998 "Indonesia's Forest Industry Dogged by Questions." *Timber and Wood Products* (28 Nov.).

Tjondronegoro, S. M. P. 1991 "The Utilization and Management of Land Resources in Indonesia, 1970–1990." In *Indonesia: Resources, Ecology, and Environment*, ed. J. Hardjono, 17–35. Singapore: Oxford University Press.

Tsing, A. 1999. "Becoming a Tribal Elder and Other Green Development Fantasies." In Transforming the Indonesian Uplands, ed. T. Li, 159–202. London: Harwood Academic Publishers.

Turner, M. 2000. "Implementing Accountability in Autonomous Regions." Technical Assistance No 3177 to the Ministry of Home Affairs, Republic of Indonesia, under the Community and Local Government Support Program (CLGSSDP), Asian Development Bank Program Loan No 1677-INO University of Canberra with SMEC International *http://www.gtzsfdm.or.id/documents/dec_ind/donor_act_re_sta/ADB_DiscussionPaper18.pdf*.

Utting, P. 2000. "An Overview of the Potential and Pitfalls of Participatory Conservation." In Forest Policy and Politics in the Philippines, ed. P. Utting. Manila: Ateneo de Manila University Press.

Vandergeest, P. 1996. "Mapping Nature: Territorialization of Forest Rights in Thailand." Society and Natural Resources 9: 159–75.

Vandergeest, P., and N. L. Peluso. 1995. "Territorialization and State Power in Thailand." *Theory and Society* 24 (3): 385–426.

Velde, J. J. van der. 1938. "De pantjang-alas in Atjeh (1931)." In *Adatrechtbundels; bezorgd door de commissie voor het adatrecht en uitgegeven door het koninklijk instituut voor de taal-, land-en volken-kunde van Nederlandsch-Indie*, 136–38. The Hague: Martinus Nijhoff.

Veth, P. J. 1873. *Atchin en Zijne Betrekkingen tot Nederland, Topographisch-historische Beschrijving*. Leiden: Gualth Kolff.

Walters, W. 2002. "Social Capital and Political Sociology: Re-Imagining Politics?" Sociology 36 (2): 377–98.

Warburton, J. 2001. "Corruption as a Social Process: From Dyads to Networks."

In Corruption and Anti-corruption, ed. P. Larmour and N. Wolanin. Canberra: Asia Pacific Press.

Warren, C. 1993. *Adat and Dinas: Balinese Communities in the Indonesian State.* Kuala Lumpur: Oxford University Press.

Warren, C., and J. F. McCarthy. 2002. "Adat Regimes and Collective Goods in the Changing Political Constellation of Indonesia." In *Shaping Common Futures: Case Studies of Collective Goods, Collective Actions in East and Southeast Asia.* London: Routledge.

Waspada. 1990. "Muncul Lagi HPH Baru di Aceh Selatan." *Waspada* (6 July).

———. 1994a. "Golkar Aceh Tenggara Dukung Rencana Pembukaan Jalan Tembus Titi Pasir-Bohorok." *Waspada* (28 Mar.).

———. 1994b. "Pencurian Kayu Dalam Party Besar Kembali Lagi di Aceh Selatan. Oknum Pengusaha Catut Nama Pejabat Teras." *Waspada* (5 May).

———. 1994c. "Seputar Rencana Pembukaan Jalan Tembus Titi Pasir—Bahorok Tokoh Masyarakat Aceh Tenggara Tolak Alasan Merusak Kawasan TN Gunung Leuser." *Waspada* (29 Mar.).

———. 1995. "Masyarakat Pasang 'Ranjau Paku' di Hutan Menggamat Aceh Selatan untuk Atasi Penebangan Liar." *Waspada* (3 Nov.).

———. 1998a. "20 Pengusaha HPH Dipanggil." *Waspada* (24 Sept.).

———. 1998b. "Seluruh HPH di Aceh Selatan Diancam Akan Dibumihanguskan." *Waspada* (26 Sept.).

Wells, M., K. Brandon, and L. Hannah. 1992. *People and Parks: Linking Protected Area Management with Local Communities.* Washington, D.C.: World Bank, World Wildlife Fund, U.S. Agency for International Development.

Wells, M., S. Guggenheim, A. Khan, W. Wardojo, and P. Jepson. 1999. *Investing in Biodiversity: A Review of Indonesia's Integrated Conservation and Development Projects.* Washington, D.C.: World Bank.

Wentzel, S. N.d. "Social Forestry Issues in Concession Areas Isu-isu Perhutanan Sosial pada Areal HPH/HPHTI." Samarinda: GTZ-SFMP.

Wessing, R. 1986. *The Soul of Ambiguity: The Tiger in Southeast Asia.* DeKalb: Northern Illinois University, Center for Southeast Asian Studies.

———. 1991. "The Last Tiger in East Java: Symbolic Continuity in Ecological Change." Royal Institute of Linguistics and Anthropology international workshop on Indonesian studies, Royal Institute of Linguistics and Anthropology, Leiden.

Western, D., and R. M. Wright. 1994. "The Background to Community-Based Conservation." In *Natural Connections: Perspectives in Community-Based Conservation,* ed. D. Western, R. M. Wright, and S. C. Strum, 1–14. Washington, D.C., and Covelo, Calif.: Island Press.

Whyte, W. F. 1984. *Learning from the Field: A Guide from Experience.* Thousand Oaks, Calif.: Sage Publications.

Wilson, R. A. 2000. "Reconciliation and Revenge in Post-Apartheid South Africa." *Current Anthropology* 41 (1): 75.

Wilshusen, P. R., S. Brechin, C. Fortwangler, and P. West. 2002. "Reinventing a Square Wheel: Critique of a Resurgent 'Protection Paradigm' in International Biodiversity Conservation." *Society and Natural Resources* 15: 17–40.

Wind, J. 1996. "Gunung Leuser National Park: History, Threats and Options."

In *Leuser A Sumatran Sanctuary,* eds. C. P. van. Shaik and J. Supratna, 4–27. Jakarta: Perdana Ciptamandiri.

WWF 1971/72. *World Wildlife Yearbook.* Gland, Switzerland: World Wildlife Fund, Public Affairs Division.

———. 1974/75. *World Wildlife Yearbook.* World Wildlife Fund, Public Affairs Division.

———. 1977/78. *World Wildlife Yearbook.* World Wildlife Fund, Public Affairs Division.

———. 1979/80. *World Wildlife Yearbook.* World Wildlife Fund, Public Affairs Division.

———. 1995. "Conservation of Gunung Leuser National Park: Wild Resources Management in National Park Interaction Areas South Aceh Regency Aceh Province." World Wildlife Fund, Directorate General Forest Protection and Nature Conservation Department of Forestry Technical Report 1.

———. 1998a. "Haze Damage from 1997 Indonesian Forest Fires Exceeds $1.3 Billion, Study Shows." http://www.worldwildlife.org/new/fires/dam.htm

———. 1998b. "New Estimates Place Damage from Fires at $4.4 Billion. Forest Networking: A Project of Ecological Enterprises." http://www.forests.org/archive/Indomalay/newindof.htm.

———. N.d. "Pelestarian Tumbuhan Rambung. Suatu Kajian Di Wilayah Interaksi Taman Nasional Gunung Leuser di Desa Batu Itam Kecamatan Tapaktuan Kabupaten Aceh Selatan." WWF ID 0106 Gunung Leuser National Park Conservation Project. Unpublished report.

World Wildlife Fund and ID 0106 Gunung Leuser National Park Conservation Project. N.d. "Profil Kemukiman Manggamat."

WWF-LP. 1995a. "Kemukiman Manggamat Melestraikan Hutan Desa." Press Release WWF/TNGL, April.

———. 1995b. "Laporan Kegiatan 28 Maret 1995–7 April 1995." Unpublished report.

———. 1995c. "Laporan Kegiatan Lapangan di Kemukiman Menggat 16–20 Maret 1995."

Young, Oran R. 2000. "Institutional Interplay: The Environmental Consequences of Cross-Scale Interactions." Paper presented at "Constituting the Commons: Crafting Sustainable Commons in the New Millennium," the eighth conference of the International Association for the Study of Common Property, Bloomington, Indiana, 31 May–4 June.

Zerner, C. 1994. "Through a Green Lens: The Construction of Customary Environmental Law and Community in Indonesia's Maluku Islands." *Law Society Review* 28 (5): 1079–1122.

———. 2000. "Towards a Broader Vision of Justice and Nature Conservation." In his *People, Plants, and Justice,* 1–20. New York: Columbia University Press.

Index

CONTEMPORARY ISSUES IN ASIA AND THE PACIFIC

A Series from Stanford University Press and the East-West Center

The authorized representative in the EU for product safety and compliance is:
Mare Nostrum Group
B.V Doelen 72
4831 GR Breda
The Netherlands

www.ingramcontent.com/pod-product-compliance
Lightning Source LLC
Chambersburg PA
CBHW021807270326
41932CB00007B/87